ADVANCES IN GAS-PHASE ION CHEMISTRY

Volume 4 • 2001

ADVANCES IN GAS-PHASE ION CHEMISTRY

Editors: NIGEL G. ADAMS
LUCIA M. BABCOCK

Volumes 1–3 were published by:

JAI PRESS INC.
100 Prospect Street
Stamford, Connecticut 06904-0811
USA

ADVANCES IN GAS-PHASE ION CHEMISTRY

VOLUME 4 • 2001

Editors: NIGEL G. ADAMS
LUCIA M. BABCOCK
Department of Chemistry
University of Georgia

2001

JAI

An Imprint of Elsevier Science
Amsterdam – London – New York – Oxford – Paris – Shannon – Tokyo

ELSEVIER SCIENCE B.V.
Sara Burgerhartstraat 25
P.O. Box 211, 1000 AE Amsterdam, The Netherlands

First edition 2001

Library of Congress Cataloging in Publication Data
A catalog record from the Library of Congress has been applied for.

ISBN: 0-444-50929-1

♾ The paper used in this publication meets the requirements of ANSI/NISO Z39.48-1992 (Permanence of Paper). Printed in The Netherlands.

CONTENTS

LIST OF CONTRIBUTORS

Robert N. Compton — Department of Chemistry
The University of Tennessee
317 Beuhler Hall
Knoxville, TN 37996
USA

Ray Fall — Department of Chemistry and Biochemistry
and Cooperative Institute for Research in
Environmental Science
University of Colorado
Boulder, CO 80309-0215
USA

Nathan I. Hammer — Department of Chemistry
The University of Tennessee
317 Beuhler Hall
Knoxville, TN 37996
USA

Thomas G. Karl — National Center for Atmospheric Research
Boulder, CO 80307
USA

Mats Larsson — Department of Physics
Stockholm University
P.O. Box 6730
S-113 85 Stockholm
SWEDEN

Thomas H. Morton — Department of Chemistry
University of California, Riverside
CA 92521-0403
USA

Lee S. Sunderlin
Department of Chemistry and Biochemistry
Northern Illinois University
DeKalb, IL 60115
USA

Masaharu Tsuji
Research Institute of Industrial Science
Kyushu University
Fukuoka, 816-8580
JAPAN

A. A. Viggiano
AFRL/VSBP
Air Force Research Laboratory
Philips Laboratory
29 Randolph Rd.
Hanscom AFB
MA 01731-3010
USA

Skip Williams
AFRL/VSBP
Air Force Research Laboratory
Philips Laboratory
29 Randolph Rd.
Hanscom AFB
MA 01731-3010
USA

DEDICATION AND EULOGY: WERNER LINDINGER 1944–2001

We dedicate this volume to the memory of our co-author Werner Lindinger who was tragically drowned in Hawaii on February 16, 2001. The scientific community lost a great friend and colleague.

Werner was born in Brixlegg/Tirol, Austria on January 25, 1944.

Following his Ph.D. degree in 1972 at Innsbruck, he began his professional career as a Max Kade Foundation postdoctoral fellow in the NOAA Aeronomy Laboratories in Boulder, Colorado from October 1973 to September 1975.

His considerable talent and exceptional energy led to an extremely productive period in Boulder. His warm and outgoing personality resulted in the formation of many deep friendships in Boulder (not only in NOAA, but also in JILA and in the Chemistry Department of the University of Colorado) and throughout the world, many persisting actively throughout his life. He vigorously exploited the newly developed Flowing Afterglow technology mainly for the measurement of thermal and low energy ion-molecule interactions and ion mobilities. In 1976, at an exceptionally young age, this outstanding research activity earned him the Fritz Kohlrausch prize, the most prestigious award from the Austrian Physical Society.

Upon his return to the Physics Faculty at Innsbruck, he was a leader instrumental in developing an atomic and ion physics program that quickly achieved international recognition, resulting also in a professorship in 1978 at the Institute of Experimental Physics. In 1987, he was elected head of the newly founded Institute of Ion Physics. Werner's group was extremely productive, making important contributions to ion-molecule reaction kinetics, and a variety of ion-molecule interaction processes as well as original contributions to thermochemistry. A notable example is the series of studies of molecular ion vibrational quenching in neutral collisions. These detailed studies led to greatly increased understanding of the reaction mechanisms.

Werner was a cofounder 23 years ago of the popular "Symposium on Atomic and Surface Physics" (SASP) held every two years, often in Tirol but also in other European countries. In recognition of this, and also for his scientific achievements, he received in 1996 the SASP Schrödinger Award and the Gold Medal of the Comenius University, Bratislava. His scientific achievements were also recognized in 1997 by the receipt of Austria's highest scientific award, the Erwin Schrödinger Prize of the Austrian Academy of Science.

In recent years, his interests broadened. His group extended the application of ion flow systems to super-sensitive detection of trace gases in real time by the development of the proton-transfer-reaction mass spectrometry (PTR-MS) technique. Werner and his colleagues pioneered its use in a variety of applications in medicine and food analysis, as well as in highly time resolved studies of the emissions into the atmosphere from vegetation and biomass burning. Many research groups around the world are now applying this technique for studies of biomass-atmosphere interactions. At the time of his death, he was in Hawaii to install one of his instruments at the NOAA Clean Air Baseline Station on Mauna Loa. His work in this area is the subject of his chapter in this volume.

In addition to his prolific publication record, Werner lectured widely in Europe and in the US for many years, being a guest professor at the Universities of Trento, Italy and Utah, Salt Lake City, USA and authoring numerous contributed and invited reviews.

Werner's interests were broad and varied. In addition to his research and teaching, he had a lively appreciation of art and music. He was a vigorous person physically, and was an avid hiker, regular tennis player and accomplished skier (like most Tiroleans). In recent years he became a serious equestrian, riding in two African safaris.

Werner's extraordinary joy of living made his friendship a rewarding and memorable experience. He will be sorely missed by us all.

Tilmann Märk, Innsbruck
Eldon Ferguson, Boulder
Paul Crutzen, Mainz
Lucia Babcock, Georgia
Nigel Adams, Georgia

PREFACE

Gas-phase ion chemistry is a broad field which has many applications and which encompasses various branches of chemistry and physics. The articles describe innovative ways of studying reactions as well as the application of unique apparatuses to problems in this field. The first chapter is a continuing theme of including applications following the precedence of earlier volumes in the series. Here, the probing of environmental, food and medical situations by proton transfer is emphasized (Lindinger, Fall and Karl). The relationship between the condensed and gaseous phases is explored by Sunderlin through the concept of hypervalent bonding. Measurements under more extreme conditions, those of very high temperatures (Viggiano and Williams), are made in a unique high temperature flowing afterglow which allows the separate effects of rotational, vibrational and electronic energy to be probed. Internal states of reaction products are probed by Tsuji in emission spectroscopic studies of ion–neutral and electron recombination reactions. Neutral products of this latter process are also investigated in sophisticated storage ring experiments (Larsson) and for ion–neutral reactions by a variety of different techniques (Morton). The above studies have been concentrated mainly on positive ions, cations, but Compton and Hammer have turned their attention to electron binding in negative ions, anions. Here the combined effect of Coulomb repulsion and polarization interactions surprisingly makes multiply charged negative ions more stable against electron loss than their singly charged parents.

The articles collected here represent only a sub-set of the advances that are being made in this rapidly developing area. Successive volumes will emphasize the progress being made in other areas of the subject. The editors are making a specific effort to include contributions from those relatively new to the field, which brings in new ideas and perspectives, as well as those more established in the field, thus bringing in their wealth of experience. Throughout each volume, a blend will be sought among experiment, theory and applications so that the reader, in addition to having up-to-date information on developments in forefront areas, will have a broad base from which to view the subject as a whole. In this volume, the medical, food and environmental applications of gas-phase proton transfer are elucidated in the chapter by Werner Lindinger to whom this volume is dedicated. Before his untimely death, Werner had set out the chapter in draft form. This chapter has been brought to the final form that is included here by the dedicated efforts of Ray Fall and Thomas Karl, two of Werner's close colleagues and collaborators. We are grateful to them for enabling Werner's latest thoughts to be made available to the scientific community. We believe it is what he would have wanted.

We feel privileged to edit a series such as this, which gives us the opportunity to look into the laboratories and the minds of the scientists who are advancing our understanding in the research area of gas-phase ion chemistry, an opportunity which, through these volumes, is passed on to our fellow scientists.

Nigel G. Adams
Lucia M. Babcock

Editors

ENVIRONMENTAL, FOOD AND MEDICAL APPLICATIONS OF PROTON-TRANSFER-REACTION MASS SPECTROMETRY (PTR-MS)

Werner Lindinger,* Ray Fall and Thomas G. Karl

OUTLINE

*Final revisions of the text were made by Ray Fall and Thomas G. Karl, who dedicate this article to the memory of Werner Lindinger, an innovative scientist, an inspiring mentor, and a loyal friend.

Advances in Gas-Phase Ion Chemistry
Volume 4, pages 1–48.

ABSTRACT

The development of proton-transfer-reaction mass spectrometry (PTR-MS) as a tool for the analysis of volatile organic compounds (VOCs) is described. PTR-MS is based on the rapid, non-dissociative transfer of protons from H_3O^+ to most common VOCs, but not to the principal gases in the air sample. Recent developments in the design of PTR-MS instruments allow detection of some VOCs in the parts per trillion by volume range. This sensitivity and the capability of PTR-MS instruments to be operated for extended periods in both laboratory and field settings has allowed exploration of many aspects of VOC analysis in environmental, food and medical applications.

I. INTRODUCTION

Numerous gas chromatography (GC) methods have been developed over the past decades for trace gas analysis to such an extent that nearly any volatile can be quantitatively analyzed with high precision even at concentrations far below the parts per trillion by volume (pptv) level. For example, Sturges et al.[1] were able to measure atmospheric concentrations of trifluoromethyl sulfur pentafluoride (CF_3SF_5) at levels of 0.1 pptv, and quite recently the group of Stuart Penkett has performed investigations on air samples from ice cores showing that the mixing ratio of halothane ($CF_3CHClBr$) in the atmosphere has risen from 5×10^{-4} pptv in the 1950s up to a maximum of 8×10^{-3} pptv in 1985, and has been declining from then on to today's level of about 5×10^{-3} pptv.[2] GC methods represent ideal tools when static or slowly changing mixtures are to be analyzed, but on-line monitoring of mixtures with fast varying concentrations—on time scales of a few seconds or minutes—has remained problematic. Mass spectrometry has an extremely fast response, but on-line gas analysis based on conventional mass spectrometry, using electron impact ionization, suffers because of considerable fragmentation of molecular ionic species. Especially when a mixture of organic compounds is to be analyzed, the complexity of break-up patterns puts severe constraints on the quantitative analysis of the concentrations of these components. For instance, electron impact on H_2O not only yields H_2O^+ ions, but also OH^+, O^+, H_2^+ and H^+, and benzene yields at least 18 different ions when ionized in this way.

The proportion of these ions also depends on electron energy. The break-up pattern of ethanol (CH_3CH_2OH) contains all the ions that also appear in electron impact ionization of methanol (CH_3OH), therefore it is nearly impossible to quantify trace amounts of methanol in commercial alcoholic products by conventional mass spectrometry.

Munson and Field[3] reported in 1966 on a technique of ionizing molecules by gas phase ion-molecule reactions, which they called chemical ionization (CI). In this way, break-up of the molecules can be greatly reduced or even avoided. Thus, measured ion currents can be correlated with the densities of the respective parent neutral compounds, allowing for on-line monitoring of rather complex gas mixtures. The fundamental principles of gas phase ion chemistry on which CI is based, as well as the instrumentation for CI, have been reviewed in great detail by Harrison.[4] The wide variety of CI methods that has been developed includes Medium Pressure Mass Spectrometry, Fourier Transform Mass Spectrometry, Quadrupole Ion Trap Mass Spectrometry, Pulsed Positive Ion-Negative Ion Chemical Ionization, and Atmospheric Pressure Ionization Mass Spectrometry (API-MS). Of these, API-MS[5] has developed into a very reliable and widely used technique for analysis of VOCs in flavor release studies and human breath.[6] A variety of API-MS applications in these fields of research has been described in a recent volume by Roberts and Taylor.[7]

A general problem in on-line monitoring is the quantification of concentrations of volatiles investigated. Usually, calibration gases are needed so that, by comparison of the respective ion signals, the actual density of the compound of interest can be evaluated. This can be avoided if the concept of swarm type experiments like that of a flow tube or flow-drift tube[8] is applied. These techniques were developed by Ferguson and his group[9,10] and similar ones by Adams and Smith[11] in order to measure rate constants for Ion-Molecule-Reactions (IMR). Ions travelling in a carrier gas containing traces of reactant gas will be depleted depending on the concentrations of these admixed gases, and from the measured ion declines, the rate constants for the specific IMR can be obtained. Turning this principle around, from the decline of a reactant ion and/or from the quantitative appearance of product ions in a flow experiment, the density of the neutral reactant can be calculated, provided that the rate constant (and reaction time) for the respective IMR is known. This procedure has been used in our laboratory to develop an on-line method for trace gas monitoring and, as the reactions on which the method is based are proton-transfer-reactions, it was named Proton-Transfer-Reaction Mass Spectrometry (PTR-MS).[12,13] In this review, a short description of the method will be given followed by results from applications of PTR-MS in the fields of environmental, food and medical research.

II. PROTON-TRANSFER-REACTION MASS SPECTROMETRY (PTR-MS)

PTR-MS combines the concept of CI with the swarm technique of the flow tube and flow-drift-tube mentioned above. In a PTR-MS instrument, we apply a CI system which is based on proton-transfer reactions, and preferentially use H_3O^+ as the primary reactant ion. As discussed earlier,[12] H_3O^+ is a most suitable primary reactant ion when air samples containing a wide variety of trace gases or VOCs are to be analyzed. H_3O^+ ions do not react with any of the natural components of air, as these have proton affinities lower than that of H_2O molecules; this is illustrated in Table 1. This table also shows that common VOCs containing a polar functional group or unsaturated bonds (e.g. alkenes, arenes) have proton affinities larger than that of H_2O and therefore proton transfer occurs between $H_3O^+$and any of these compounds (see Equation 4). The measured thermal rate constants for proton transfer to VOCs are nearly identical to calculated thermal, collisional limiting values (Table 1), illustrating that proton transfer occurs on every collision.

From investigations of the ion-neutral induced (or permanent) dipole interactions (using the strongly polar ArH_3^+ and KrH_3^+ ions),[14] and from other highly accurate measurements on proton transfer reactions, it has been shown that there is excellent agreement between measured rate constants for exoergic proton transfer reactions and respective calculated values k_c.[14,15] But how do we obtain values for these latter rate constants?

A. Rate Constants for Proton-Transfer Reactions

Wherever an ion approaches a neutral (molecule or atom), which does not have a permanent dipole, its Coulomb field induces a dipole within this neutral which results in an attractive force. This leads to the formation of an ion-neutral collision complex when the impact parameter is below a critical value, and, as has been shown by Gioumousis and Stevenson,[16] the rate constant for formation of such complexes is independent of temperature, and has the value

$$k = 2\pi e \left(\frac{\alpha}{m_r} \right)^{1/2} = k_L, \tag{1}$$

where α is the polarizability of the neutral collision partner, m_r is the reduced mass of the ion and the neutral, and k_L is called the Langevin limiting value and can be seen as a capture rate constant; that is, the value indicates the rate at which the reactants are captured into spiralling orbits.

Table 1. Examples of Volatile Substances Typically Present in Air Mixtures to be Analyzed by PTR-MS, their Molecular Formulas, Protonated Masses, and Proton Affinities.[a] Water and Ammonia (bold) can Serve as Primary Ions.

Compound	*Formula*	*Protonated mass*	*PA[b] [kcal/ mol^{-1}]*	*Measured thermal rate constant[c] [10^{-9} cm^3 s^{-1}]*	*Calculated thermal rate constant[d] [10^{-9} cm^3 s^{-1}]*
Helium	He	—	42.5	—	—
Neon	Ne	—	48.6	—	—
Argon	Ar	—	88.2	—	—
Oxygen	O_2	—	100.6	—	—
Hydrogen	H_2	—	100.9	—	—
Krypton	Kr	—	101.5	—	—
Nitrogen	N_2	—	118.0	—	—
Xenon	Xe	—	118.6	—	—
Carbon dioxide	CO_2	—	129.2	—	—
Carbon monoxide	CO	—	141.7	—	—
Water	$\mathbf{H_2O}$	**19**	**165.2**	—	—
Hydrogen sulfide	H_2S	35	168.5	1.4	1.9
Formaldehyde	CH_2O	31	170.4	3.0	3.3
Formic acid	CH_2O_2	47	177.3	2.7	2.2
Benzene	C_6H_6	79	179.3	2.1	1.9
Propene	C_3H_6	43	179.6	1.5	1.7
Methanol	CH_4O	33	180.3	2.2	2.7
Acetaldehyde	C_2H_4O	45	183.8	3.6	3.7
Ethanol	C_2H_6O	47	185.6	2.8	2.7
Acetonitrile	C_2H_3N	42	186.2	4.7	5.1
Toluene	C_7H_8	93	187.4	2.1	2.2
Propanal	C_3H_6O	59	187.6	—	3.6
1-Propanol	C_3H_8O	61	188.2	2.3	2.7
Butanal	C_4H_8O	73	189.5	—	3.8
Xylene	C_8H_{10}	107	190.0	—	2.2
2-Propanol	C_3H_8O	61	190.1	2.8	2.8
Acetic acid	$C_2H_4O_2$	61	190.2	3.0	2.7
Methylpropanal	C_4H_8O	73	190.7	—	—
1,4-Dioxane	$C_4H_8O_2$	89	190.7	—	1.9
2,3-Butanedione	$C_4H_6O_2$	87	192.1	—	—
Acetone	C_3H_6O	59	194.1	3.9	3.9
Phenol	C_6H_6O	95	195.0	—	2.7
Butanone	C_4H_8O	73	197.8	—	—
Dimethylsulfide	C_2H_6S	63	198.6	2.1	2.6
Isoprene	C_5H_8	69	198.9	1.3	2.0
Ammonia	$\mathbf{NH_3}$	**18**	**204.1**	2.2	2.6

(continued)

Table 1. Continued

Compound	*Formula*	*Protonated mass*	*PA*[b] *[kcal/mol^{-1}]*	*Measured thermal rate constant*[c] *[10^{-9} cm^3 s^{-1}]*	*Calculated thermal rate constant*[d] *[10^{-9} cm^3 s^{-1}]*
Diethylsulfide	$C_4H_{10}S$	91	204.5	—	—
3-Penten-2-one	C_5H_8O	85	206.6	—	—
2-Methylfuran	C_5H_6O	83	206.9	—	—
Pyrrole	C_4H_5N	68	209.2	—	—
Pyrazine	$C_4H_4N_2$	81	209.5	—	—

[a]The last two columns show the measured and calculated rate constants for reaction of some volatile substances with H_3O^+. Updated from data presented in Ref. [12].
[b]Data from NIST Standard Reference Database Number 69 (August 1997 release).
[c]SIFDT measurements from the University of Innsbruck.
[d]Upper collisional limit values obtained from Ref. [93].

In cases where neutral reactants already possess a permanent dipole moment, μ_D, the capture rate coefficient is larger than k_L. Su and Bowers[17] have derived the expression

$$k_{ADO} = 2\pi e\left(\frac{\alpha}{m_r}\right)^{1/2} + C\left(\frac{2\pi e\mu_D}{m_r}\right)\left(\frac{2}{\pi kT}\right)^{1/2}, \tag{2}$$

which is called the average dipole orientation (ADO) limiting rate constant k_{ADO}. The factor C is a weighting coefficient depending on the degree of orientation of the permanent dipole and depends on the ratio $\mu_D/\alpha^{1/2}$. Values of C are listed in Ref. [17]. The rotational motion of the molecule is hindered by the presence of permanent dipoles, and the effect is more pronounced in systems having strongly anisotropic potentials. Thus a variety of more complex theories have been developed to account not only for permanent dipole, but also for quadrupole moments.[18] A new computational technique involving a combination of adiabatic capture and centrifugal sudden approximations (ACCSA) was applied by Clary.[19] This theory predicts sharply increasing rate constants as the temperature decreases. Parameterization of the ion-polar molecule collision rate constant by trajectory calculations was done by Su and Chesnavich[20] leading to the temperature dependent expression $k_c(T) = k_L K_{cap}$, where

$$K_{cap} = \begin{cases} 0.476x + 0.6200; & x \geq 2, \\ \dfrac{(x + 0.5090)^2}{10.526} + 0.9754; & x \leq 2. \end{cases} \tag{3}$$

with $x = 1/\sqrt{T_R} = \mu_D/(2\alpha k_B T)^{1/2}$. Except at low temperatures, K_{cap} leads to values similar to k_{ADO}, and at elevated temperatures the two often differ only by a few percent. The calculated rate constants k_L, k_{ADO}, $k_c(T)$ are usually called "collisional limiting values", k_c, and the above discussion infers that the values of k_c represent upper limits for rate constants of actual ion molecule reactions (IMR).

It should be stressed that in the case of exoergic proton transfer reactions involving small reactant neutrals (masses up to 100 Dalton), the measured values of k invariably agree with k_c to within a few percent. When rate constants are needed for measuring densities of VOCs using PTR-MS, it is recommended that calculated values of k_c be used unless very reliable experimental data are available.

An additional advantage of using primary H_3O^+ ions, besides that they do not react with the natural compounds of air, is that many of their proton transfer processes are non-dissociative, so that only one product ion species occurs for each neutral reactant with a mass one Dalton greater than the neutral. Table 1 (updated from Ref. 12) shows a variety of such cases. There are, however, compounds where dissociation does occur. These quite often follow a straightforward pattern like in the cases of the reactions of H_3O^+ with alcohols, where proton transfer followed by the ejection of an H_2O molecule is the predominant reaction pathway.[21]

B. The Concept of PTR-MS

PTR-MS uses the concept of swarm type experiments. In order to allow for an accurate quantification of the densities of neutral compounds from data on primary and product ion signals, the reactions of H_3O^+ ions with neutrals must proceed under well-defined conditions. Such ideal conditions in combination with long and adjustable reaction times are obtained in plasma and swarm type experiments using instruments like the Flowing Afterglow, the Flow Drift Tube (FDT) or the Selected Ion Flow (Drift) Tube (SIF(D)T), details of which have been discussed in several review articles.[22–25] For the development of the PTR-MS system we have used a FDT type which had been developed by Ferguson and his colleagues[8,22] to be used for the investigation of IMR, their energy dependencies and their thermodynamic properties.

Descriptions of the technical details and operation of the PTR-MS instrument have been published elsewhere,[12,13,26] therefore we will describe only the salient features of the system and will add new aspects of its operation that have since then turned out to be important. In swarm type experiments, and especially in drift experiments using PTR-MS (Figure 1 shows a schematic representation of the system), primary (reactant) ions travel through a buffer/carrier gas B to which the reactant gas R is added in small amounts, so that the density [B] is much larger than the density [R].

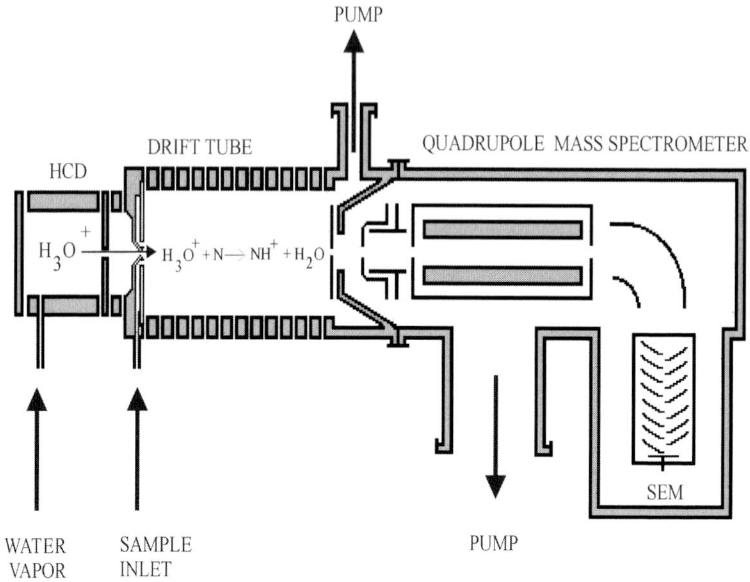

Figure 1. A schematic representation of the PTR-MS system. Abbreviations: HCD, hollow cathode discharge; N, neutral gas species; SEM, secondary electron multiplier.

On the way through the reaction region, the ions have many non-reactive collisions with buffer gas atoms or molecules; however, once they collide with a reactant gas particle they may undergo a reaction and specifically in the case of reactant ions H_3O^+, they perform a proton transfer reaction (if energetically allowed),

$$H_3O^+ + R \xrightarrow{k} RH^+ + H_2O. \tag{4}$$

If only trace components are present to react with H_3O^+, as it is usually the case in PTR-MS applications, the H_3O^+ ion signal does not decline significantly (about one to a few percent), so that $[RH^+] \ll [H_3O^+]$ is always valid. Therefore, by analogy with the detailed description in Ref. [24] at the end of the reaction section, the density of product ions $[RH^+]$ is given by

$$[RH^+] = [H_3O^+]_o(1 - e^{-k[R]t}) \approx [H_3O^+]_o[R]kt \tag{5}$$

where $[H_3O^+]_o$ is the density of H_3O^+ ions in absence of reactant neutrals in the buffer gas, k is the reaction rate constant for the proton transfer reaction, and t is the average time or "reaction time" the ions spend in the reaction region. As [R] denotes small densities of trace constituents, then

$[RH^+] \ll [H_3O^+] \approx [H_3O^+]_o = \text{constant}$. Equation (5) shows that without the use of any calibration gas, the density of a neutral component R can be measured by obtaining the primary and product ion signals, $[H_3O^+]$ and $[RH^+]$ respectively. Only the value of the respective reaction rate constant, which is obtained as discussed above, and the reaction time t need to be known. The reaction time, however, does not change since the pressure and electric field strength in the drift tube are kept constant at all times.

To reach a high sensitivity requires a high ion count rate $i(RH^+)$ per unit density [R] in the gas to be analyzed. This obviously can be achieved by keeping the $[H_3O^+]$ density high, and by not diluting the gas to be analyzed in an additional buffer gas like helium as is done in conventional Flow-Drift-Tube experiments, but by using the air itself (which contains the trace constituents to be analyzed) as the buffer gas. This can especially be done when H_3O^+ ions are used as the ionic reactant species because, as discussed above, these ions do not react with the major components of air.

The required high density of primary ions, H_3O^+, is provided by means of a hollow-cathode-discharge ion source, which provides H_3O^+ ions with a purity of about 99.5% or better. This situation has two advantages. High concentrations and therefore high count rates of primary H_3O^+ ions are obtained in the ion detection system (typical count rates are $\sim 10^6$ counts s^{-1}) and no quadrupole system needs to be installed to preselect the reactant ions H_3O^+ before entering the reaction region of the system. The only significant impurity ions observed are O_2^+ ions, which are produced within the "source drift region" due to the charge transfer from H_2O^+ ions to O_2 diffusing from the reaction region toward the ion source system, or by direct electron impact ionization of O_2. As O_2^+ does not react with H_2O,[27] it is not converted into another ionic form once it is produced in an H_2O environment. In contrast, N_2^+ or N^+ are rapidly converted into H_3O^+ in successive reactions with H_2O.

From the hollow-cathode source, ions are extracted into a short "source drift region" filled with water vapor. After passing this small drift section, the H_3O^+ ions reach a reaction region which is in the form of a drift section of about 20 cm length and 5 cm inner diameter, filled with the air (pressure 2–3 mbar) containing the trace constituents to be analyzed. No further buffer gas is needed and therefore the original mole fraction of R in air is retained in the reaction region. On the way from the Venturi type air inlet to the downstream end of the drift section, H_3O^+ ions undergo non-reactive collisions with any of the common components in air (see Table 1), but a small fraction (typically in the order of a percent) react with trace constituents.

The last important quantity we have to obtain in order to calculate the density [R] according to equation (5) is the reaction time t. This is the time the H_3O^+ ions take to transverse the length of the drift tube, which can be measured directly, either by pulsing the entrance and the exit slits of the drift tube and monitoring the arrival time spectrum, or by calculating the time from mobility values, μ, of H_3O^+ in air, reported in the literature.[28] The drift velocity v_d of the ions, which is obtained from the relation $v_d = \mu E$, where E is the electric field strength, is always kept high enough to ensure that it is large compared to the flow velocity of the air through the drift tube.

Usually, throughout the measurements, E/N is kept at values of 120–140 Td (1 Td = 1 Townsend = 10^{-17} cm^2 V^{-1} s^{-1} and N is the buffer gas number density) resulting in mean collision energies (KE_{cm}) of about 0.25 eV. This is a good compromise between avoiding formation of too many cluster ions, $H_3O^+ \bullet (H_2O)_n$, which would obscure the data due to switching reactions, and collision-induced break-up of product ions in the drift region;[28,29] the latter fragmentation would complicate the identification of neutrals. Under these conditions, typical count rates $i(H_3O^+)$ are in the order of a few 10^6 s^{-1} and $i(RH^+)$ of 20 to 60 s^{-1} per ppbv are reached. For many masses, the background is of the order of 0.1–0.5 counts s^{-1} (in the case of aromatic compounds even lower), so that concentrations of 50 to a few hundred pptv and in some cases even significantly lower are measurable with good accuracy. An important advantage of PTR-MS compared to an API technique is that any instrumental drift in ion counts is compensated through application of equation (5).

C. Identification of Volatiles

PTR-MS measures the density of a neutral compound by monitoring the mass-analyzed signal of an ion, which is usually the protonated compound. From that, we obtain information about its mass, but there are large numbers of compounds that have the same mass. Whenever qualitatively unknown mixtures of compounds are to be investigated, the problem of identification is a crucial one. PTR-MS is a method for on-line monitoring of compounds and not primarily for gas analysis. Quite often, however, the number of possible compounds of the same mass is drastically limited due to the origin of the mixture to be analyzed. If we monitor the $M59^+$ ion, investigating for example the emissions of a chemical plant, this could be indicative of propanal (CH_3CH_2CHO) or acetone ($CH_3C(O)CH_3$) both having the same unprotonated mass (58 Da). However when human breath is investigated, the $M59^+$ ion signal would be primarily related to acetone. This is due to information from the literature, where the presence of acetone, but rarely propanal, has been reported in breath. But even if several distinct compounds of the same mass must be considered as possible candidates to be present, we do have a variety of methods to distinguish between these compounds using PTR-MS. Several of these possibilities have been discussed earlier[12,13,26] and thus are only mentioned briefly:

- Changing the mean collision energy between H_3O^+ ions and the neutral reactants by variation of the electric field strength, E, in the drift tube (and thus changing E/N, N being the buffer gas number density) allows product ions to be distinguished by competition between the usual binary proton transfer reactions and association processes which are quite sensitive to E/N. Thus, a strong increase in $M55^+$ with decreasing E/N identifies it as the cluster $H_3O^+ \bullet (H_2O)_2$ which is produced in association processes at low E/N. Conversely, the protonated acetone signal at $M59^+$ does not increase strongly, being produced in the binary proton transfer from H_3O^+ to acetone.
- Isotopic abundances often allow the number of carbon atoms in a compound to be identified. For example, $M73^+$ which is observed in ambient air could be a protonated oxygenated hydrocarbon ($C_4H_8OH^+$) or the third water cluster ($CH_3OH\,(H_2O)_3H^+$). In typical analyses at $E/N = 120$ Td, average isotopic mass ratios for M73/M74 indicate a ^{13}C abundance of 4.6%. This is close to the expected value of 4.4% for a 4-carbon compound, suggesting that most of the signal can be related to the hydrocarbon.
- Collision-induced dissociation of product ions is another method for identification. Increasing E/N at the very downstream end of the drift tube by applying a high voltage between the last two drift rings and the end plate of the drift section leads to collision-induced breakup of the product ions. This breakup is strongly dependent on the type of isobaric ions and therefore can be readily used for identification purposes. For example, acetic acid (CH_3COOH) and *n*-propanol ($CH_3CH_2CH_2OH$) both give protonated $M61^+$, but exhibit a very different pattern of positive ions as the breakup voltage is varied from 10 to 50 V (see data in ref. 12).
- Use of NH_4^+ as the primary reactant ion can also aid compound identification. A hollow cathode ion source operating with H_2O vapor produces nearly exclusively H_3O^+ and some H_3O^+-water cluster ions. Similarly, when NH_3 is used, only NH_4^+ ions emerge from the source and can thus be used as primary reactant ions. While H_3O^+ ions perform proton transfer to all VOCs having a proton affinity (PA) higher than 165.2 kcal mol^{-1}, NH_4^+ only performs proton transfer to compounds with proton affinities in excess of 204.1 kcal mol^{-1} (Table 1). When air to be analyzed contains traces of two compounds with the

same mass, but one having a PA between that of H_2O and NH_3 and the other one having a PA higher than 204.1 kcal mol^{-1}, these compounds can be distinguished from each other by using the two different primary ions. Unfortunately, most other vapors used in a hollow cathode discharge would produce more than one terminating reactant ion species, and thus are not practical for PTR-MS operation.

- Use of Henry's Law partitioning from solution can be informative. The partitioning of a volatile compound from solution to the headspace is governed by a specific Henry's Law Constant (HLC), and PTR-MS can be used to monitor this process, calculate the HLC, and use this information for VOC identification. By bubbling air through a solution containing a VOC or VOCs, the decline of the head space concentration(s) due to mass transport by the bubbling air can be observed. In the case of only one volatile component, a single exponential decline is seen, as in the case of furfural ($C_5H_5O_2$) detected at M97$^+$ (Figure 2a). However, when two different exponential declines of a particular mass are observed, two different compounds with different HLCs are suspected. Figure 2b shows an example obtained from investigation of coffee,[30] where the two compounds, pyrazine ($C_4H_4N_2$) and furfuryl alcohol ($C_5H_6O_2$), both detected at M81$^+$, are clearly distinguishable due to their strongly different HLCs, represented by the different slopes in the figure. With the combined information of the protonated mass and the HLC, many volatiles can be identified this way. It is also noteworthy that PTR-MS provides a facile method for determination of HLCs.

While identification of compounds in many cases can be done unambiguously, it should be emphasized that PTR-MS primarily has its strength in monitoring fast concentration changes of compounds rather than in compound identification.

D. GC-PTR-MS Coupling

For identification of compounds in complex mixtures, gas chromatography-mass spectrometry (GC-MS) methods undoubtedly will remain the techniques of choice. In typical GC-MS applications, analytes emerging from a capillary GC are coupled to an electron impact source, a mass separator and a mass detector, generally allowing compound identification by analysis of the parent ion and its positive ion fragments.[31] This has the same problems discussed earlier for electron impact ionization. Similarly, we have demonstrated that it is possible to identify volatiles by coupling a PTR-MS instrument to a conventional capillary GC system. In this case, the

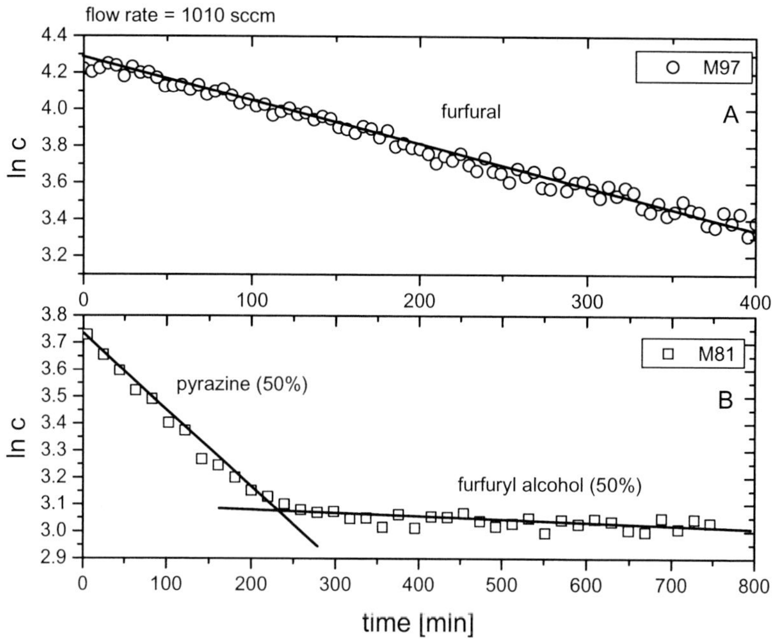

Figure 2. PTR-MS can be used to distinguish volatiles detected at the same mass by differences in their Henry's Law partitioning. (**A**) Measurement of the water:air partitioning of a single VOC, furfural ($M97^+$), by PTR-MS. (**B**) In this example, two coffee VOCs, pyrazine and furfuryl alcohol, both appearing at $M81^+$,[30] are clearly distinguishable due to their strongly different Henry's Law constants, represented by the different slopes in the figure. VOCs were added at time zero to a gas stripping bottle containing water, and partitioning of VOCs to the air stream was measured by PTR-MS.[34]

effluent of the GC column is continuously pulled into the PTR-MS inlet (Figure 1), and selective mass scanning is used to detect emerging analytes. The value of PTR-MS-GC coupling is illustrated in the following example.

In the performance of on-line PTR-MS measurements of VOCs in ambient air at the Sonnblick Observatory in Austria, we observed significant intensities at masses that are indicative of volatile leaf wound compounds (more details are presented in Section IV A).[32] Such compounds are released very shortly after wounding of leaves has occurred, and in the case of their observation at the Sonnblick, it was suspected that freezing of living plants might be the cause for their appearance. To test this idea, we needed to confirm the identity of detected masses, so a PTR-MS instrument was interfaced with a gas chromatograph equipped with a capillary column. Details of the chromatography can be found in references [33] and [34], and the interface was accomplished as follows. The flow was split after the separation column, with half of it directed through a conventional FID detector and the other half (1 sccm) leading into a 20 cm heated (100°C) deactivated glass capillary. The capillary was interfaced with the PTR-MS by adding 9 sccm of ultrapure air to its helium flow. This maintained the overall flow rate of 10 sccm through the PTR-MS and helped to suppress different mobility values, which would be caused by a pure helium buffer. The resulting dilution of 1:9 was compensated by longer VOC sampling times.

Figures 3a and 3b show GC-PTR-MS analysis of VOCs released from plants treated by a freeze-thaw method in a leaf cuvette. Mass scans revealed the presence of most major leaf-wound VOCs (discussed in Section IV A), but, for simplicity, only the VOCs detected at $M69^+$ and $M83^+$ are shown. The main VOC at $M69^+$ in all the tested plants (Figure 3a) was 1-penten-3-ol with smaller amounts of methylbutanals in clover and *cis*-2-penten-1-ol in each. The main VOC at $M83^+$ was hexanal in most freeze-thawed leaves; *trans*-2-hexenol, also detected at $M83^+$, was most abundant in freeze-thawed larch needles (Figure 3b). Although not as extensive, GC-PTR-MS analysis of Sonnblick air samples demonstrated that the major VOCs detected following a major freeze in nearby forests were 1-penten-3-ol ($M69^+$) and hexanal ($M83^+$), strongly supporting their origin from freeze-damaged vegetation.[33]

Work is in progress in Joost de Gouw's laboratory to investigate the specificity of the PTR-MS response to many different VOCs using GC-PTR-MS analyses of samples from various origins. The first results are promising, and indicate that even in urban air, the response to many VOCs (e.g. methanol, acetonitrile, acetaldehyde, benzene, toluene) is free of interference from other compounds.[35]

III. HISTORICAL

The path for development of PTR-MS was not a simple one; it was blocked by many obstacles and was time consuming. Thus a brief summary of the main cornerstones may be of interest to some readers.

PTR-MS is based on knowledge about IMR and mass spectrometric diagnostics obtained by the primary author and his coworkers over the past three decades. When mass spectrometric and Langmuir probe diagnostics were applied to Hollow Cathode Discharges (HCD) operating with rare gases in 1970, it was recognized that the presence of rather small amounts of H_2O vapor converted nearly all of the primary rare gas ions, R^+, into RH^+, H_2O^+ and H_3O^+ ions. As the negative glow of a HCD is essentially a field free plasma region, primary ions spend their time diffusing within this region

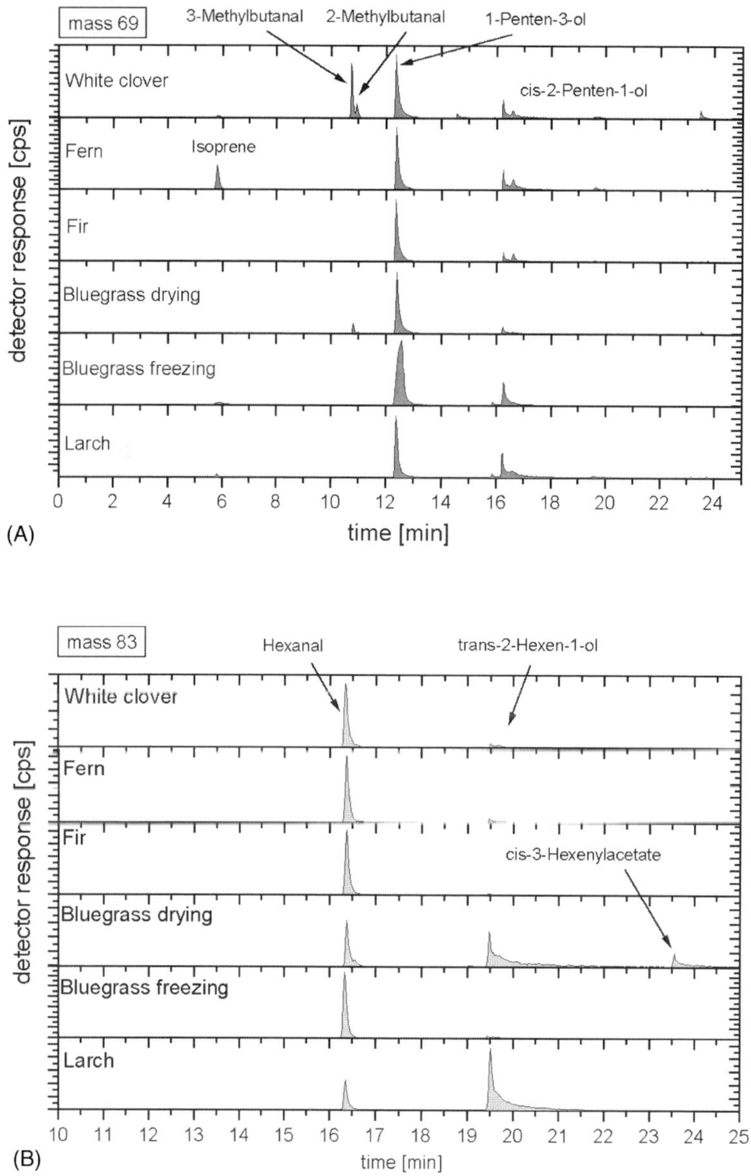

Figure 3. GC-PTR-MS coupling can be used to determine the identity of VOCs. GC-PTR-MS analysis of VOCs at (**A**) $M69^+$ and (**B**) $M83^+$ released during leaf freeze-thaw experiments with different plants sampled in the Innsbruck area. Retention times and PTR-MS profiles of indicated standard compounds were obtained with reagent grade chemicals. Replotted from data shown in Ref. [32].

performing many elastic collisions with rare gas atoms. However, when they undergo a collision with H_2O, every such collision is a reactive one, and, in the case of Ar being the rare gas, it only takes two such collisions until all the primary ions are converted into H_3O^+ due to the following reaction scheme. This shows how the different ions are created:

$$\left.\begin{array}{llll}
(a)\,Ar^+: & Ar+e_f & \rightarrow Ar^+ + 2e & k_e \\
(b)\,ArH^+: & Ar^+ + H_2O & \rightarrow ArH^+ + OH & k_1 \\
(c)\,H_2O^+: & Ar^+ + H_2O & \rightarrow H_2O^+ + Ar & k_2 \\
(d)\,H_3O^+: & ArH^+ + H_2O & \rightarrow H_3O^+ + Ar & k_3 \\
(e)\,H_3O^+: & H_2O^+ + H_2O & \rightarrow H_3O^+ + OH & k_4
\end{array}\right\} \quad (6)$$

Radial profiles of ion densities (such as shown in Figure 4) were obtained by mass spectrometric sampling of each type of ion, X^+, that was present,

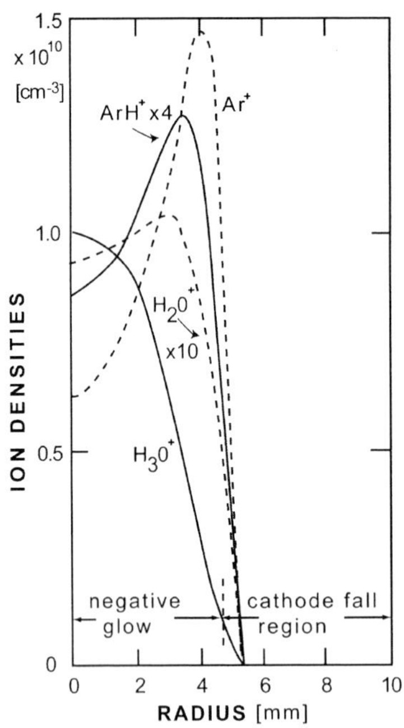

Figure 4. Radial profiles of ion densities in a hollow cathode discharge (HCD). The main ions, Ar^+, ArH^+, H_2O^+, and H_3O^+, in a hollow cathode discharge were measured by mass spectrometric analysis. Conditions: 2 cm diameter, 3 cm length, 0.34 torr pressure (Ar with 0.15% H_2O), and discharge current of 3 mA. Redrawn from data presented in Ref. [36].

application of the steady state equation

$$d[X^+]/dt = D_a\Delta[X^+] + k_e[e_f][R] + \nu_p - \nu_l - \alpha[X^+][e_s] \qquad (7)$$

and using approximate values for all the rate constants, k_e, $k_1, \ldots k_5$, involved. In equation (7), $D_a\,\Delta[X^+]$ is a diffusion term, and $[e_f]$ denotes the density of fast electrons with energies high enough for production of ions X^+ by electron impact ionization on neutrals R with a rate constant k_e, whereas $[e_s]$ denotes the density of slow plasma electrons which usually is much higher than $[e_f]$. These slow plasma electrons recombine with ions X^+ with the recombination coefficient α. ν_p and ν_l are terms representing reactions leading to the production and loss of X^+ ions which generally occurs via IMR.

From many such investigations,[36,37] the reaction kinetics proceeding in HCDs became well-known. Therefore it was quite natural to choose this kind of discharge when an efficient source for H_3O^+ ions was needed in the development of PTR-MS in the early 1990s.

Initial attempts were made in 1977 to combine a HCD with a drift tube for gas analysis using Kr^+ and Xe^+ as primary reactant ions. These failed because of the high impurity levels due to the vacuum equipment in use at that time. Oil was present in all the vacuum pumps, from roughing to high vacuum and nearly all masses in the spectrum showed background count rates originating from this source. Therefore this combination,[38,39] which was already quite similar to that used in today's PTR-MS instruments, was used only for measuring of reaction rate constants of IMR where much higher densities of reactant gases are present so that impurity effects are not very important. It appeared that swarm type experiments were not usable for gas analysis purposes, because the residence time of the buffer gas in the drift tube (on the order of a second) was too long, allowing impurities desorbing from walls to build up to intolerable concentrations.

The next step was a type of low energy single-collision system, where the primary reactant ions (again Kr^+ and Xe^+ were used, because charge transfer to nearly any volatile caused little fragmentation due to the low ionization potentials of Xe and Kr) were enclosed in a multipole RF-field region. This acted as a reaction chamber into which the gas mixture to be analyzed was introduced. This IMR-MS-system was quite efficiently used[40,41] for on-line monitoring of car exhaust and for emissions from cement plants, especially since compounds like NO, CO and CO_2 could be monitored. A further development of this system is now commercially available from a company run by former students of the primary author.[42] However, the dynamic range of the IMR-MS was limited by reactions of the primary ions with the oxygen of the air to be analyzed, and because depletion of Kr^+ and Xe^+ was in competition with their reactions with volatile components of interest.

From the above considerations, it is obvious that for measurement of trace compounds in air, a primary reactant ion is desirable which does not react with any of the natural components of air. In the early 1990s, a wealth of data on IMR was available as well as information on PA of molecules from which one could infer that H_3O^+ was the reactant ion of choice if VOCs in air require monitoring. Lane had earlier tried using $H_3O^+ \bullet (H_2O)_n$ as the reagent ion in API mass spectrometry, but this approach was hindered by the problem of deconvolution of cluster ion data.[43] Rather than a cluster distribution of primary ions, we proposed using H_3O^+ as the primary ion in a paper on IMR-MS[41] in 1993, and used a SIFDT system to prove that this can be done by measuring the concentrations of methanol, ethanol and acetone in human breath.[44] By that time, vacuum technology had also advanced so far that both roughing pumps and high vacuum turbo molecular pumps were working without oil, so that the background pressures of VOCs originating from this source were reduced by orders of magnitude.

Thus, it was time to revive the old combination of a HCD source, now producing H_3O^+ ions, with a small drift tube, small enough that the air to be analyzed, and which flows through it, only has a residence time of a few tenths of a second. A first, small and transportable PTR-MS system, weighing only about 100 kg, already allowed VOC concentrations of 1 ppbv to be measured with a time resolution of less than 1 s. Results from this system were reported in 1995.[26] Two years later, the sensitivity of PTR-MS had been increased to a few pptv. This was mainly achieved by changing critical orifices at source and detection lenses and increasing the drift pressure up to 2.3 mbar. This improved system was configured for operation on aircraft to measure height profiles of VOCs in ambient air, and has also been deployed for continuous air analysis in other field experiments described below. More recently, a test-variant of the PTR-MS has been designed for faster response in order to undertake direct eddy covariance. With the help of Alfons Jordan, we redesigned the original drift-tube segment and minimized the exchange time of the buffer gas (down to ~0.12 s), which makes this new instrument suitable for eddy covariance sampling at frequencies up to 8 Hz. Additionally, a fast inlet system was designed to minimize delay times and achieve highly turbulent flow conditions.[45] Viton gaskets in the hollow cathode discharge and the drift-tube segment were changed to PFA-Teflon. This helped to reduce the instruments background down below 10 pptv. Versions of these instruments are commercially available (www.ptrms.com), and thus have found their way into industrial companies as well as research institutes and are being used for environmental, food and medical research. Examples of some of these applications with higher sensitivity PTR-MS instruments are described in the next section.

IV. APPLICATIONS OF PTR-MS

The development of a chemical–physical method like PTR-MS is an advancement in gas phase ion chemistry, the subject of this volume, and more experimental details and basic background could be given here. However, as this has been done in several recent reviews,[12,13] it may be more useful for the reader of this Chapter to be given information on recent applications of PTR-MS. Papers on these applications are widely scattered in the literature, and therefore this is a good chance to bring this information together in a short summary of recent results. Again, as the reader is most likely to be educated in various areas of gas phase ion chemistry, rather than being an expert in any of the specific fields of PTR-MS application, the examples are presented in a way that relates to everyday life and thus fall into the category of general education.

A. Environmental Applications

The use of PTR-MS for environmental applications was strongly influenced by Paul Crutzen, who in 1997 urged us to take part in the Large Scale Biosphere–Atmosphere Experiment in Amazonia (LBA-CLAIRE), during which for the first time a PTR-MS instrument was operated in flight, measuring VOCs above the rain forest in Surinam. The main interest in atmospheric VOCs above biogenic sources such as forests is that these reactive trace gases can have significant impacts on levels of oxidants such as ozone (O_3) and the hydroxyl radical (OH), as well as on carbon monoxide (CO), precipitation acidity, and aerosol formation.[46] In addition, biogenic VOCs like acetone can be an important source of $HO_x(OH+HO_2)$ radicals in the upper troposphere,[47] and contribute to formation of peroxyacetic nitric anhydride (PAN; $CH_3C(O)OONO_2$), a compound that can transport reactive nitrogen oxide equivalents over long distances.[48] The examples below illustrate how the use of PTR-MS technology has enhanced our understanding of these biosphere–atmosphere exchange processes and photochemical processing of biogenic VOCs.

Atmospheric Profiles of Biogenic VOCs and Their Oxidation Products

The photochemical reactions of VOCs in the atmosphere are complex, depending on the presence of OH, nitrogen oxides ($NO_x = NO + NO_2$) and ultraviolet light. The generally accepted net reaction of a simple hydrocarbon (RH) is the following:[46]

$$RH + 4O_2 + 2h\nu \rightarrow 2O_3 + \text{carbonyl(s)} + 2H_2O. \tag{8}$$

OH and NO_x do not appear in the net reaction as they are regenerated in other reactions. Notably, atmospheric oxidation of RH (at sufficient NO_x) results in the production of ozone and a carbonyl that can undergo further photochemistry. If RH is the slowly reacting methane (typical lifetime, about 8 y), the carbonyl product is formaldehyde (HCHO). If RH is the more reactive hydrocarbon isoprene ($CH_2{=}C(CH_3)CH{=}CH_2$; typical lifetime, 1–2 h) the net reaction produces two carbonyl products: formaldehyde and a 4-carbon carbonyl, either methacrolein (MAC; $CH_2{=}C(CH_3)CHO$), or methylvinyl ketone (MVK; $CH_2{=}CHC(O)CH_3$) depending on the site of OH attack.

From these considerations, we conclude that a VOC generally contributes more to ozone formation the higher its density is and the faster its reaction with OH radicals proceeds. Thus, in order to understand quantitatively tropospheric ozone chemistry, it is a necessary prerequisite to know the VOC distribution within the lower troposphere as well as VOC sources. In both these areas of research, PTR-MS has been used extensively. Profiles of VOCs in the troposphere have been investigated by PTR-MS in flight experiments during LBA-CLAIRE mentioned above, on shipboard in the Indian Ocean Experiment (INDOEX) in 1999, at a ground site in the Houston area during the Texas Air Quality Study in 2000 (TexAQS 2000), and on a mountain top at the Sonnblick Observatory in Austria. Here, some results obtained in the Surinam experiment will be described.[49–51]

In Surinam, a PTR-MS instrument was configured for aircraft operation, making it possible to perform on-line analyses of a variety of ambient VOCs up to altitudes of 12 km. For example, Figure 5 shows profiles of isoprene (at the protonated mass $M69^+$) and the sum of its primary oxidation products MAC plus MVK (both at $M71^+$).[51] The lifetimes of these VOCs are too short to allow for ready mixing within the troposphere, therefore all three compounds have strong vertical gradients as shown in the figure. The data were obtained during one of several flights performed on three different days, but along nearly the same route and at different times of the day. There was a general trend of increasing mixing ratios (i.e. the relative number of molecules of a given type in an air sample) from morning to late afternoon reflecting the increase of isoprene emission with increasing

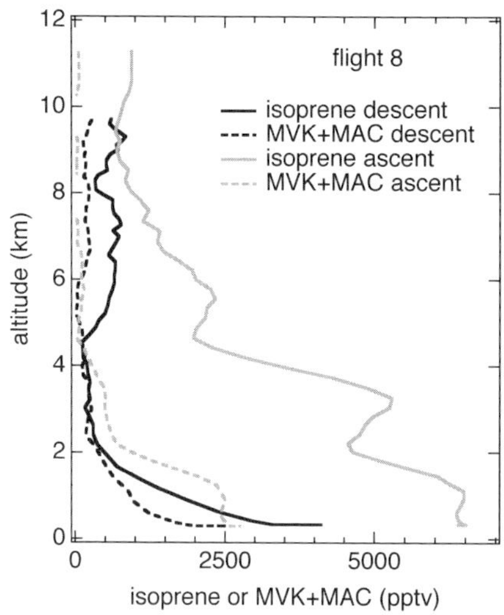

Figure 5. Isoprene and its oxidation products, methacrolein (MAC) plus methyl-vinyl ketone (MVK), have been measured by an airborne PTR-MS instrument. Data are from an aircraft flight (about 16:30 to 19:30 local time) over a tropical rainforest in Surinam,[51] and show the vertical gradients of isoprene (detected at $M69^+$), and the sum of MAC and MVK (both detected at $M71^+$).

temperature. Biogenic VOC mixing ratios close to the ground (e.g. the forest source) are higher than those at high altitudes, this being the consequence of reactive losses due to reactions with OH as well as convective transport to higher altitudes. In contrast to this pattern, the mixing ratios of longer lived compounds like acetone and acetonitrile showed no significant dependencies on altitude throughout the troposphere, due to strong convective transport and, in the case of acetone, in situ photochemical production.[49]

This extensive data set showed good agreement with a Master Chemical Mechanism[52] for isoprene oxidation. In addition, one of the predictions of the mechanism is that at low NO_x levels, like those seen in Surinam, isoprene hydroperoxides (six isomers, e.g. $HOCH_2C(OOH)(CH_3)CH{=}CH_2$) will accumulate. It was noted that correlations between isoprene and other VOCs (different times of day and altitude) were greatest with $M101^+$, which could be indicative of isoprene hydroperoxides.[50] This result is an example where PTR-MS analysis can detect previously unmeasured VOCs, although as mentioned above, verification of the identity of unknown positive ions requires complementary methods (e.g. GC-MS).

In addition to measurements of isoprene and its oxidation products, the flights in Surinam illustrate the versatility of PTR-MS. In the mass scan mode the instrument was able to obtain vertical profiles of a large range of VOCs in addition to those mentioned above, including a variety of oxygenated VOCs (alcohols, carbonyls and acids), and indicators of biomass burning (acetonitrile and acetone) and marine air masses (dimethyl sulfide).[49,50] Analysis of this range of oxygenated and organosulfur VOCs by other methods would usually require more than one instrument.

On-line Analysis of Organic Nitrates

Besides monitoring the common partially oxygenated VOCs like the ones discussed so far in this chapter, PTR-MS also can be used to detect PAN (mentioned above) which originates from the atmospheric degradation of acetaldehyde and acetone that produce acetylperoxy radicals.[47,48] These radicals in turn associate with NO_2 to form PAN, which acts as a relatively unreactive temporary reservoir for NO_x. It is in this form that NO_x equivalents are transported over wide distances, e.g. from urban to rural areas, where they can contribute to photochemical ozone formation with biogenic VOCs. This explains why the highest ozone concentrations often do not occur within large cities or along highways, both being strong sources of anthropogenic NO_2 and CO, but rather at distances of 50 to 100 km away, or—in the Alps—at altitudes of about 2000 m, where the optimum combination of PAN-related NO_2, OH and biogenic VOCs exists.[53]

Using the known thermal instability of PAN to decompose to peroxyacetic acid ($CH_3C(O)OOH$) and NO_2, Hansel and Wisthaler[54] let air entering the PTR-MS inlet pass through a heated stainless steel tube and were able to separate the signal originating from PAN from most of the background associated with $M77^+$, such as the acetone–water cluster $[C_3H_6OH^+ \bullet (H_2O)]$. The rapid response time of the instrument allowed rapid transitions between signals from unheated and heated air. Similarly $M91^+$ and $M103^+$ were used as indicators of PPN (peroxypropionic nitric anhydride; $CH_3CH_2C(O)OONO_2$) and MPAN (peroxymethacrylic nitric anhydride; $CH_2{=}CH(CH_3)C(O)OONO_2$, respectively, known peroxynitric anhydrides arising from oxidation of anthropogenic hydrocarbons (PPN) or isoprene (MPAN).[55] On-line aircraft analyses of PAN were performed in summer 2000 in the Texas Air Quality Study, and may allow unprecedented time resolution of NO_x product partitioning in power plant and refinery plumes.[56]

Sources of Tropospheric Biogenic VOCs

In addition to its utility in measuring ambient levels of VOCs in air, PTR-MS technology has been applied to investigations of biogenic

VOC emissions from living and dead vegetation. Because this method allows on-line analysis of VOCs, not readily achieved with traditional GC or GC-MS methods, it has provided new insights into biological processes.

As mentioned above, living vegetation releases a large number of VOCs, and many of these can significantly alter the chemistry of the atmosphere,[57–59] as in the case of biogenic isoprene, which is probably the most important VOC for tropospheric ozone. It is worth noting that the first observations of isoprene emission from plants were made around 1960 by Rei Rasmussen during his Ph.D. work[60] and independently by Sanadze and Dolidze,[61] yet the biological reason for its production is still a great mystery to scientists. The pathways along which it is created in plants have been uncovered, and we know that its production in the light occurs very fast.[59] After uptake of CO_2 by plants it only takes a few minutes until the same carbon atoms of those CO_2 molecules appear in the emitted isoprene. This was impressively demonstrated in a recent PTR-MS experiment where the natural CO_2 of air (with the fraction of $^{13}CO_2$ being ca. 1.1%) was replaced by 100% $^{13}CO_2$.[62] Figure 6 shows that after exposure of an oak plant in a leaf cuvette to $^{13}CO_2$ in a rapid succession, reaching steady state after about 20 min, all five ^{12}C atoms are replaced by ^{13}C in about 60% of the isoprene emitted. Detection of the ^{13}C-labeled intermediates was achieved by PTR-MS mass scanning at $M69^+$ to $M74^+$. After about 10 min of labeling, all of the emitted isoprene molecules contain at least one ^{13}C atom. Similar reversal of labeling patterns and kinetics were observed following a return to $^{12}CO_2$ (Figure 6). These types of on-line results have provided a very large ^{13}C-labeling data set that can be compared to results from traditional GC-MC methods,[63] and should provide new insights into the pool sizes of isoprene precursors in leaves.

Other significant biogenic sources of VOCs that have been investigated by PTR-MS are the emissions from plants initiated by "wounding", from decaying biomass, and from biomass burning.

Plant Wounding

When plants are wounded due to attack by insects, bacteria and fungi, they react by producing wound compounds,[59] which are thought to repel the intruders due to their unpleasant taste, smell and/or antibiotic properties. That wounding induces VOC release can be readily sensed in the odor of newly mown grass. As reviewed in detail by Gardner[64] and Hatanaka,[65] leaves of most wounded plants have the potential to produce and emit a series of C_6 aldehydes and C_6 alcohols and their derivatives, referred to here

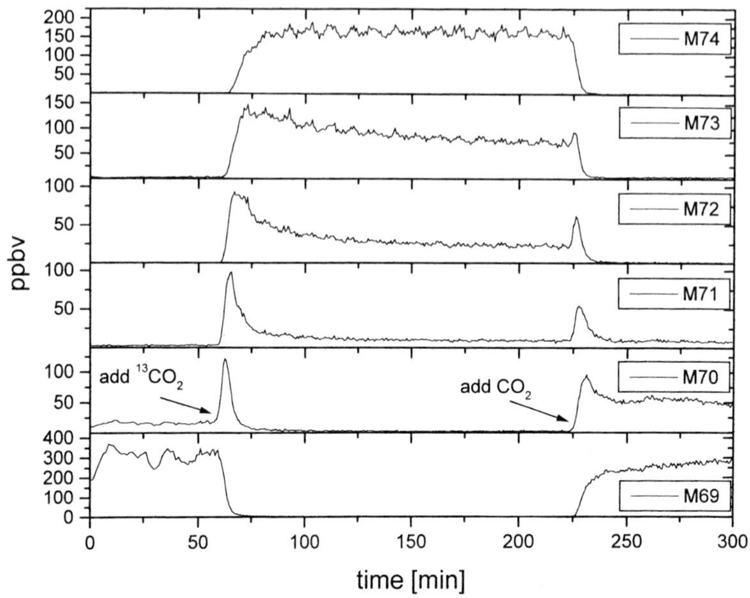

Figure 6. Incorporation of $^{13}CO_2$ into isoprene emitted from intact leaves can be monitored by on-line PTR-MS. After exposure of an oak leaf (*Quercus agrifolia*) in a cuvette to $^{13}CO_2$, in rapid succession all five ^{12}C atoms were replaced in about 60% of the isoprene emitted, and isoprene molecules with five ^{13}C atoms ($M74^+$) were dominant. The labeling pattern returned to predominantly ^{12}C-isoprene ($M69^+$), after a return to normal $^{12}CO_2$. (Unpublished data from Thomas Karl, Peter Prazeller, and Guenther Seufert.)

as the hexanal and hexenal families of VOCs. As Figure 7 illustrates, following leaf wounding, the formation of hexanal and hexenal family VOCs results from the oxidative cleavage of membrane fatty acids, linoleic and α-linolenic acid, respectively. Analysis of this complex mixture of wound VOCs is a serious challenge for the analytical chemist, made even more daunting if one wishes to observe the kinetics of VOC release following wounding. We used a PTR-MS instrument to measure leaf wound VOCs, first establishing the fragmentation patterns with standard compounds.[66] The structures and major positive ions for wound VOCs seen by PTR-MS are indicated in Figure 7, indicating that some produce unique ions (e.g. (*Z*)-3-hexenal at $M81^+$) and some have overlapping fragments (e.g. hexanal and hexenals each give $M83^+$ as the major ion). Using the unique ions as tracers, we were able to demonstrate for the first time: (a) the detailed kinetics of appearance and disappearance of C_6 metabolites; (b) that leaf wound VOCs are mainly formed within seconds of wounding and

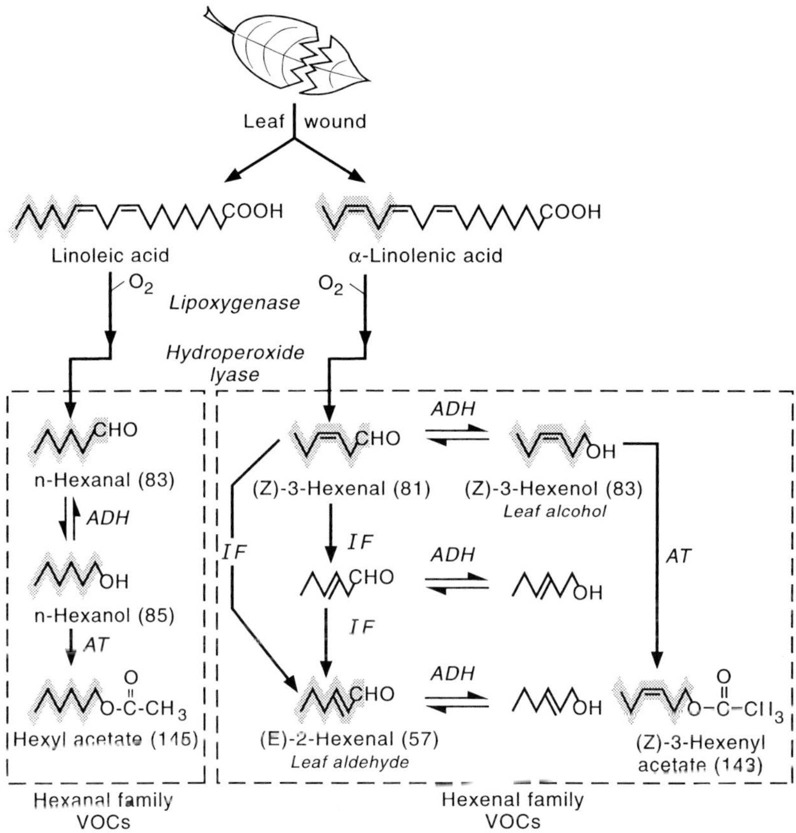

Figure 7. Scheme for VOC formation from leaf fatty acids following wounding. The enzymatic origins of hexanal and hexenal family VOCs are shown, and the unique or major positive ions detected by PTR-MS are indicated in parentheses for many of these VOCs. Abbreviations: ADH, alcohol dehydrogenase; AT, acetyl transferase; IF, isomerization factor. Reprinted with permission of the American Geophysical Union from Ref. [66].

do not arise from pre-existing leaf pools; and (c) that emissions of hexenal family VOCs are greatly enhanced as detached leaves dry out.

These laboratory results may help explain field observations. First, Helmig et al.[67,68] have recently reported rather high emissions of (*Z*)-3-hexenyl acetate from natural vegetation (i.e., 20–25 μg C g^{-1}(dry weight) h^{-1} in oak and raspberry), and have monitored ambient levels of (*Z*)-3-hexenyl acetate in an oak canopy. Since it is estimated that a large fraction of all leaves in a forest are wounded by herbivores at any time, it is likely that wound compounds must be considered a significant source of tropospheric VOCs. Second, cut lawns or hay crops which are left to dry, as well as savanna grasses which dry due to lack of rain, are strong emitters of volatiles that are easily smelled. We speculate that during the drying process, cell structures break down, thus initiating the wound compound production described above. We observed this recently on the campus of the University of Innsbruck.[69] When the grass was cut by a lawn mower, initial bursts of wound compounds (e.g. (*Z*)-3-hexenal) were observed, reaching ambient levels of 10–20 ppbv and then declining. However one to two hours later, when the radiation of the sun dried some uncollected grass, a continuous emission of various reactive VOCs was observed.

Several methods have been used to measure biogenic VOC fluxes near vegetated surfaces, such as the surface layer gradient, the mixed layer gradient, relaxed eddy accumulation, and disjunct eddy covariance systems (reviewed in Ref. 45). Eddy covariance is a statistical tool used to relate high frequency wind and scalar atmospheric data, and true eddy covariance measurements for VOC flux determination require sampling rates in the order of several Hz. There are few VOC analyzers with this capability. However, we have shown that PTR-MS can be used quite successfully for this kind of investigation, and have evaluated a standard PTR-MS instrument and a fast response instrument with a redesigned inlet system (see Section III) for assessing fluxes of VOCs during hay harvesting.[45,70] Eddy covariance methods are based on measurement of the time-dependent densities of VOCs at a constant height above ground and on simultaneous measurement of the vertical component of the movement of the air at the same spot. When there exists a source of VOCs on the ground, as is the case when grass is drying, upward moving eddies will have higher concentrations of VOCs than those moving downward. Therefore we expect a pronounced positive correlation between the wind vector in the z-direction and the measured densities. Figure 8 shows raw data on the concentration of methanol measured at a height of 450 cm above ground and the corresponding vertical wind speed. From these data and information on micrometeorology, the fluxes of VOCs can be calculated. Emission fluxes of methanol at a hay field site in western Austria were measured for two days. As shown in Figure 9, it can be seen that even on the second day after cutting, significant methanol fluxes were still detectable as the hay dried. The fluxes generally declined in the afternoon with decreasing latent heat flux, approaching zero when the hay was harvested (starting at 15:45). These results are notable, since we estimated the potential area average reactive VOC emission for this site to be $\sim 1.6 \times 10^5\ \mathrm{g\,km^{-2}\,d^{-1}}$. This emission is higher than the estimated area average reactive VOC emission ($\sim 1 \times 10^5\ \mathrm{g\,km^{-2}d^{-1}}$) for the South Coast Air Basin (SOCAB, Los Angeles).[71] Thus, VOC emissions

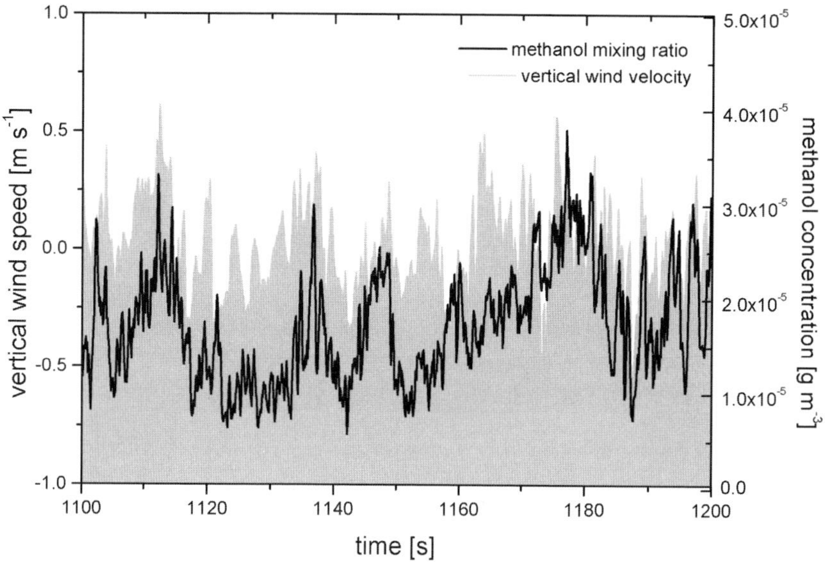

Figure 8. PTR-MS can be used for eddy covariance measurements of VOC fluxes. A fast-response PTR-MS instrument, described in the text, was used to measure the concentration of methanol at a height of 450 cm above ground in a hay field following mowing; the corresponding vertical wind speed at this height was also measured. Data redrawn from Ref. [45].

associated with crop harvesting may have a significant influence on local tropospheric ozone formation.

High levels of wound compounds are also produced by frost damage to plants, as discussed in Section II D. It is likely that ice crystals created by frost within the cells of plants lead to their destruction, thus initiating the enzymatic processes like those shown in Figure 7.[33]

VOC Signatures

Following deployment of a PTR-MS instrument at the Sonnblick Observatory atop of the central ridge of the Austrian Alps (3106 m above sea level), we were able to make long-term measurements (e.g. several months) of ambient VOCs in the free troposphere (but with occasional intrusions of boundary layer air moving over the Alps or of stratospheric air). A unique feature of the experiment is that the instrument was operated remotely, and data was downloaded by phone line. The instrument was placed at this high mountain site to monitor long-range transport of VOCs,

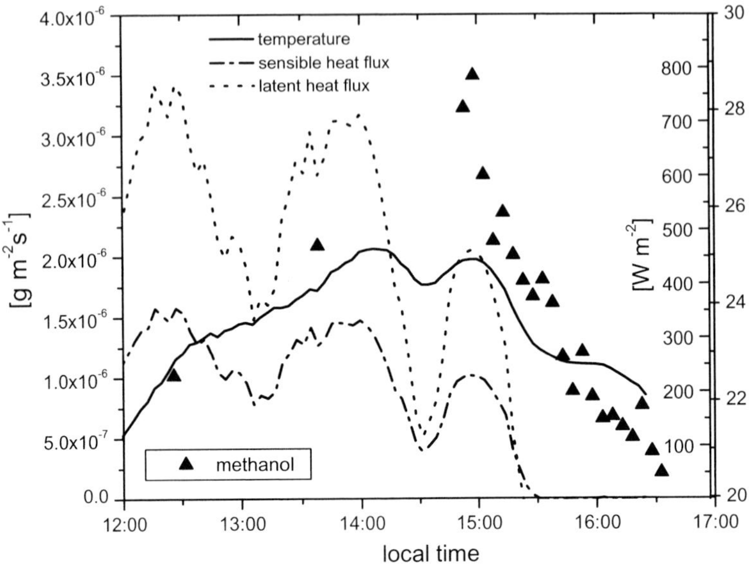

Figure 9. Fluxes of methanol measured with a fast response PTR-MS instrument during hay harvesting at a field site in western Austria. The data shown are for methanol fluxes on the second day after hay cutting, and for periods when the prevailing wind was suitable. Methanol fluxes largely correlated with air temperature, declined sharply in the afternoon, and approached zero as the hay harvest began after 15:45. Also plotted are sensible heat flux (i.e. transfer of heat due to conduction and convection) and latent heat flux (i.e. heat loss due to evaporation of liquid water). Data redrawn from Ref. [45].

primarily those such as aromatic hydrocarbons that are derived from distant urban centers. However, it soon became apparent that at certain periods, high levels of biogenic VOCs were present (discussed in more detail in refs. 32–34). This large PTR-MS data set allowed a test of the origins of these VOCs, using conventional principal component analysis[33] and by the recently proposed variability-lifetime approach.[72] Figure 10 shows the source profiles for 13 VOCs measured at the Sonnblick, which, when analyzed by factor analysis, revealed 3 distinct sources that accounted for more than 90% of the variance in the dataset.[33] Source profile 1 (i.e. anthropogenic signature) contains aromatic compounds originating from fuel combustion processes. The second distinct source profile (i.e. biomass burning signature) contained acetonitrile (CH_3CN) a marker for wood fires and biomass burning.[73] Four groups of compounds, including pentenols, hexanal, ethylvinyl ketone plus pentenals, and hexenals, clustered as a third source (i.e. biogenic signature). Our laboratory experiments demonstrated

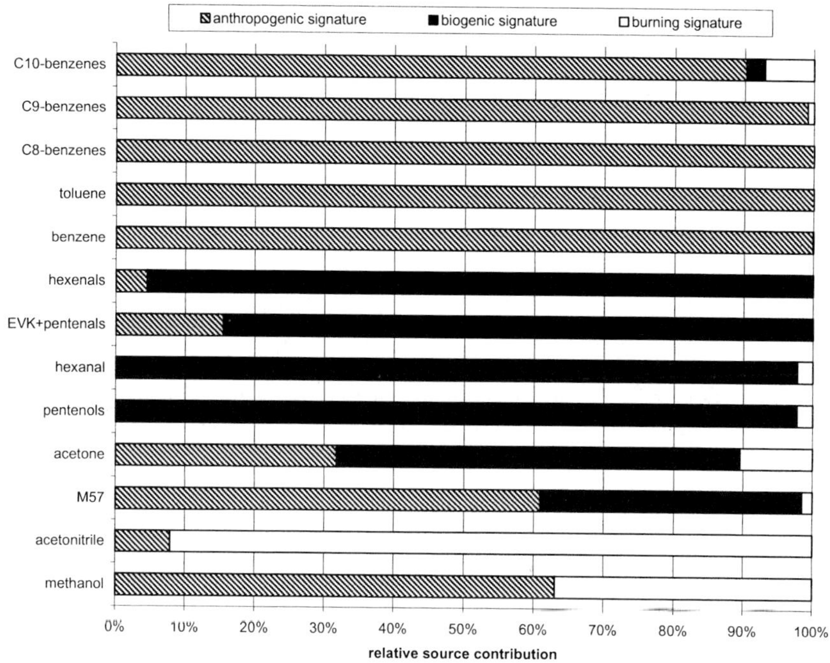

Figure 10. Principal component analysis of signature VOCs observed by PTR-MS monitoring of ambient air at the Sonnblick Observatory from 11 to 16 November 1999. Details are discussed in the text. The component listed as M57 is the PTR-MS signal at m/z 57, which can be derived from both anthropogenic and biogenic VOCs. Data replotted from Ref. [33].

that these latter VOCs are produced from damaged leaves, so we assign the source profile 3 to a similar biogenic origin. Acetone and methanol showed mixed factor loadings which is in agreement with emission from various sources.[74]

The variability-lifetime relationship is based on theoretical considerations (see Ref. [72]) in the form,

$$S_{\ln x} = A\tau^{-b}, \tag{9}$$

where $S_{\ln x}$ is the relative standard deviation of the natural logarithm of mixing ratios, τ is the atmospheric lifetime and A and b are empirical fitting parameters. If a statistically significant dataset is subjected to this treatment, trends between VOCs from similar sources follow relation (9) giving characteristic parameters A and b. The data set for late November showed that biogenic VOCs followed a different trend than VOCs mainly derived from anthropogenic activities, such as aromatic compounds. Using estimated reaction rates with OH radicals for several long-chained aldehydes we could also show that these compounds closely followed the biogenic trend.[75]

Decaying Biomass

Once not only grass and leaves, but also conifer needles, have become dry, this biomass still may remain a strong source of tropospheric VOCs. As in coffee roasting or cooking, non-enzymatic browning reactions (i.e. Maillard reactions) occur between reducing sugars and amino groups in biomass, leading to the formation of a wide variety of VOCs even at temperatures of 30–60°C.[76] Such temperatures are easily reached when sun shines onto leaves lying on the ground or onto dry grasses and grain crops still standing upright. These VOCs stick to the surfaces within the cell structures, but once water comes into contact with this "roasted" biomaterial, VOCs become dissolved in the water and are consequently released to the atmosphere due to the action of Henry's law, in much the same way as aroma compounds in coffee are released only after we add water to the coffee powder. It is in this way that, after a hot summer day when the first drops of a rainstorm fall onto the ground of a meadow or forest, a strong smell is produced. The same occurs when after a hot day dew brings moisture to the biomaterial on the ground during the night. Warneke et al.[76] used a combination of heating/wetting cycles to estimate the total VOC release to the atmosphere. These data obtained from a variety of biomaterial show that the relative emission of acetone and methanol can be at least 10^{-4} and $3–5 \times 10^{-4}$ g/g of decaying dry plant matter, respectively. If these results may be extrapolated, global annual emissions of 6–8 Tg of acetone and 18–40 Tg of methanol would result, adding strongly to the total annual emissions of these compounds to the atmosphere, estimated to be 56 Tg and 122 Tg, respectively.[74] But "roasting" of biomatter by hot air and the sun is not the only biomass-source of tropospheric VOCs.

Biomass Burning

Every year in the tropics, large amounts of biomass in the order of $1.8–4.7 \times 10^{15}$ g C are burning.[77] Bushfires and burning of savannah grass are important sources of local to global air pollution. Besides carbon dioxide, the main product of burning, many chemically and radiatively active gases as well as particulate matter are released to the atmosphere, as summarized by Crutzen and Andreae.[73] In biomass burning experiments,

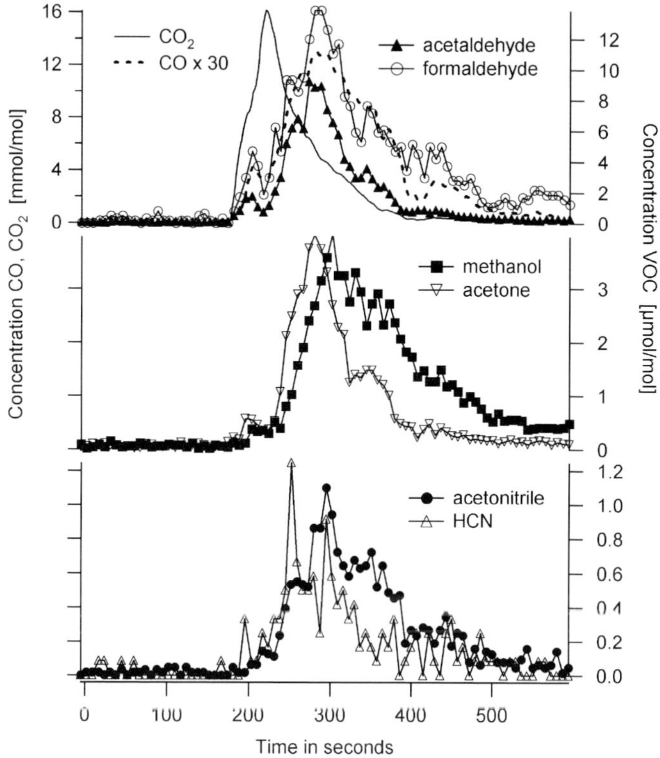

Figure 11. Biomass burning experiments using PTR-MS to monitor emissions of VOCs. The data illustrate a typical biomass burning experiment where VOCs and some known combustion products (HCN, CO and CO_2) were analyzed during the burning of tropical plant matter. Data replotted from Ref. [77].

using PTR-MS data, Holzinger et al.[77] have shown that the emissions of VOCs strongly correlate with the appearance of CO rather than CO_2, as shown in Figure 11. By comparing the ratios of CO versus various VOCs and using estimates on global CO release due to biomass burning, these data allow calculation of the global source strengths from biomass burning (all in $Tg\,y^{-1}$): formaldehyde (5–13), acetaldehyde (3.8–10), methanol (1.5–4), acetone (2.3–6.1) and acetonitrile (0.4–1.0). Online analysis of VOC formation in rapidly changing, burning processes can be easily accomplished by PTR-MS.

Biogenic VOCs also originate due to stresses on plants such as flooding,[78] and due to the action of bacteria and fungi in dead biomatter and in soil. Analysis of these processes represents future additional applications of

PTR-MS. As shown in a recent field experiment during the TexAQS 2000 campaign, PTR-MS methods are also well suited to measurements of anthropogenic sources of VOCs from industrial plants and traffic.[79]

B. Food Research

Monitoring the emissions of VOCs has many potential applications in the food industry, since biological materials that are aged, cooked, treated, etc. often release characteristic "signature" volatiles that can be indicators of the extent of changes in the food product. There has been considerable development of chemical sensors (i.e. "electronic noses") to detect such volatiles.[80] The following examples illustrate how PTR-MS methods can also be used to detect signature volatiles in food research.

Why Do We Put Oil on the Salad?

If the readers of this text ask themselves which part of a head of lettuce they like most, many will state that they prefer the inner, usually somewhat yellow leaves. Why is this the case? We have used PTR-MS to measure the volatiles released when lettuce is eaten, placing the inlet to the PTR-MS in the nose space of human volunteers. In this way, volatiles appearing in nose space air when salad is eaten can be monitored as shown in Figure 12.[81] Here the densities of hexenal compounds ($M81^+$), originating from wounding of salad leaves by processes like those in Figure 7, are compared to acetone ($M59^+$), an endogenous compound in human breath, which is not influenced by eating salad. The data are taken with high enough time resolution so that each breath cycle is resolved, as can be seen from the acetone traces. While chewing an outer (green) leaf, concentrations of hexenal increase much more than while chewing an inner leaf (Figure 12a); this is consistent with the higher production of wound compounds in leaves with more chloroplasts.[65] As wound compounds alter the taste of the leaf, this may explain why the inner leaves are preferred.

In another experiment, a test person chewed green (outer) leaves of endive with and without a film of salad oil; in this, one leaf was split through the center and the test person ate the first half of it without oil and afterwards the second half with a film of oil. As can be seen from Figure 12b, the concentration of emitted hexenal from the leaf chewed without oil is much higher than from the leaf that covered by oil. There are two likely explanations. First, the film of oil partly prevents oxygen from reaching the tissue of the leaf during chewing and therefore only smaller amounts of "wound compounds" are produced in the oxygen-dependent lipoxygenase reaction (Figure 7). Second, the hexenals are quite soluble in oil, and thus do

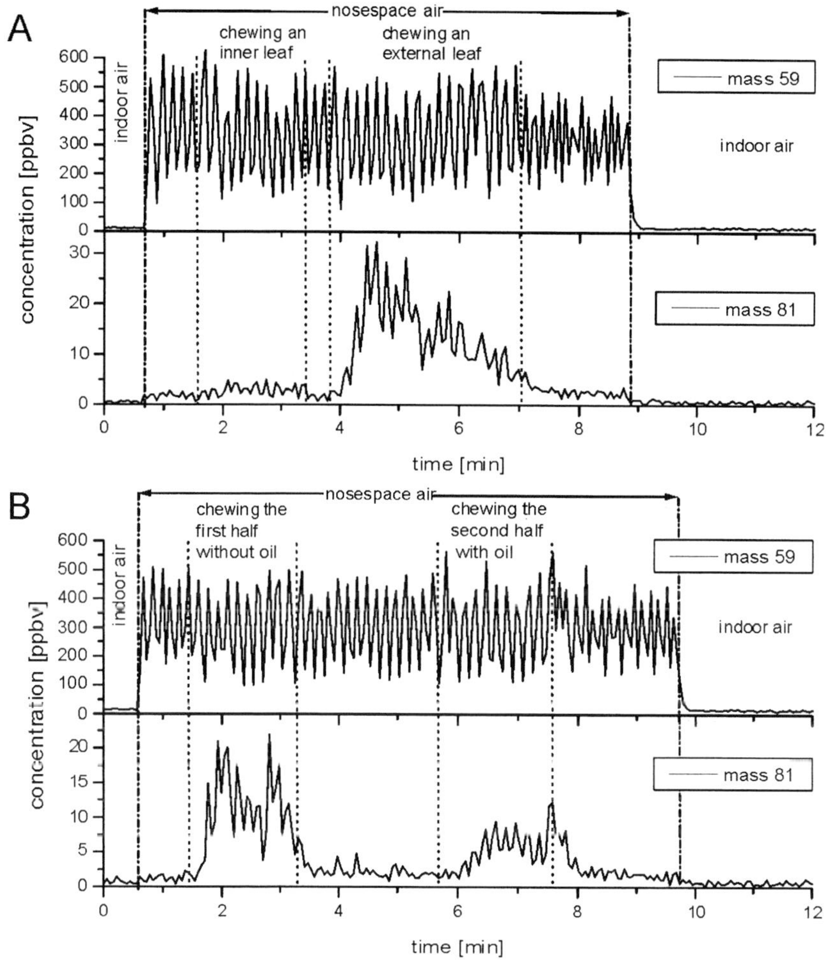

Figure 12. PTR-MS measurements of nose space concentrations of leaf-wound VOCs appearing when endive is eaten. (A) More hexenal ($M81^+$) is released from the external, green leaves of endive. (B) The concentrations of emitted hexenal from a green endive leaf chewed without salad oil are much higher than the concentrations from the leaf that was covered by oil. In each case the concentration of acetone ($M59^+$) in nose space air indicated individual breaths. Data replotted from Ref. [81].

not partition as readily to the nose space air. These observations probably do not resolve why salad oil improves flavor, but they do show how fast-response, on-line PTR-MS methods can be used to monitor dynamic biological processes.

Ripening of Fruit

There is considerable interest in being able to monitor the ripening of fruit, and previously we showed some examples of the use of PTR-MS for this purpose.[12] The following is a recent additional example of this application. It is known that strawberry fruit, depending on the degree of ripening, contains hundreds of volatile components, and the most important aroma-producing compounds have been determined by human "sniffers" who literally smell GC column effluents and assign aromas to strawberry volatiles.[82] The analytical capability of these human noses is impressive, but there is continued interest in developing reliable instruments that can discriminate these aroma compounds and be used to monitor large scale food processing. With this background, we tested samples of strawberries raised in open field and in greenhouse cultivation, each harvested at the same time. During the following two weeks, their emissions of several aroma-related and other compounds were investigated.[83] Every day during the first week, and occasionally later, a test person ate one or two berries in order to check on their taste. Most of the aroma-related compounds (four are shown in Figure 13) reached a maximum after about five days, and at that time the berries also had the best taste. During the following days, methanol, acetaldehyde and ethanol, typical fermentation products, increased strongly and simultaneously the berries visually started to decay. Similarly, raspberries, blueberries, red currants and white currants all showed a significant increase in the emission of methanol, acetaldehyde and ethanol slightly before or after starting to decay. Thus PTR-MS can be used to discriminate optimal flavor formation during fruit ripening and detect the onset of decay processes.

Coffee Roasting

As with fruit ripening, where complex patterns of volatiles are signatures for processes within a tissue, optimizing coffee flavors during roasting is an important commercial process that is more art than science. Together with Chahan Yeretzian from Nestlé R&D, we have been interested in using PTR-MS methods to help in coffee volatile analysis. Roasted coffee contains hundreds of volatile compounds, some of which contribute to the flavor of a fresh cup of coffee. Most of these volatiles are produced during the roasting via non-enzymatic, thermochemical Maillard processes mentioned above.[30] Some of the VOCs that are generated during the roasting can be observed in the off-gas of the roaster by PTR-MS as shown in Figure 14. The data were obtained by sampling the headspace from six Arabica coffee beans roasted at 185°C. The occurrence of coinciding maxima in the concentrations of

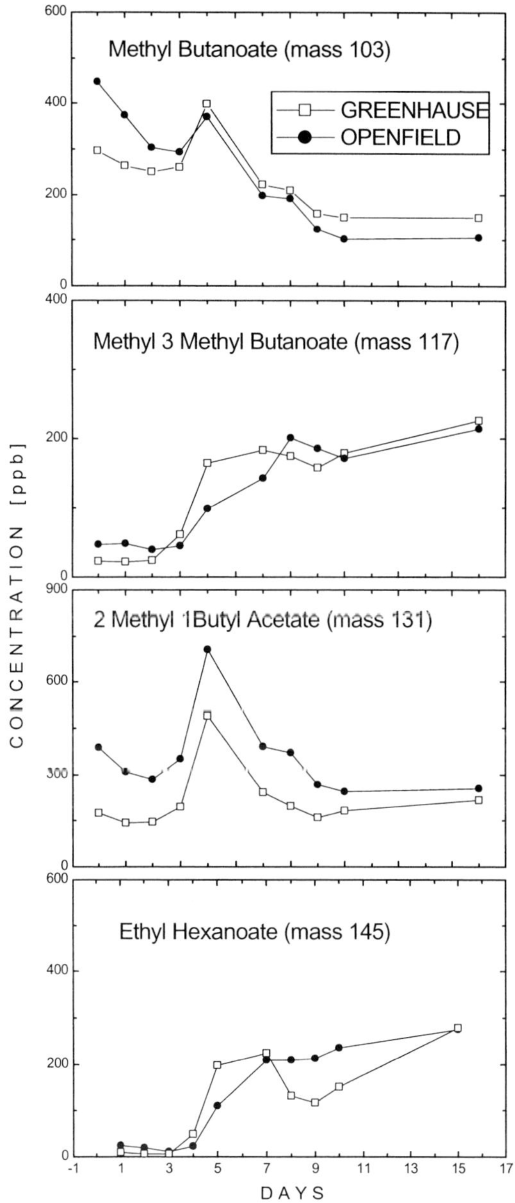

Figure 13. VOCs released during the ripening of strawberries. Strawberries were obtained from an open field or a greenhouse as indicated, and each day during the first week, and occasionally later, a test person ate one or two berries in order to check on their taste, and PTR-MS was used to assess the headspace volatiles. Data replotted from Ref. [83].

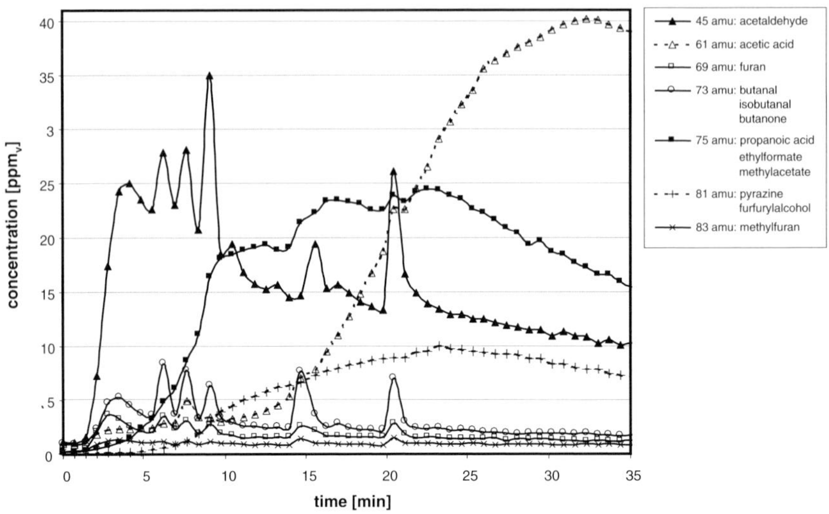

Figure 14. Detection of some volatiles emitted during roasting of 6 coffee beans. Six green Arabica coffee beans were roasted at 185°C, and headspace VOCs were monitored by PTR-MS; the ions monitored and the corresponding VOCs are indicated. Data replotted from Ref. [30].

several compounds at certain times are due to "popping" of single beans. Despite the observed emissions, many of the VOCs created during roasting are polar and remain chemically or physically associated with the polar cell material of the beans. Even if one grinds roasted coffee beans, the release of VOCs from dry coffee powder is relatively low. However, after water is added to the coffee powder, strong emission of aroma compounds occurs within a few seconds. The highly polar water molecules replace the VOC molecules attached to the cell material and the VOCs are either directly released into the gas phase or dissolve in the water from which they are released according to their liquid-gas partition coefficients. The resulting coffee brew is a mixture of dilute solutions of VOCs in water, and if we keep the head space above the brew enclosed, the partial pressures in the gas phase in contact with the liquid tend towards values that are governed by Henry's Law. By measuring the concentrations in the head space and using known values of Henry's Law Constants (HLCs), the amounts of individual VOCs dissolved in the brew and thus the amounts present in the coffee powder can be calculated. Figure 15 presents measured head space concentrations of coffee volatiles as dependent on the roasting time, showing that under the given roasting conditions most of the aroma compounds reach a maximum for a medium roasted coffee after about

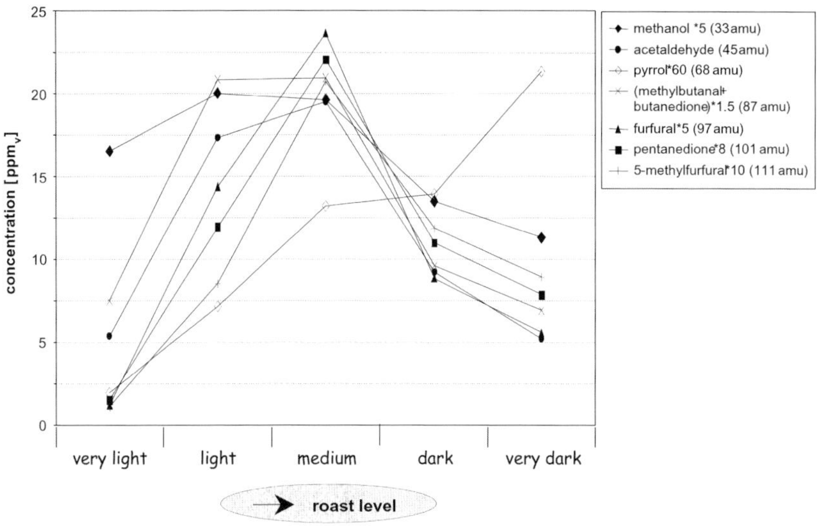

Figure 15. Measured head space concentrations of coffee VOCs are dependent on roasting time. Protonated masses detected by PTR-MS are indicated in parentheses. Data replotted from Ref. [30].

15 min ("medium" roast), and are depleted and/or destroyed if roasting continues.[30] Pyrrole (C_4H_5N), however, which is one of the compounds giving the coffee a bitter taste, accumulates with increasing roasting time.

Measurements of this kind provide important information towards elucidating the kinetics of formation and release of coffee flavor compounds during roasting, and their dependence on process parameters. Ultimately, such data can be used to optimize roasting conditions with respect to aroma intensity and composition of the roasted coffee.

Quality Control of Meat

There is increasing concern over the presence of pathogenic bacteria in food products such as meat.[84] PTR-MS provides a simple and fast working tool for monitoring changes in volatiles released from meat that might be associated with bacterially-induced spoilage. Figure 16 (left) shows the emission of various compounds from a beef sample of meat purchased from a local supermarket, and then kept at room temperature for about 40 h. The data show large increases of methanethiol (CH_3SH), dimethylsulfide (CH_3SCH_3) and dimethyl disulfide (CH_3SSCH_3) from about 10–30 ppbv at 22 h to more than 1000 ppbv at 39 h. It is common practice for butchers

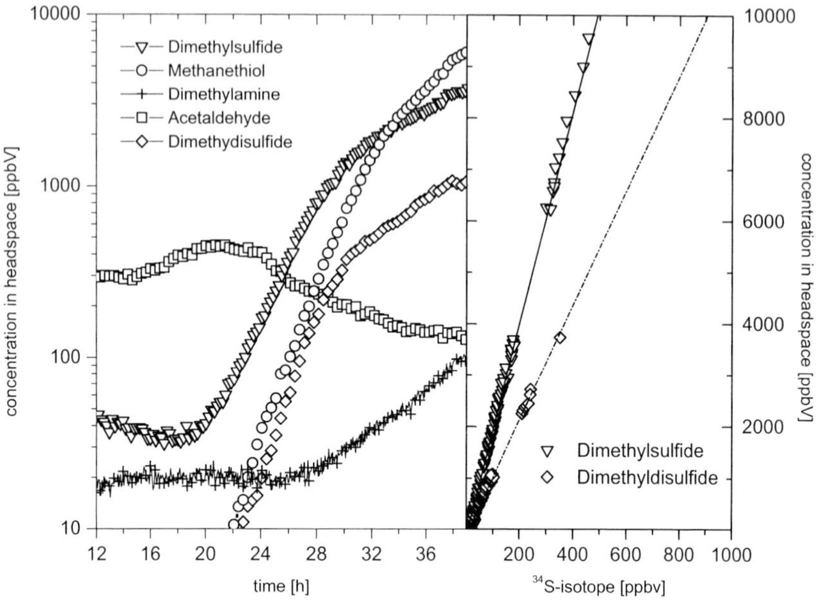

Figure 16. (*Left*) The emissions of various compounds from a sample of beef purchased and kept at room temperature for about 40 h were monitored by headspace PTR-MS. (*Right*) The appearance of sulfur gases around 22 h (left panel) can be confirmed by the abundance of ^{34}S isotopes measured at the $(M+2)^+$ masses.

and consumers to sniff meat for the presence of "off" odors, which can include thiols that arise from sulfur amino acid breakdown.[85] It is obvious that the head space concentration of the sulfur compounds in Figure 16 are a clear indicator of the onset of degradation of meat. PTR-MS tests are especially useful as they can be performed within a few minutes, while results of bacteriological tests are available only after several days. Figure 16 also illustrates the ability of PTR-MS to verify the presence of S-containing VOCs using natural ^{34}S isotopic abundance (4.1%). The large release of VOCs identified as dimethylsulfide ($M63^+$) and dimethyl disulfide ($M95^+$), is paralleled by increases in $M65^+$ and $M97^+$, respectively (Figure 16, right). These latter ion abundances are close to that expected for ^{34}S abundance in compounds with 1 S atom (4.46%) or 2 S atoms (8.92%).

C. Medical Applications

Exhaled human breath contains not only the natural constituents of air, but also a variety of endogenous VOCs,[86] such as acetone, methanol, and isoprene. Acetone is normally present in concentrations of ~ 1 ppmv, the others have concentrations of typically one hundred to a few hundred ppbv, and there are other less abundant compounds at concentrations of a few ppbv. Many VOCs are produced within the human body in metabolic processes. If these processes are influenced by intake of unusual amounts of specific kinds of food or chemicals, but also by illness, VOCs in the body and thus in the breath can show concentrations deviating significantly from the "normal" values. Previously reviewed PTR-MS experiments[12] provided detailed analysis of such human breath VOCs, including examples like the conversion of ingested isopropanol into acetone, formation of "garlic breath" volatiles, isoprene levels in adults and humans, and production of methanol after fruit consumption. In the following, we will show recent PTR-MS data on two medically-related investigations, (1) quantification of passive smoking, and (2) the relation between endogenous isoprene and cholesterol levels in the human body, both of which are of wide spread interest. In each case, the use of PTR-MS instruments allow sensitive, non-invasive analysis of human breath samples.

Quantification of Passive Smoking

Prazeller et al.[87] have shown that the relative abundance of acetonitrile and other nitriles in exhaled breath might provide good markers for smoking and passive smoking. Nearly all the VOCs contained in tobacco smoke are removed from the body quite rapidly, presumably via enzymatic reactions and excretion. Figure 17 shows how the concentrations of acetonitrile (CH_3CN) and acrylonitrile ($CH_2{=}CHCN$) increase rapidly in the breath of a test person smoking three cigarettes. Following cigarette smoking, breath concentrations of acrylonitrile decline rapidly to initial levels, but acetonitrile is removed very slowly from the body, so it will accumulate in proportion to the amount of this compound inhaled by a test person. For this reason acetonitrile can be used for quantification of passive smoking.

In several experiments, test persons were put in a room where other persons smoked cigarettes causing air contaminations similar to that found in public bars where heavy smoking occurs. Breath acetonitrile concentrations were measured before, during and after the experiment, and the concentration of acetonitrile in the room was monitored. Figure 18 shows typical results obtained with two test persons. Also shown in the figure is the increase of the breath acetonitrile concentration calculated by using the measured acetonitrile concentration in the room, assuming a breath rate of 5 and 7.5 l/min (typical for persons in a relaxed state) and using Henry's Law to obtain the increase of the acetonitrile concentration in the blood under

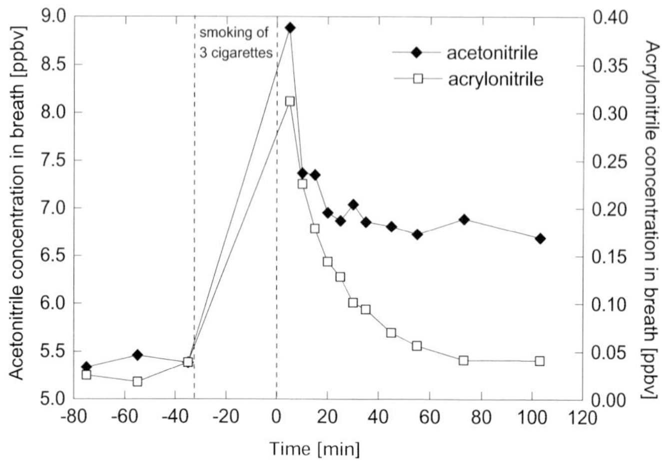

Figure 17. The concentration of acrylonitrile and acetonitrile in the breath of a test person increases due to smoking of cigarettes, as determined by PTR-MS. Breath acrylonitrile and acetonitrile concentrations were measured before, during and after smoking three cigarettes; other details are described in the text. Data replotted from Ref. [87].

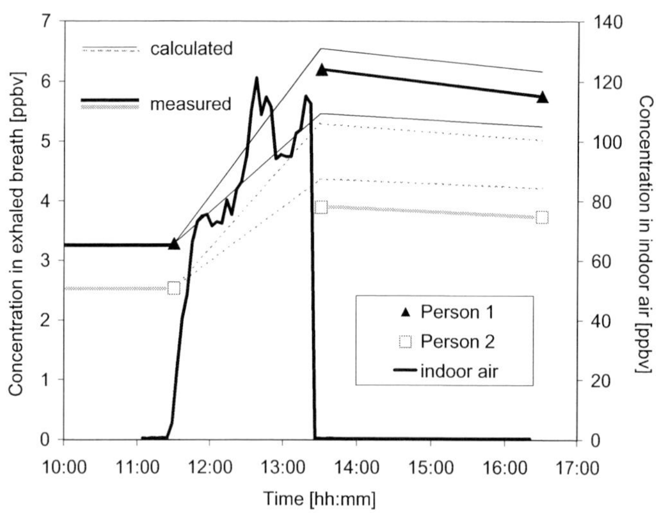

Figure 18. Breath acetonitrile concentrations before and after exposure to second hand smoke. PTR-MS was used to measure breath acetonitrile concentrations in two individuals before and after exposure to second hand cigarette smoke in a room where acetonitrile concentrations in air were also measured. The calculated levels of breath acetonitrile (see text) are also shown. Data replotted from Ref. [87].

the pretext that all the inhaled acetonitrile is reabsorbed. The measured value for the first test person lies ideally between the calculated values for the two breath rates. In the second case, the measured value is on the low side, but the test person was sitting with little movement at a computer during the experiment, thus the value still can be regarded as consistent. From nine passive smoking experiments like this and 40 measurements during direct smoking, we conclude that someone staying for a whole working day in a typically smoke contaminated room is passively inhaling the equivalent of about 1–4 cigarettes.

These types of experiments demonstrate the ability of PTR-MS methods to analyze the exposure of individuals to VOCs in the environment. A similar approach has recently been used to analyze exposure of health care workers to anesthetic gases in hospitals.[88]

Physiology of Breath Isoprene and Its Relation to Blood Cholesterol

An interesting application of human breath analysis by PTR-MS is the exploration of the physiology of isoprene. The origin of human isoprene is probably related to the isoprenoid biosynthetic pathway, but whether its formation is enzymatic or non-enzymatic is still uncertain.[89] Pronounced diurnal changes in isoprene concentration in breath were reported by De Master and Nagasawa[90] peaking between the hours of 02:00 and 07:00 a.m. to a level nearly four times greater than their daytime levels. Cailleux and Allain[91] showed that this diurnal variation is associated with the state of sleep and wakefulness rather than an intrinsic circadian rhythm. Earlier, we had also investigated this phenomenon by PTR-MS,[92] and in agreement with previous results, we found an increase by a factor of 2–4 in isoprene during the night for the adult participants in our study. However, as described below, recent analysis has revealed that this increase is not due to a true diurnal variation in isoprene formation.

Insights into this problem were obtained from breath analysis of individuals undergoing exercise on a stationary bicycle. As shown in Figure 19, before the start of physical exercise the test person had a breath isoprene concentration of about 75 ppbv. A few minutes after the start of the exercise, the isoprene concentration rose rapidly to a maximum of ~ 275 ppbv, and a few minutes later it had dropped to ~ 50 ppbv. When the exercise was decreased and then stopped, the breath isoprene concentration rose to values similar to those before the exercise had begun. At first sight, one might be tempted to interpret the variation in breath isoprene concentration as indicative of the variation of endogenous isoprene production in the body. However, using a simple two compartment model (details in Ref. 89) we showed that this is not the case. Briefly, our

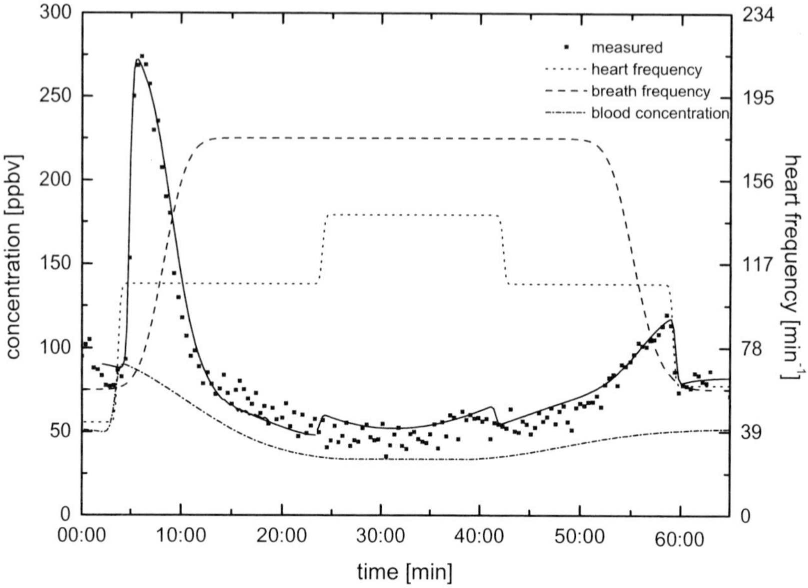

Figure 19. Isoprene concentrations in human breath during exercise on a stationary bicycle. Breath isoprene was measured by PTR-MS, and the heart beat rate was varied by changing the exercise rate in steps; breathing frequency was also monitored. The exercise started at 2.5 min and ended at 60 min. A curve (solid line) derived from a two-compartment isoprene model (see the text) shows an excellent fit to the measured breath isoprene data. Replotted from data presented in Ref. [89].

explanation is the following. In contrast to most VOCs present in human breath, isoprene has a very low solubility, i.e. it has a small Henry's law constant, so that, when isoprene produced in the body is transported via the blood stream to the lungs, it evaporates quite efficiently. The actual concentration of the isoprene in the breath is governed by the production term (which we assume is constant), by the velocity of the blood stream pumped through the lungs (which is proportional to the heart beat frequency), and by the breathing rate. As seen in Figure 19, as soon as exercise starts the heart beat rate increases within seconds, with a corresponding increase in isoprene in breath. Then, the enhanced rate of evaporation leads to a decline in the blood isoprene concentration and thus of the evaporation rate, reaching a steady state after about 10–15 min. Breath isoprene returns to normal as the exercise is ended, and breath rate and heart beat rate again reach normal values. Similarly, the observed diurnal variations of the breath isoprene concentrations mentioned above can be explained on the basis of a (nearly) constant endogenous isoprene

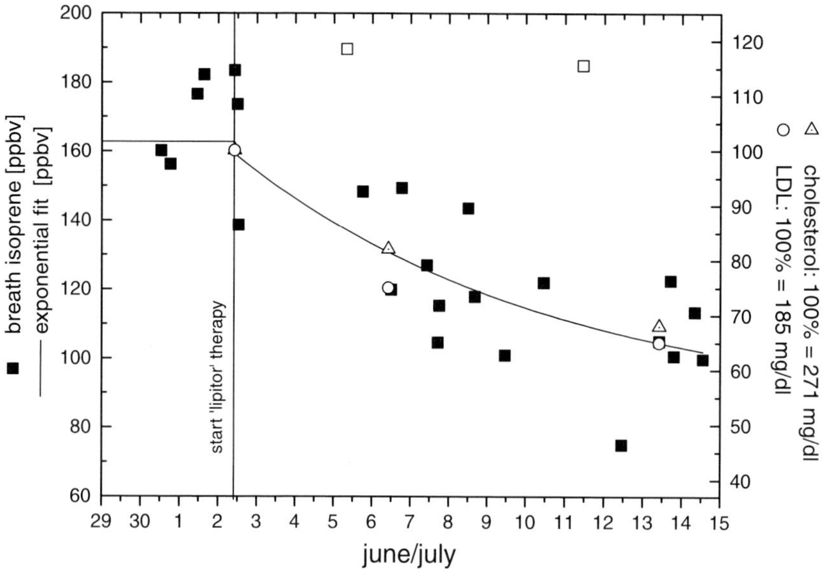

Figure 20. A decline in serum cholesterol and low density lipoprotein (LDL) levels induced by the cholesterol-lowering drug, Lipitor, is paralleled by a decline in breath isoprene concentrations. Breath isoprene was measured by PTR-MS under carefully controlled conditions as described in the text. Replotted from data presented in Ref. [89], where two atypical samples (□) are discussed.

production, but note that a rapid increase in breath isoprene is induced by the increase in heart beat caused by awakening an individual suddenly (for breath sampling).

These data clearly demonstrate that whenever the breath isoprene concentration or other breath volatiles are to be used for diagnostic purposes, these concentrations have to be measured under well-defined conditions. An example of this was used in checking the average breath isoprene concentrations of a test person undergoing medical treatment for lowering of blood cholesterol level by intake of the drug, Lipitor, over a period of about two weeks.[89] Figure 20 shows that a decline in cholesterol levels by about 35% over this period of time is paralleled by a decline of the breath isoprene concentration of about the same relative amount. These data infer that measurements of breath isoprene could be used for a screening of the population in search for those who have elevated cholesterol levels and therefore should undergo medical treatment.

It is quite stunning that up to now VOC analysis of the human breath has hardly been used for medical diagnostic purposes, but from the data mentioned above, medical applications of breath analysis will be very rewarding in the future.

V. CONCLUSIONS

As a technique, the use of proton transfer reactions together with mass spectrometry (PTR-MS), emerged from the development of instruments to study ion molecule reactions (IMR), and has now become a very versatile analytical tool for measuring VOCs. Its development has relied on an understanding of IMR and particularly proton transfer from H_3O^+ to the trace VOC species to be detected. As reviewed here, the applicability of PTR-MS for on-line measurements of trace constituents has been demonstrated by examples in the fields of environmental research, food processing, and medicine. The high sensitivity of the system that now has been reached, allows for continuous monitoring of ambient VOCs not only in urban air, but also in clean air in remote environments. Fluxes of VOCs can now be measured by direct eddy covariance with fast-response PTR-MS designs or by disjunct sampling strategies (i.e. intermittent sample accumulation and analysis) with slower-response instruments. In many cases, detected VOCs can be verified by GC-PTR-MS methods. The on-line capability of PTR-MS, and its ability to detect signature VOCs arising from food products, make it an especially promising analytical tool for food process monitoring. In addition, food spoilage by pathogenic bacteria might be detectable with these instruments. Further exploitation of the method in medicine might include non-invasive medical diagnostics, investigations of metabolic processes and drug detection, as well as monitoring of harmful VOC emissions in and from industrial plants. Other promising applications may also involve the monitoring of catalytic processes and of materials production. It is also noteworthy that PTR-MS instruments have been deployed in a variety of settings, including university and hospital laboratories, on ships, aircraft, mountain tops and other field sites, and it is likely that the next decade will see continued applications of the method in many locations. Such diverse uses are sure to bring new impetus to increasing the sensitivity and selectivity of PTR-MS instruments, and one can anticipate interesting developments in this field.

ACKNOWLEDGMENTS (FROM W.L.)

Only due to the intense cooperation with many colleagues in different fields of research was it possible to obtain the large body of information mentioned in this chapter. The development of PTR-MS could not have been done without the possibility to learn about the use of swarm type experiments during my post-doctoral years from 1973–1975 at NOAA Boulder with Eldon Ferguson, Dan Albritton, Fred Fehsenfeld, Art Schmeltekopf and Carl Howard. Special thanks go to Paul Crutzen, MPI Mainz, for his continuous encouraging support over the past years paving our way to tropospheric VOC monitoring. Ray Fall, University of Colorado, Boulder, made it possible to perform many breathtaking investigations on biogenic VOCs and Alex Guenther, NCAR, Boulder, gave us insight into the techniques of VOC flux measurements. Especially gratifying is the recent revival of the cooperation with Fred Fehsenfeld, which has led to joint projects on tropospheric VOCs. Research on coffee roasting, done jointly with Chahan Yeretzian at Nestlé (Lausanne), gave valuable insight in processes related to VOCs in environmental as well as food related areas. Herwig Paretzke, GSF (Munich), and Guenther Seufert, JRC (Ispra), provided valuable support and ideas for investigations of biogenic sources of VOCs. This chapter is also dedicated to Patrizia Jilg. Her book on "Planetenkräuter" is a valuable source of information on properties of plants related to research described in this chapter. Many more colleagues, not mentioned by name, have provided ideas, financial and psychological support which always have been highly appreciated.

Most of the actual developmental work on PTR-MS was done together with my coworkers Armin Hansel and Alfons Jordan. Based on the technical skills of Alfons Jordan, PTR-MS has become a reliable scientific tool working uninterruptedly for months in the lab and field. Special thanks go to my (former) students and coworkers Johann Taucher, Peter Prazeller, Rupert Holzinger, Helmut Judmair, Carsten Warneke, Armin Wisthaler, Martin Graus, Dagmar Mayr, Elena Boscaini and Thomas Karl for their enthusiastic work using PTR-MS and to my colleague Tilmann Märk as well as to our secretary Monika Heigl for their continuous support. Finally I want to thank my son Christian who is consistently improving the software for PTR-MS as well as the remote control of PTR-MS instruments.

Financial support for this work from the "Fonds zur Förderung der wissenschaftlichen Forschung" under Project P 14130 is appreciated. R.F. was also supported by National Science Foundation grant ATM-9805191.

REFERENCES

[1] Sturges, W. T.; Wallington, T. J.; Hurley, M. D.; Shine, K. P.; Sihra, K.; Engel, A.; Oram, D. E.; Penkett, S. A.; Mulvaney, R.; Brenninkmeijer, C. A. M. *Science* **2000**, *289*, 611.

[2] Stuart Penkett. University of East Anglia, 2000, private communication.

[3] Munson, M. S. B.; Field, F. H. *J. Am. Chem. Soc.* **1966**, *88*, 2621.

[4] Harrison, A. G. *Chemical Ionization Mass Spectrometry, 2nd ed.*; CRC Press: Boca Raton, FL, 1992.

[5] Bruins, A. P. *Mass Spectrom. Rev.* **1991**, *10*, 53.

[6] Taylor, A. J.; Linforth, R. S. T.;Harvey, B. A.; Blake, A. *Food Chemistry* **2000**, *71*, 327.

[7] Roberts, D. D.; Taylor, A., Eds.; *Flavor Release*; ACS Symposium Series; American Chemical Society, Washington, DC, 2000.

[8] Mc Farland, M.; Albritton, D. L.; Fehsenfeld, F. C.; Ferguson, E. E.; Schmeltekopf, A. L. *J. Chem. Phys.* **1973**, *59*, 6620.

[9] Ferguson, E. E.; Fehsenfeld, F. C.; Schmeltekopf, A. L. In *Advances in Atomic and Molecular Physics, Vol. 5*, Bates, D. R.; Estermann, I., Eds.; Academic Press, New York, 1969, p.1.
[10] Fehsenfeld, F. C.; Ferguson, E. E.; Schmeltekopf, A. L. *J. Chem. Phys.* **1966**, *44*, 3022.
[11] Adams, N. G.; Smith, D. *Int. J. Mass Spectrom. Ion Phys.* **1976**, *21*, 349.
[12] Lindinger, W.; Hansel, A.; Jordan, A. *Int. J. Mass Spectrom. Ion Processes* **1998**, *173*, 191.
[13] Lindinger, W.; Hansel, A.; Jordan, A. *Chem. Soc. Rev.* **1998**, *27*, 347.
[14] Praxmarer, C.; Hansel, A.; Lindinger, W. *J. Chem. Phys.* **1994**, *100*, 8884.
[15] Praxmarer, C.; Hansel, A.; Lindinger, W. *Int. J. Mass Spectrom. Ion Processes* **1996**, *156*, 189.
[16] Gioumousis, G.; Stevenson, D. P. *J. Chem. Phys.* **1958**, *29*, 294.
[17] Su, T.; Bowers, M. T. In *Gas Phase Ion Chemistry, Vol. 1*; Bowers, M. T., Ed.; Academic Press, New York, 1979, p. 84.
[18] Troe, J. *Chem. Phys. Lett.* **1985**, *122*, 425.
[19] Clary, D. C. *Molecular Physics* **1984**, *53*, 3.
[20] Su, T.; Chesnavich, W. J. *J. Chem. Phys.* **1982**, *76*, 5183.
[21] Spanel, P.; Smith, D. *Int. J. Mass Spectrom. Ion Processes* **1997**, *167*, 375.
[22] Ferguson, E. E. *J. Am. Soc. Mass Spectrom.* **1992**, *3*, 479.
[23] Lindinger, W. In *Gaseous Ion Chemistry and Mass Spectrometry*; Futrell, J. H., Ed.; John Wiley & Sons: New York, 1986, p. 141.
[24] Lindinger, W. In *Gaseous Ion Chemistry and Mass Spectrometry*; Futrell, J. H., Ed.; John Wiley & Sons: New York, 1986, p. 237.
[25] Lindinger, W.; Smith, D. In *Reactions of Small Transient Species*, Chapter 7; Fontijn, A.; Clyne, M. A. A., Eds.; Academic Press: London, 1983, p. 387.
[26] Hansel, A.; Jordan, A.; Holzinger, R.; Paretzke, P.; Vogel , W.; Lindinger, W. *Int. J. Mass Spectrom. Ion Processes* **1995**, *149/150*, 605.
[27] Ikezoe, Y.; Matsuoka, S.; Viggiano, A. *Gas Phase Ion-Molecule Reaction Rate Constants through 1986*; Maruzen Company, Ltd., Tokyo, 1987.
[28] Ellis, H.; Pai, R.; McDaniel, E.; Mason, E.; Viehland, L. A. *Atom. Data Nucl. Data Tables* **1976**, *17*, 77.
[29] Praxmarer, C.; Hansel, A.; Jordan, A.; Kraus, H.; Lindinger, W. *Int. J. Mass Spectrom. Ion Processes* **1993**, *129*, 121.
[30] Yeretzian, C.; Jordan, A.; Brevard, H.; Lindinger, W. In *ACS Symposium Series 763: Flavor Release;* Roberts, D. D.; Taylor, A. J., Eds.; American Chemical Society: Washington, DC, 2000; p. 112.
[31] Hübschmann, H-J. *Handbook of GC/MS. Fundamentals and Applications*; Wiley-VCH: Berlin, 2001.
[32] Karl, T.; Fall, R.; Crutzen, P. J.; Jordan, A.; Lindinger, W. *Geophys. Res. Lett.* **2001**, *28*, 507.
[33] Fall, R.; Karl, T.; Jordan, A.; Lindinger, W. *Atmos. Environ.* **2001**, *35*, 3905.
[34] Karl, T. G. Ph.D thesis, University of Innsbruck, 2000.
[35] Joost de Gouw. Utrecht University, private communication, 2001.
[36] Lindinger, W. *Phys. Rev. A* **1973**, *7*, 238.
[37] Howorka, F.; Lindinger, W.; Pahl, M. *Int. J. Mass Spectrom. Ion Phys.* **1973**, *12*, 67.
[38] Lindinger, W.; Alge, E.; Störi, H.; Varney, R. N.; Helm, H.; Holzmann, P.; Pahl, M. *Int. J. Mass Spectrom. Ion Phys.* **1979**, *30*, 251.
[39] Störi, H.; Alge, E.; Villinger, H.; Egger, F.; Lindinger, W. *Int. J. Mass Spectrom. Ion Phys.* **1979**, *30*, 263.
[40] Lindinger, W.; Leiter, K.; Andriollo, M. *Chemie-Technik*, **1991**, *7*, 1.
[41] Lindinger, W.; Hirber, J.; Paretzke, H. *Int. J. Mass Spectrom. Ion Phys.* **1993**, *129*, 79.

[42] Villinger, H.; Federer, W. V+F Analysetechnik, A-6067 Absam, Austria (www.vandf .com).

[43] Lane, D. A. *Environ. Sci. Technol.* **1982**, *16*, 38A.

[44] Lagg, A.; Taucher, J.; Hansel, A.; Lindinger, W. *Int. J. Mass Spectrom. Ion Processes* **1994**, *134*, 55.

[45] Karl, T. G.; Guenther, A.; Lindinger, C.; Jordan, A.; Fall, R.; Lindinger, W. *J. Geophys. Res.* **2001**, in press.

[46] Atkinson, R. *Atmos. Environ.* **2000**, *34*, 2063.

[47] Singh, H. B.; Kanakidou, M.; Crutzen, P. J.; Jacob, D. J. *Nature* **1995**, *378*, 50.

[48] Singh, H. B. *Environ,. Sci. Technol.* **1987**, *21*, 320.

[49] Pöschl, U.; Williams, J.; Hoor, P.; Fischer, H.; Crutzen, P. J.; Warneke, C.; Holzinger, R.; Hansel, A.; Jordan, A.; Lindinger, W.; Scheeren, H. A.; Peters, W.; Lelieveld, J. J. *Atmos. Chem.* **2001**, *38*, 115.

[50] Williams, J.; Pöschl, U.; Crutzen, P. J.; Hansel, A.; Holzinger, R.; Warneke, C.; Lindinger, W.; Lelieveld, J. J. *Atmos. Chem.* **2001**, *38*, 133.

[51] Warneke, C.; Holzinger, R.; Hansel, A.; Jordan, A.; Lindinger, W.; Williams, J.; Pöschl, U.; Hoor, P.; Fischer, H.; Crutzen, P.J.; Scheeren, B.; Lelieveld, J. *J. Atmos. Chem.* **2001**, *38*, 167.

[52] Saunders, S. M.; Jenkin, M. E.; Derwent, R. G.; Pilling, M. J. *Atmos. Environ.* **1997**, *31*, 1249.

[53] Wotawa, G.; Kröger, H.; Stohl, A. *Atmos. Environ.* **2000**, *34*, 1367.

[54] Hansel, A.; Wisthaler, A. *Geophys. Res. Lett.* **2000**, *27*, 895.

[55] Roberts, J. M. et al. (18 co-authors) *J. Geophys. Res.* **1998**, *103*, 22, 473.

[56] Armin Hansel. University of Innsbruck, private communication, 2000.

[57] Fehsenfeld, F. et al., *Global Biogeochem. Cycles* **1992**, *6*, 389.

[58] Helas, G.; Slanina , J.; Steinbrecher, R., Eds. *Biogenic Volatile Organic Compounds in the Atmosphere*, SPB Academic Publishing, Amsterdam, 1997.

[59] Fall, R. In *Reactive hydrocarbons in the Atmosphere;* Hewitt, C. N., Ed.; Academic Press: San Diego, California, 1999, p. 43.

[60] Rasmussen, R. A. *Progress Report*, Gas Chromatography Laboratory, Monsanto, Co., St. Louis, Missouri, 1961.

[61] Sanadze, G. A.; Dolidze, G. M. *Soobshch. Akad. Nauk Gruz. SSR* **1961**, *27*, 747.

[62] Karl, T.; Prazeller, P.; Seufert, G. unpublished data, 2000.

[63] Delwiche, C. F.; Sharkey, T. D. *Plant Cell Environ.* **1993**, *16*, 587.

[64] Gardner, H. W. *Biochim. Biophys. Acta* **1991**, *1084*, 221.

[65] Hatanaka, A. *Phytochemistry* **1993**, *34*, 1201.

[66] Fall, R.; Karl, T.; Hansel, A.; Jordan, A.; Lindinger, W. *J. Geophys. Res.* **1999**, *104*, 15,963.

[67] Helmig, D.; Greenberg, J.; Guenther, A.; Zimmerman, P.; Geron, G. *J. Geophys. Res.* **1998**, *103*, 22,397.

[68] Helmig, D.; Klinger, L. F.; Guenther, A.; Vierling, L.; Geron, C.; Zimmerman, P. *Chemosphere*, **1999**, *38*, 2163.

[69] Karl, T.; Fall, R.; Jordan, A.; Lindinger, W. *Environ. Sci. Technol.* **2001**, *35*, 2926.

[70] Karl, T.; Guenther, A.; Jordan, A.; Fall, R.; Lindinger, W. *Atmos. Environ.* **2001**, *35*, 491.

[71] Harley, R.; Hannigan, M.; Cass, G.; *Environ. Sci. Technol.* **1992**, *26*, 2395.

[72] Jobson, B. T.; McKeen, S. A.; Parrish, D. D.; Fehsenfeld, F. C.; Blake, D. R.; Goldstein, A. H.; Schauffler, S. M.; Elkins, J. W. *J. Geophys. Res.* **1999**, *104*, 16,090.

[73] Crutzen, P. J.; Andreae, M. O. *Science* **1990**, *250*, 1669.

[74] Singh, H.; Chen, Y.; Tabazadeh, A.; Fukui, Y.; Bey, I.; Yantosca, R.; Jacob, D.; Arnold, F.; Wohlfrom, K.; Atlas, E.; Flocke, F.; Blake, D.; Blake, N.; Heikes, B.; Snow, J.; Talbot, R.; Gregory, G.; Sachse, G.; Vay, S.; Kondo, Y. *J. Geophys. Res.* **2000**, *105*, 3795.

[75] Karl, T. G.; Crutzen, P. J.; Mandl, M.; Staudinger, M.; Guenther, A.; Jordan, A.; Fall, R.; Lindinger, W. *Atmos. Environ.* **2001**, in press.
[76] Warneke, C.; Karl, T.; Judmaier, H.; Hansel, A.; Jordan, A.; Lindinger, W.; Crutzen, P. J. *Global Biogeochem. Cycles* **1999**, *13*, 9.
[77] Holzinger, R.; Warneke, C.; Hansel, A.; Jordan, A.; Lindinger, W.; Scharffe, D, H.; Schade, G.; Crutzen, P. J. *Geophys. Res. Lett.* **1999**, *26*, 1161.
[78] Holzinger, R.; Sandoval-Soto, L.; Rottenberger, S.; Crutzen, P.J.; Kesselmeier, J. *J. Geophys. Res.* **2000**, *105*, 20579.
[79] Thomas Karl, National Center for Atmospheric Research, unpublished data, 2001.
[80] Gardner, J. W.; Bartlett, P. N. *Electronic Noses. Principles and Applications*; Oxford University Press: Oxford, 1999.
[81] Mayr, D.; Graus, M.; Karl, T.; Jordan, A.; Prazeller, P.; Lindinger, W. *Ber. Nat.-med. Verein. Innsbruck* **2000**, *87*, 7.
[82] Gomes da Silva, M. D. R.; Chavel das Neves, H. J. *J. Agric. Food Chem.* **1999**, *47*, 4568.
[83] Boscaini, E.; Boschetti, A.; Biasioli, F.; Gallerani, G.; Gasperi, F.; Jordan, A.; Lindinger, W.; Iannotta, S. *Symposium on Atomic and Surface Physics and Related Topics*, Trento, Italy, Jan. 30–Feb. 5, 2000 (www.science.unitn.it/sasp/index.html).
[84] Hui, Y. H.; Pierson, M. D.; Gorham, J. R., Eds.; *Foodborne Disease Handbook, Volume 1: Bacterial Pathogen*; Marcel Dekker: New York, 2001.
[85] Mussinan, C. J.; Keelan, M. E., Eds.; *Sulfur Compounds in Foods*; American Chemical Society: Washington, D.C., 1994.
[86] Fenske, J. D.; Paulson, S. E. *J. Air Waste Manag. Assoc.* **1999**, *49*, 594.
[87] Prazeller, P.; Karl, T.; Jordan, A.; Holzinger, R.; Hansel, A.; Lindinger, W. *Int. J. Mass Spectrom.* **1998**, *178*, L1.
[88] Rieder, J.; Prazeller, P.; Boehler, M.; Lirk, P.; Lindinger, W.; Amann, A. *Anesth. Analg.* **2001**, *92*, 389.
[89] Karl, T.; Prazeller, P.; Mayr, D.; Jordan, A.; Rieder, J.; Fall, R.; Lindinger, W. *J. Appl. Physiol.* **2001**, *91*, 762.
[90] DeMaster, E. G.; Nagasawa, H. T. *Life Sci.* **1978**, *22*, 91.
[91] Cailleux, A.; Allain, P. *Life Sci.* **1989**, *44*, 1877.
[92] Taucher, J.; Hansel, A.; Jordan, A.; Fall, R.; Futrell, J. H.; Lindinger, W. *Rapid Commun. Mass Spectrom.* **1997**, *11*, 1230.
[93] Hunter, E. P. L.; Lias, S. G. *J. Phys. Chem. Ref. Data* **1998**, *27*, 413.

HYPERVALENT BONDING IN GAS-PHASE ANIONS

Lee S. Sunderlin

OUTLINE

Advances in Gas-Phase Ion Chemistry
Volume 4, pages 49–84.

ABSTRACT

A flowing afterglow-tandem mass spectrometer has been used to derive bond energies in hypervalent systems AX_n^-, where A is a central atom and X is a halogen atom. Periodic trends in the bond energies as a function of A, X, and n are discussed. The gas-phase bond energies in some cases are over 200 kJ/mol, substantially stronger than in solution. The differences can be readily explained by solvation effects. The bond energies support the three-center, four-electron model rather than the expanded octet model for hypervalent bonding. Efforts to derive additional thermochemistry for odd-electron systems by modeling competing dissociation reactions are described.

I. INTRODUCTION TO HYPERVALENT BONDING

A common justification for the study of chemistry in the gas phase is that comparison of gas- and solution-phase chemistry defines the effect of the solvent on a chemical reaction or property.[1,2] Ion–solvent interactions are generally substantially stronger than neutral-solvent interactions, so the above statement is even more true about gas-phase ion chemistry. One would expect chemists who study gas-phase ions to be less susceptible to preconceived notions about chemical behavior in the absence of solvents. One of the most interesting aspects of the study of gas-phase hypervalent ions is that the experimental results challenge ideas arising from the study of solvated systems that have been almost universally taught to chemistry students (including myself) starting in first-semester introductory chemistry.

Dirac wrote soon after the development of quantum mechanics that "the underlying physical laws necessary for the mathematical theory of a large part of physics and the whole of chemistry are thus completely known, and the difficulty is only that the exact application of these laws leads to equations much too complicated to be soluble." This statement certainly applies to chemical bonding, and models of varying degrees of accuracy and complexity have been developed to describe bonding in systems of interest. The simplicity and the accuracy of the predictions of a model are key indicators of its usefulness, and are generally apparent to users of the model. Limits to the range of applicability of a model, however, are often much less obvious.

One of the most important chemical models is the use of Lewis dot structures to predict the structure and bonding of molecules. Although this model predates modern quantum mechanics,[3,4] it is still the primary means most chemists use to form a mental picture of a molecule. The tremendous success of this model, however, can blind chemists to its limitations. Studies of the thermochemistry of gas-phase ions demonstrate dramatically that this model can be misleading in some circumstances.

A. The Octet Rule

The Lewis dot structure essentially simplifies electronic structure by limiting electrons to orbitals involving two nuclei (not counting resonance structures).[4] The Lewis dot structure model is strongly tied to the Octet Rule, which indicates that main-group elements in their lowest-energy configuration have eight electrons in their valence shell. Atoms are deemed hypervalent[5] if they have an apparent electron count greater than eight, and hypovalent if the electron count is less than eight. Hypovalent compounds support the octet rule by being reactive Lewis acids (electron pair acceptors). Hypervalent compounds, on the other hand, should dissociate according to the octet rule. Several explanations have been proposed to explain the stability of hypervalent compounds.

B. Models for Hypervalent Bonding

Expanded Octets

The expanded octet model includes one or more *d* orbitals in the valence shell, allowing an electron count greater than eight.[4] The energy cost of promoting electrons from an *p* orbital to a *d* orbital (with the same principle quantum number) is presumed to be made up by the additional covalent bonding. The electron hybridization of a ten-electron system, for example, is sp^3d, which can be broken down into a *pd* pair of axial orbitals and a sp^2 trio of equatorial orbitals. Expansion of the octet is not feasible for the second-row elements, such as C, N, O, and F, because there are no 2*d* orbitals.

Three-Center, Four-Electron (3c-4e) Bonding

A second model for hypervalent bonding was developed by Hach and Rundle[6] and Pimentel.[7] This model involves use of three *p* orbitals, which are aligned along the axis of three atoms A, B, and C in a linear arrangement. These can be combined to form three molecular orbitals, as shown in Figure 1. Similar schemes can be used to describe the bonding in related 3c-4e bonding situations, such as the allyl anion.[8,9] The lowest molecular orbital is bonding between A and B and between B and C. The next highest orbital is nonbonding. The third orbital is antibonding between both pairs of neighboring atoms. In hypervalent systems, there are four electrons in these orbitals. Thus, the bonding and the nonbonding orbitals are filled and the antibonding orbital is empty. If the number of bonds is equated with the number of electron pairs in bonding orbitals,[10] then this scheme shows no increase in bonding over that in Figure 2, which also has two bonding electrons. Simply counting electrons, however, is not sufficient.

A more accurate way to determine bond order is to look at the atomic constituents of the molecular orbitals, as sketched in Figures 1 and 2. For simplicity, the degree of orbital overlap is neglected in the following discussion. It should be noted that for optimized bonding, electrons in the bonding orbital in Figure 1 spend 50% of their time on the central atoms and 25% of their time on each terminal atom,[11] not 33% on each atom.

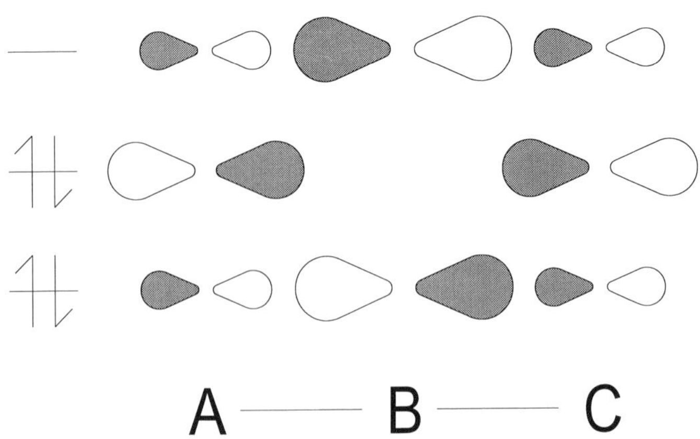

Figure 1. Molecular orbital scheme for hypervalent bonding using a three-center, four-electron bond. In this scheme, only *p* orbitals are used.

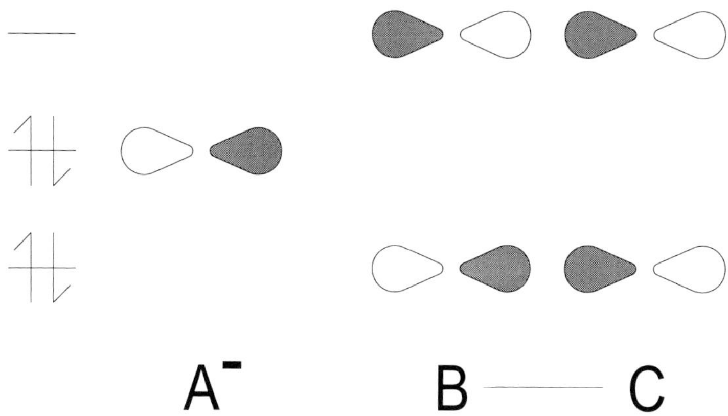

Figure 2. Molecular orbital scheme for an asymmetric $A^{-}\bullet B\text{–}C$ complex. The two filled orbitals have a repulsive interaction.

For each electron in an orbital, the contribution to the bond order between adjacent atoms is equal to the product of the two orbital coefficients.[12] Thus, for electrons in the bonding orbital in Figure 2, the contribution is $1/\sqrt{2} \times 1/\sqrt{2} = 1/2$. For electrons in the antibonding orbital in Figure 2, the contribution is $1/\sqrt{2} \times (-1/\sqrt{2}) = -1/2$. Two electrons in these orbitals give the expected bond orders of 1 and -1, respectively. The situation in Figure 1 is different: each electron in the bonding orbital gives a bond order of $1/2 \times 1/\sqrt{2} = 1/2\sqrt{2}$ between *both* sets of adjacent atoms. Two electrons in this orbital gives a bond order of $1/\sqrt{2}$ between both pairs; thus, one pair of electrons results in a total molecular bond order of $\sqrt{2}$ rather than 1.

In summary, a simple molecular orbital model predicts that formation of a 3c-4e bond between A^{-} and BC increases the total bond order from 1 to $\sqrt{2}$. While this calculation ignores issues of overlap and bond length, it indicates that the A^{-}–BC bond in ABC should be roughly 0.4 times as strong as the B–C bond in BC.

These two models are in fact end points on a continuum with varying amounts of *d* orbital occupancy on the central atom. This is illustrated in Figure 3, which shows that *d* orbital occupancy is readily accommodated (although not essential) in the 3c-4e scheme.[13] It has been noted that the issue of *d* orbital occupancy is not strictly resolvable because... atomic orbitals presumed to be eigenfunctions of the angular momentum operator cannot be discerned in molecules that possess no spherical symmetry..."[14] This is indeed correct, but models that consider the atomic basis of molecular orbitals are essential for human comprehension. Thus, this work

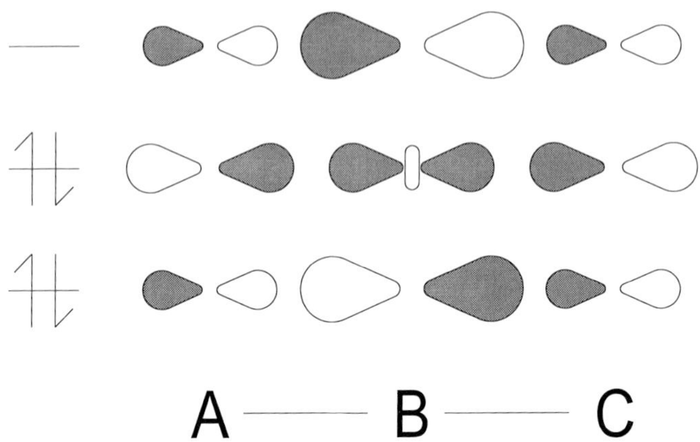

Figure 3. Molecular orbital scheme for hypervalent bonding with an expanded octet on the central atom. The middle molecular orbital uses a *d* orbital on the central atom, and bonds all three atoms together.

will use the expanded octet and 3c-4e models as needed, with the understanding that reality can lie between.

Resonant Ionic Bonding

A third description of hypervalent compounds is given by Pauling.[15] It treats a molecule like ABC^- as a combination of two resonance structures, $A^- \bullet B\text{–}C \longleftrightarrow A\text{–}B \bullet C^-$. This is a description of three-center molecular orbitals as resonance structures of two-center orbitals. As such, it can be considered as a variant of the 3c-4e model that avoids the question of *d* orbital occupancy. Some theoreticians prefer to view hypervalent bonding in this manner;[14,16] this model and the 3c-4e model will be considered equivalent in this chapter.

Donor–Acceptor Interactions

One more conceptual framework for understanding hypervalent bonding is to consider a molecule ABC^- to be held together by the interaction between an electron-pair donor A^- and an electron-pair acceptor BC, as seen in Figure 2. If the acceptor orbital is the $d\sigma$ orbital on the central B atom, this is the expanded octet model. If the acceptor orbital is a σ^* orbital of BC, then this is the 3c-4e model. This model is useful for association/dissociation processes, and it also can be effective for considering asymmetric systems (A $\neq$ C), where each center may have different electronic properties.

C. Comparing the 3c-4e and Expanded Octet Models

Condensed-Phase Experiments

Several predictions from the models above can be compared to experimental results. Crystallographic results in several systems, primarily trihalides, are consistent with both the predictions of Valence Shell Electron Pair Repulsion and with linear 3c-4e bonds. More telling are the bond strengths. Figure 4 shows the bond energies in aqueous solution for the homonuclear trihalide anions X_3^-. Two features stand out in this data. One is that the bond energies are much weaker for elements higher in the periodic table (in fact, F_3^- has not been observed in solution). The other is that the X^-–X_2 bond energies are very weak, 17 kJ/mol or less. In contrast, the gas-phase bond energies in F_2 Cl_2, Br_2, and I_2 are 156, 240, 190, and 149 kJ/mol, respectively.[17] The very weak hypervalent bonds in solution are not consistent with the prediction of the 3c-4e bond model of a bond strength roughly half that of a covalent bond; the weak bonds are consistent with high promotion energy costs. Furthermore, the periodic trend in hypervalent bond strength is much more consistent with the expanded octet model. In particular, the nonexistence of F_3^- (where promotion of an electron from a $2p$ orbital to a $3d$ orbital is energetically prohibitive)

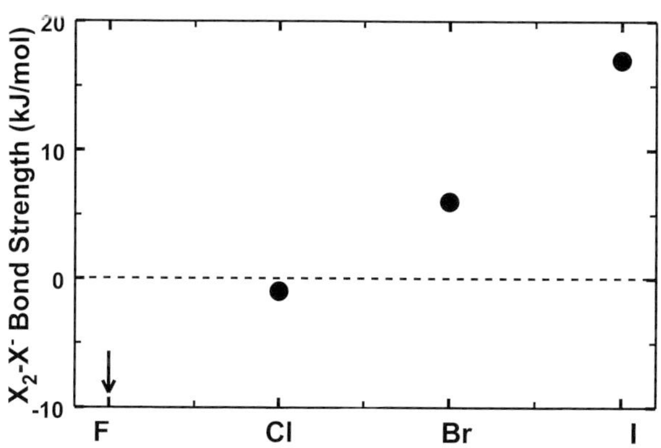

Figure 4. Trihalide bond enthalpies in aqueous solution. The arrow is an upper limit on the F_2–F^- bond strength, since aqueous F_3^- has not been observed. For details see Ref. [58].

is consistent with an expanded octet. Similar results are found for other main group hypervalent anions.[18]

Computational Results

Early computational work on hypervalent compounds indicated that inclusion of *d functions* in the basis set is important in these systems, suggesting that *d* orbitals were occupied. It is now generally believed that the *d* functions in ab initio calculations act as polarization functions.[19–24] Indeed, similar *d* function occupancy is seen in both hypervalent and nonhypervalent systems.

The 3c-4e scheme results in a formal charge of 0 on the central atom and $-1/2$ on each of the terminal atoms, while the expanded octet model places the formal negative charge on the central atom. The 3c-4e model is in better agreement with results for trihalide systems $X\text{–}X\text{–}X^-$, which are not biased by different electronegativities on the central and terminal atoms. For these systems, Atoms In Molecules (AIM)[14,25,26] calculations at the B3LYP/6–31 + G(d) level give charges on the central atom of -0.08 for X = F, -0.06 for X = Cl, and -0.04 for X = Br. Calculations on other hypervalent systems suggest that the terminal atoms are anionic as well.[14] Thus, most recent theoretical and computational work has supported the 3c-4e bonding scheme.[8–13,19–27]

Introductory Textbooks

Surprisingly, these developments have not been reflected in chemistry pedagogy. Although some advanced texts give a careful account of hypervalent bonding,[18,28] chemistry textbooks at the introductory level still follow the expanded octet model.[29] Thus, a strange situation has arisen where nearly all chemistry students are taught a model that is rejected by most specialists active in the field. Lewis said in 1923 that "... the chemical bond is always a pair of electrons which lies between two atomic centers and is held jointly in the shells of the two atoms,"[4] and most textbooks leave it at that.

Importance of Hypervalent Bonding

This lack of discussion does not mean that hypervalent bonding is uncommon. It occurs for nearly all of the *p*-block elements.[30] Noble gas compounds, of course, have 3c-4e bonds. The central atom in trihalide anions such as I_3^- is hypervalent. Similar 3c-4e interactions occur for halides of group 13–17 elements; examples include anions such as AlF_6^{-3},

SiF_6^{-2}, PF_6^-, as well as neutral compounds such as PF_5, SF_6, and IF_5.[11] Silicon forms a wide variety of hypervalent compounds and the gas-phase chemistry of these has been explored in some detail.[31,32] Hypervalent compounds of such unlikely elements as C[33] and He[34] have been detected, or predicted using high-level ab initio calculations. Hydrogen bonding is also 3c-4e bonding: for hydrogen or helium as the central atom, the 1*s* orbital is substituted as the middle orbital in the schemes discussed above.

Another important example of hypervalent bonding is the transition state in the S_N2 nucleophilic substitution reaction,[2,35] which has the same orbital configuration and electron count as a 10-electron hypervalent system. For nucleophilic substitution, the hypervalent species is a transition state rather than a stable species, as shown schematically in Figure 5. Experimental and computational studies on transition states are generally difficult. Stable hypervalent systems can serve as more tractable benchmarks to test the accuracy of computational techniques used on 10-electron systems.[36]

Gas-Phase Ion Chemistry to the Rescue

There is an obvious discrepancy between the experimental solution thermochemistry, which seems to support the expanded octet model, and recent computational results, which are more consistent with the 3c-4e

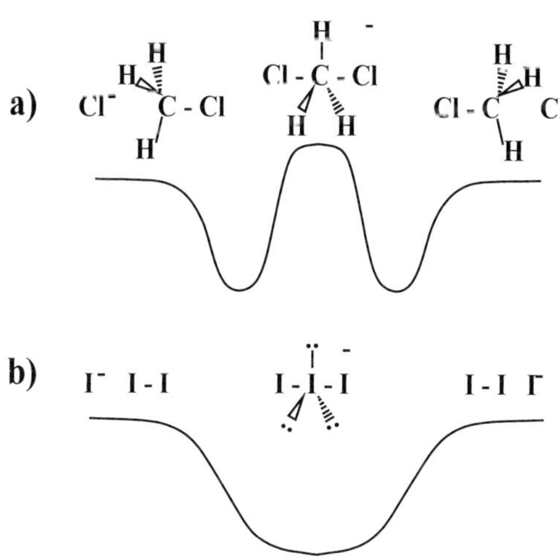

Figure 5. Qualitative potential energy surfaces comparing two ten-electron systems. (a) An unstable nucleophilic substitution (S_N2) intermediate. (b) A stable hypervalent molecule.

model. Gas-phase experiments demonstrate that the recent work of theoreticians is generally correct, and that the discrepancy can be readily explained by the effects of solvation. These experiments, which involve hypervalent main-group halide anions, are the subject of this chapter.

II. EXPERIMENTAL SECTION

The experiments discussed in this chapter were carried out using a flowing afterglow-tandem mass spectrometer (MS) described in detail previously,[37] and shown in Figure 6. The basic procedure involves making the ions, cooling them to a known temperature, and then adding energy by collisional activation. Analysis of the resulting energy-dependent decomposition gives the desired thermochemistry. These steps are detailed below. Related experiments have been described in Volume 1 of this series.[38]

A. Making Hypervalent Ions

Making the hypervalent ions is often the most difficult step in their study. Ion precursors are introduced into the ion source as a vapor. A flow of unreactive carrier gas over the sample, or if necessary a heated solids inlet probe, can be used to increase the flow of precursor. The ion source used in these experiments is a DC discharge that is typically set at 1200 V with 1 mA

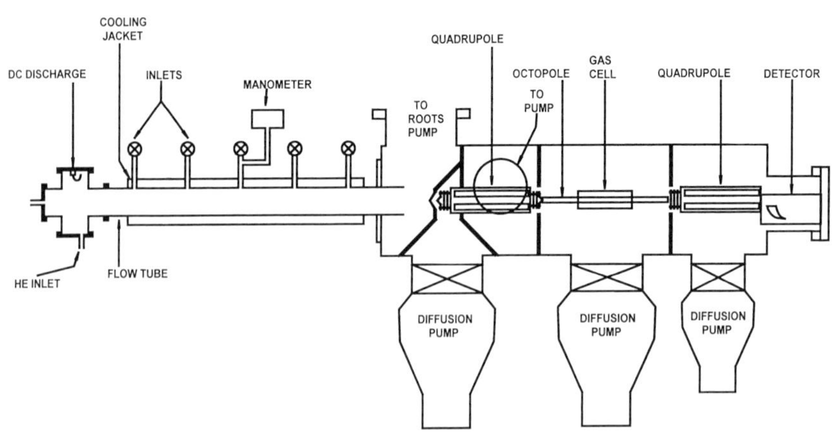

Figure 6. The flowing afterglow-tandem mass spectrometer. Hypervalent anions are formed and cooled in the flow tube, and collision-induced dissociation occurs in the quadrupole–octopole–quadrupole tandem mass spectrometer.

of emission current. This high-energy electron source generally makes ions through dissociative ionization, usually producing halide ions from any halogenated precursor (reaction 1).

$$AX_n + e^- \rightarrow AX_{n-1} + X^- \tag{1}$$

Therefore, secondary reactions are necessary to form the desired ions. In this situation, the halide ion and a second precursor molecule form a complex, and then the complex is stabilized by further collisions with either a third precursor molecule or an inert buffer gas M (reactions 2 and 3).

$$AX_n + X^- \rightleftharpoons AX_{n+1}^{-*} \tag{2}$$

$$AX_{n+1}^{-*} + M \rightarrow AX_{n+1}^{-} + M \tag{3}$$

Amazingly, the third-order kinetics of this type of reaction were first measured for I_3^- in 1928;[39] this was the first study of anion reaction chemistry performed using a mass spectrometer. Some electrons are thermalized rather than captured by the precursor. This can lead to electron attachment to form odd-electron ions, while halide ion association forms even-electron ions.

The clustering reactions necessary for producing the desired ions are not typically very efficient, since the reverse of reaction 2 can be very fast. Therefore, in order to make an adequate number of ions for study, the pressures of the precursor and the buffer gas pressure must be kept high and the residence time must be kept long. These conditions can be met by using either a static system (such as a high pressure mass spectrometer)[40] or a flow tube. A static system has the advantage of lower precursor flow rates. Flow systems have the advantage that the ion source plasma is physically remote from where reactions 2 and 3 take place in the flow tube. This assures that the ions emanating from the flow tube are truly thermalized to ambient temperature, and allows more flexibility in performing sequential reactions. In the present experiments, a 92 cm long, 7.3 cm i.d. stainless steel flow tube operating at a carrier buffer gas pressure of 0.4 Torr and flow rate of 200 standard $cm^3 s^{-1}$ was used. The buffer gas is He with up to 5% Ar added to stabilize the DC discharge source. Approximately 10^5 collisions with the buffer gas (and precursor molecules) in the flow tube ensure that the ions are cooled to room temperature.[41]

B. Collision-Induced Dissociation

The disadvantage of the high pressure used in the ion source is incompatibility with the high vacuum needed to operate the tandem MS. In these experiments, ions are gently extracted from the flow tube through a

0.5 mm orifice by applying a 0–4 V potential to the nose cone at the end of the flow tube. The small orifice ensures that there is a drastic (ca. 10^4) drop in the buffer gas pressure between the flow tube and the mass spectrometer, at the cost of a significant decrease in the ion population. The tandem MS is contained in a stainless steel box that is divided by interior partitions into five chambers. Further differential pumping on the five chambers ensures that further collisions of the ions with the buffer gas after extraction from the flow tube are unlikely.

The tandem MS includes a quadrupole mass filter, an octopole ion guide,[42] a second quadrupole mass filter, and an ion detector. The ions from the flow tube are focused through electrostatic lenses into the first quadrupole, where a particular reactant ion is selected. These ions are then focused into the octopole, which passes through a cell that contains the collision gas. From the octopole, the dissociated and unreacted ions are focused into a second quadrupole for mass analysis. The detector is an electron multiplier operating in pulse-counting mode.

Lighter collision gases minimize the effect of the spread in ion energies, but heavier collision gases improve energy transfer during collisions. The collision gases used in these experiments were the rare gases Ne (for F_3^-), Ar (for ions with a mass of less than 200 u), and Xe (for heavier ions). These choices were found experimentally to give the best results with the present instrument.

C. Data Analysis

The threshold energy for collision-induced dissociation (CID) of a reactant ion is determined by modeling the intensity of product ions as a function of the reactant ion kinetic energy in the center-of-mass (CM) frame, E_{CM}. The reactant ion beam energy zero is measured using the octopole as a retarding field analyzer.[42,43] The first derivative of the beam intensity as a function of ion energy is approximately Gaussian, with a typical full-width at half-maximum of 1.0 electron-volt (1 eV = 96.49 kJ/mol). The laboratory energy E_{lab} in eV is given by the octopole rod offset voltage measured with respect to the center of the Gaussian fit. This energy is corrected at low offset energies to account for truncation of the ion beam.[43]

Conversion to the CM frame is accomplished by use of $E_{CM} = E_{lab}m/(m + M)$, where m and M are the masses of the neutral and ionic reactants, respectively. The isotopic distribution of compounds containing multiple chlorine or bromine atoms gives a relatively large number of reactant peaks in the mass spectrum. To improve signal, the mass filters were generally operated at the lowest possible resolution, where all of the isotopic peaks were simultaneously transmitted. Operating at low resolution also tends to minimize broadening of the translational energy distribution of the reactant ions. For data collected in this manner, the weighted average of the isotopic masses is used as the ion mass. Similarly, the average neon and xenon masses are used. For several ions, some data sets were collected at higher resolution on an individual isotopic peak. The results are indistinguishable from the low mass resolution results.

Total cross sections for reaction, σ_{total}, were calculated using equation 4, which is Beer's Law rewritten in terms of the variables appropriate for the CID experiments.[43]

$$I = I_0 \exp(-\sigma_{total} nl) \quad (4)$$

In this equation, I is the intensity of the outgoing reactant ion beam, I_0 is the intensity of the incoming ion beam ($I_0 = I + \Sigma I_i$), and I_i are the intensities for each product ion. The number density of the neutral collision gas is n, and l is the effective collision cell length, 13 ± 2 cm.[37] Individual product cross sections σ_i are equal to $\sigma_{total}(I_i/\Sigma I_i)$.

To derive CID threshold energies, the threshold region of the data is fitted to the model function given in equation 5,[44]

$$\sigma(E) = \sigma_o \Sigma_i g_i (E + E_i - E_T)^n / E \quad (5)$$

where $\sigma(E)$ is the cross section for formation of the product ion at center-of-mass energy E, E_T is the desired threshold energy, σ_o is a scaling factor, n is an adjustable parameter, and i denotes rovibrational states having energy E_i and population g_i ($\Sigma g_i = 1$). The CRUNCH computer program written by P.B. Armentrout and coworkers is used in the threshold analysis described above (see Chapter 3, Volume 1 of the present series).

To fit the CID data, a model cross section is first chosen. The CRUNCH program then accounts for the effects of ion internal energy, broadening due to the thermal motion of the collision gas (Doppler broadening), and the kinetic energy distribution of the reactant ion by convoluting the model cross section with these energy distributions. The convoluted cross section is then compared to the experimental data, and the fitting parameters are adjusted to give the best least-squares fit. The broadening effect of the various energy distributions cause the apparent reaction threshold to be generally lower than the reaction threshold derived from the fitting procedure; this is particularly noticeable on a logarithmic cross section scale. In this chapter, the unconvoluted fits are shown as dashed lines, while the convoluted fits are shown as solid lines.

Hypervalent species tend to contain heavy atoms, and have generally low vibrational frequencies. Thus, even at room temperature most modes are active, and the internal energy can be significant (6–10 kJ/mol for the triatomic molecules, ca. 20 kJ/mol for penta-atomic molecules, and even more for larger species). Fortunately, the average internal energy content is only moderately sensitive to the vibrational frequencies. In some cases, experimental values for the frequencies are available; otherwise values are calculated using commercially available software,[45,46] either in-house or in collaboration with computational chemists.

D. Competing Dissociation Pathways

Recently, Armentrout, Ervin, and Rodgers have developed software for predicting the dissociation kinetics of collisionally activated ions using the RRKM model.[47] This modeling requires that the transition states for dissociation be known. Fortunately, the relatively long-range ion-induced dipole attraction between separating ionic and neutral products tends to dominate any kinetic bottlenecks during the dissociation process for most ionic systems. Thus, the transition state for simple bond cleavage is essentially the separated products. This greatly simplifies the task of obtaining the necessary rotational and vibrational constants for the activated complex and the dissociation transition states.

One important aspect of the kinetics is that ion dissociation may not occur during the time between ion activation near the middle of the octopole, and ion detection in the final quadrupole, ca. 30 μs. The effect of incomplete dissociation on the experimental time scale (the kinetic shift) is negligible for the triatomic atoms and is 0–1 kJ/mol for a larger molecule like Br_5^-,[58] the magnitude depending on whether the transition state for dissociation is loose (product-like) or tight (reactant-like). Therefore, the derived thresholds are only slightly sensitive to the vibrational and rotational constants. Because the reactant and product internal energy distributions are taken into account, the reaction thresholds correspond to bond energies (or enthalpies) at 0 K.

The collision gas pressure can influence the observed cross sections because an ion that is not sufficiently energized by one collision with the target gas may gain enough energy in a second collision to be above the dissociation threshold. Such collisions can lead to a measured threshold that is too low. This is accounted for by linearly extrapolating data taken at several pressures to a zero pressure cross section,[48] which is then fit with the method described above.

E. Other Experimental Methods

Several other methods have been used to derive thermochemistry for hypervalent species. The most important of these is high pressure mass spectrometry (HPMS),[40] where measurement of equilibrium constants for association or ligand-exchange reactions can be used to derive not only bond dissociation enthalpies, but also entropies and free energies. Some hypervalent bonds are sufficiently strong that dissociation equilibria cannot be measured and therefore, measurements of a series of relative halide affinities must be made to derive a "ladder" of halide affinities. This is a painstaking task, made more difficult for the molecules discussed in this work by their corrosive properties and, in some cases, instability. The technique, however, has the advantage of very high precision in favorable circumstances.

Ion cyclotron resonance (ICR)[49] and flowing afterglow experiments[24] can also be used to derive relative affinities. Neutral beam experiments, where a beam of alkali atoms such as Cs is crossed with a beam of molecules such as PCl_3 or $(Cl_2)_2$ have been used to derive thermochemistry for anions such as PCl_3^-[50] and Cl_3^-[51], but proper analysis of this type of data is difficult.[51] High-resolution negative ion photoelectron spectroscopy (NIPES) experiments can provide otherwise unobtainable information on hypervalent anions, including precise electron affinities and vibrational frequencies.[52,53] This technique has limited applicability to hypervalent species with more than three atoms because of vibrational congestion from low-frequency modes.

One final method for deriving gas-phase thermochemistry of hypervalent ions does not involve gas-phase experiments at all. It requires measuring solid-state heats of formation of ionic salts and correcting to gas-phase values using calculated lattice energies for these salts.[54] Since the lattice energies are in the vicinity of 1000 kJ/mol, 1% accuracy on these values is necessary to derive gas-phase thermochemistry with error limits of ca. 10 kJ/mol. This accuracy has yet to be met in practice.[54–56]

III. TRIHALIDES

A. Homonuclear Trihalides

The four homonuclear trihalides X_3^- (X = F, Cl, Br, and I) are the simplest hypervalent systems. It should be remembered that X_3^- is an AX_2^- system, where the central atom A is the same as the terminal atoms X. The aqueous bond energies have been discussed above. Figures 7–10 show the collision-induced dissociation results for these ions in the gas phase. Reactions 6a and 6b were the observed products for all four halogens studied.

$$X_3^- \rightarrow X^- + X_2 \tag{6a}$$

$$\rightarrow X + X_2^- \tag{6b}$$

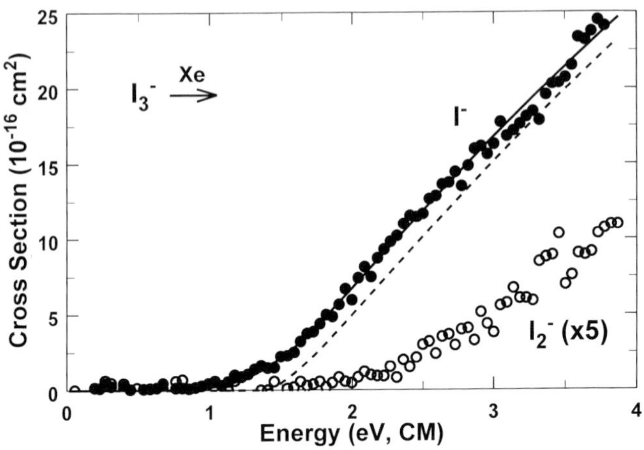

Figure 7. Appearance curves for collision-induced dissociation of I_3^- as a function of translational energy in the center-of-mass frame. At 3.5 eV, I^- is the dominant product by a factor of 10. The solid and dashed lines represent convoluted and unconvoluted fits to the I^- data, respectively.

The results for iodine are shown in Figure 7.[37] Although a moderate amount of I_2^- was observed experimentally, the predominant product at all energies was I^-. This is consistent with the relative thermodynamics of these two products: the difference in energy between these two channels is the difference between the electron affinities (EAs) of X and X_2. In the case of iodine, the difference is 0.535 eV,[57] ensuring that I^- is favored.

The CID threshold is the most striking aspect of the data. Although the aqueous bond energy is only 17 kJ/mol, the bond energy in the gas phase is 126 ± 6 kJ/mol. Thus, solvation has a massive effect on this energy; the origin of this effect will be discussed below.

Figure 8 shows the CID appearance curves for tribromide.[58] In comparison to the iodine data, the overall threshold is essentially the same, while the amount of X_2^- is somewhat less. For trichloride (Figure 9),[58] the overall threshold is now somewhat less, and the amount of Cl_2^- observed is negligible. The trend in the relative amounts of X_2^- is consistent with the differences between the X and X_2 electron affinities; a higher ΔEA results in a greater X^-/X_2^- ratio.

These three ions were successively harder to form in sufficient quantities for CID experiments. Early attempts to make F_3^- were unsuccessful, leading us to assume that F_3^- is less strongly bound than the other three trihalides. Recently, however, Tuinman et al. successfully synthesized F_3^- in the gas

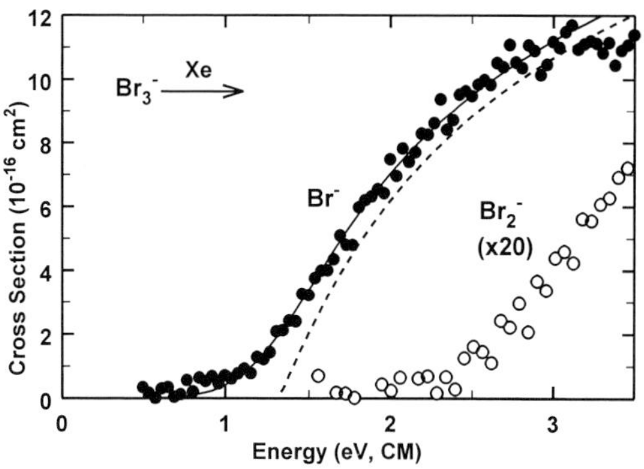

Figure 8. Appearance curves for collision-induced dissociation of Br_3^- as a function of translational energy in the center-of-mass frame. At 3.5 eV, Br^- is the dominant product by a factor of 30. The solid and dashed lines represent convoluted and unconvoluted fits to the Br^- data, respectively.

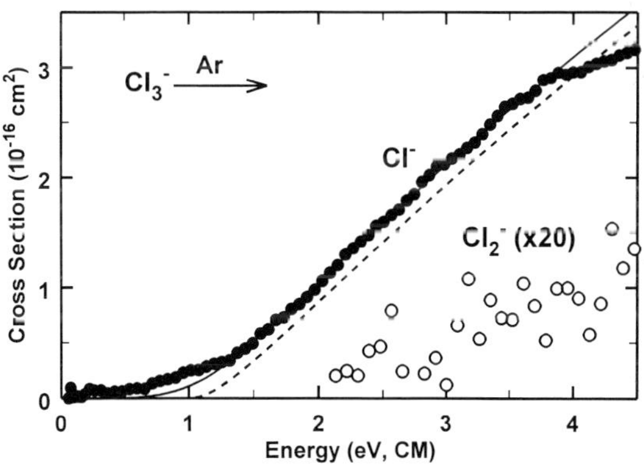

Figure 9. Appearance curves for collision-induced dissociation of Cl_3^- as a function of translational energy in the center-of-mass frame. At 3.5 eV, Cl^- is the dominant product by a factor of 60. The solid and dashed lines represent convoluted and unconvoluted fits to the Cl^- data, respectively.

phase.[59] Subsequently, the Wenthold group at Purdue and our group at NIU made sufficient quantities of F_3^- as well.[60] Apparently, the difficulty is not in the thermodynamics of F_3^-, since the bond strength is very similar to that of Cl_3^-. Rather, the kinetics of formation of F_3^- are not favorable: the metastable F_3^{-*} formed by reaction 2 decays too rapidly back to the reactants. There are two reasons why this should be a particular problem for F_3^-, both of which are due to the higher vibrational frequencies in the lighter trihalide. One is that higher frequencies lead to a lower density of states and thus a shorter lifetime for the metastable F_3^-. The other is that collisional cooling by helium is most effective for species with low vibrational frequencies, where vibrational-translational energy coupling is greater. Addition of more effective polyatomic cooling agents to the flow tube or ion source is the solution. F_2,[59] N_2O,[60] and NF_3 [60] have all been found effective for this purpose. Fortunately, ions with more than three atoms have longer metastable lifetimes such that addition of additional cooling agents is not necessary.

CID of the trifluoride ion gives surprising results; at high collision energies, F_2^- is the dominant product. This contrasts with the other trihalides, and seems to contradict the fact that F has a higher electron affinity than F_2.[61] Energy-resolved CID data are given in Figure 10, and show a transition between the favored channels at a collision energy near 3 eV, not seen in the other systems. This effect is nevertheless in agreement with the predictions of the statistical model for decomposition, as seen by the fits in Figure 10. The rotational and vibrational constants for F_2^- are significantly smaller than those for F_2, increasing the density of states (and the dissociation probability) for the F_2^- channel. The difference between the thresholds of the two channels (0.28 ± 0.06 eV) gives a new determination of 3.12 ± 0.06 eV for the electron affinity of F_2. Previous

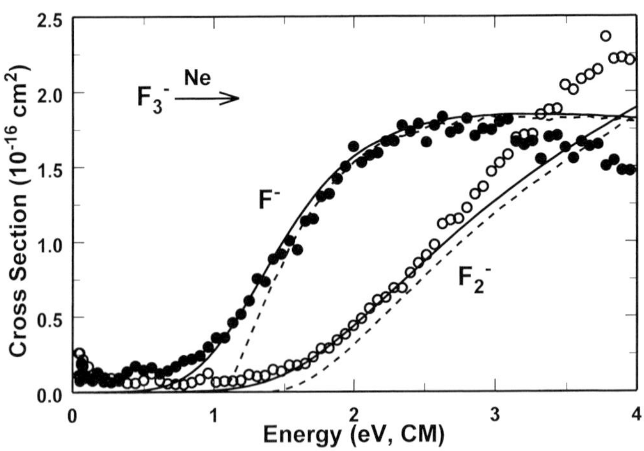

Figure 10. Appearance curves for collision-induced dissociation of F_3^- as a function of translational energy in the center-of-mass frame. At 3.5 eV, the cross sections for F^- and F_2^- are nearly equal. The solid and dashed lines represent convoluted and unconvoluted fits to both product channels, respectively.

values of 3.01 ± 0.07 eV[61] and 3.08 ± 0.10 eV[62] are consistent with this measurement.

It should be kept in mind that at any given collision energy, the efficiency of energy deposition into internal energy of the reactant ion can range from 0 to 100%. Thus, the calculation of branching ratios at a given collision energy involves numerical integration of the dissociation probabilities at a given energy over the internal energy probability distributions.[47] In the F_3^- reaction, for example, ions that have less than 1.0 eV of energy after collision do not dissociate. Ions with between 1.0 and 2.1 eV of energy dissociate primarily to form $F^- + F_2$, and ions with more than 2.1 eV of energy are most likely to dissociate to form $F + F_2^-$.

B. Analysis of X_2^- as a Minor Channel

A similar analysis of the triiodide dissociation[63] gives $EA(I) - EA(I_2) = 0.55 \pm 0.04$ eV, in excellent agreement with the 0.535 ± 0.005 eV difference determined by Neumark and coworkers.[57] This measurement is in many ways an ideal case. Many of the sources of error in measuring thresholds for individual reaction channels cancel out in comparisons of data for two channels. The spectroscopic constants are sufficiently well known. Also, there is a reasonably large energy range in the threshold region where both channels have a significant cross section.

It would be useful to apply the competitive modeling procedure to derive the electron affinities of large numbers of closed-shell molecules. Such values can be difficult to obtain in many cases, primarily because large geometry differences between neutral and anionic species makes photoelectron spectroscopy measurements of adiabatic EAs (electron affinities) difficult or impossible. Unfortunately, the mechanism of energy deposition at collision energies beyond the threshold region is not well understood. In the present context, the threshold region is taken to be the energy range where the data can be fit with the cross section form given in equation 5. For Cl_3^- and Br_3^-, the cross sections for the X_2^- channels do not become sufficiently large for accurate data analysis until collision energies beyond the threshold region. This gives analysis of competing CID reactions a limited dynamic range of $\Delta EA \leq$ ca. 0.6 eV, which is too small to allow the measurement of EAs for most closed-shell AX_{n-1} species by competitive CID of AX_n^-.

Cross-Section Ratio Analysis

An alternative fitting technique can potentially be used to fit data where the difference in the energetics of the two channels is large. This method involves calculating the *ratio* of the competing cross sections rather than the individual cross sections. It has the advantage of allowing data analysis at energies higher than the threshold region. Several factors limit the effectiveness of this technique. The energy deposition at collision energies in the threshold region is partially determined by the derived fitting parameters. The energy deposition function becomes less well known with increasing energy. Similarly, the density of states calculations are less accurate at energies farther from the rovibrational ground state, such that the calculated dissociation probabilities are more subject to error. Finally, secondary dissociation, reaction 7, can skew the observed ion product ratio by depleting the signal for X_2^-.

$$X_3^- \rightarrow X_2^- + X \rightarrow X + X^- + X \tag{7}$$

A more fundamental question is whether the products should be statistically distributed at all. The reactant X_3^- initially is in the ground singlet electronic state. Collisional activation is expected to cause rotational and vibrational excitation, but not electronic excitation. Dissociation on the same surface should lead to the ground state singlet products $X_2 + X^-$. The other product channel, $X + X_2^-$, correlates to singlet and triplet excited electronic states of the activated complex. Electronic state correlations during ion dissociation have been discussed by Armentrout and Simons,[64] and a sketch of the relevant potential energy surfaces is shown in Figure 11.

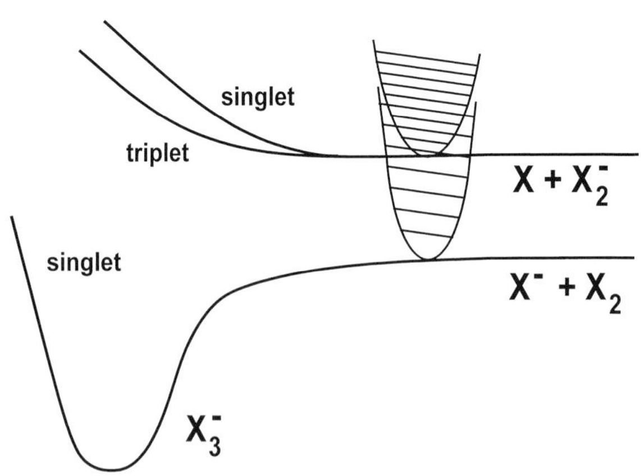

Figure 11. Potential energy surfaces for dissociation of X_3^- showing the lack of crossings of the electronic states, but with potential coupling in the vibrational manifolds. For details on the potential energy surfaces, see Ref. [64].

The ratio technique gives EA values for F_2 and I_2 in good agreement with the values discussed above. The fact that statistical modeling of the CID data gives results in agreement with known thermochemistry indicates that vibrationally mediated coupling between different electronic states is efficient, at least when the energies of the two dissociation channels are reasonably close. Cross-section ratio analysis of the CID data for Br_3^- gives $EA(Br_2) = 2.53 \pm 0.07$ eV, in agreement with recent literature values (2.42 to 2.62 eV).[17] Here, the difference between EA(Br) and $EA(Br_2)$ is 0.74–0.94 eV. The difference between EA(Cl) and $EA(Cl_2)$ is larger still, 1.09–1.29 eV.[17] This is apparently too great a difference for quantitative thermochemistry to be derived from competitive CID experiments, since fitting the data for Cl_3^- gives $EA(Cl) - EA(Cl_2) = 1.8$ eV. This discrepancy is presumably due to a breakdown of the statistical dissociation assumption, which would lead to an overestimate of the difference in the energetics of the two channels. This suggests that the ratio technique is only applicable in cases where the electron affinity difference between the two products is roughly 1 eV or less.

Larger molecules, with more degrees of freedom and longer activated complex lifetimes, should be more amenable to this type of analysis. However, as seen in the case of trifluoride, the vibrational and rotational constants of all product species have a significant influence and must be reasonably well known to derive useful electron affinity values. Few experimental frequencies are known for the larger anionic products, and theoretical methods are not well developed for frequency calculations on halide anions. Testing of various computational techniques to provide the necessary information is underway in collaboration with Professor T. M. Gilbert, also at Northern Illinois University.

IV. MORE COMPLICATED EXAMPLES: PCl_4^-, PCl_5^-, AND PCl_6^-

While collisionally activated trihalides lose halide anions or halogen atoms, larger systems undergo more complicated reactions. For example, Figures 12–14 show CID data for PCl_4^-, PCl_5^-, and PCl_6^-; the reactions observed[65] are listed in equations 8–10. $[PCl_4]$ may be a weakly bound PCl_4 radical, or the dissociated form $[PCl_3 + Cl]$.

These plots are put on a semi-log scale to show more clearly the competition between different product channels. It should be kept in mind that by emphasizing relatively small cross sections at low

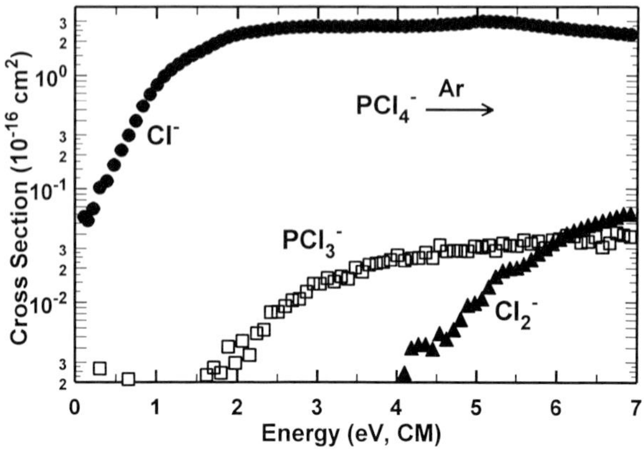

Figure 12. Product ion appearance curves for collision-induced dissociation of PCl_4^- as a function of translational energy in the center-of-mass frame. Chloride loss is the dominant product.

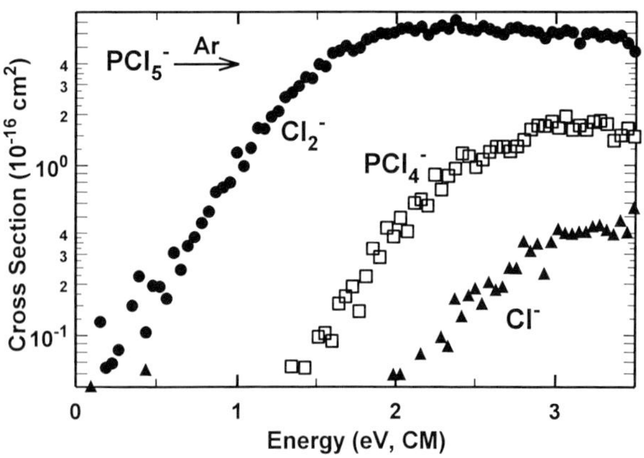

Figure 13. Product ion appearance curves for collision-induced dissociation of PCl_5^- as a function of translational energy in the center-of-mass frame. For this reaction, the (PCl_3 + Cl_2^-) channel is dominant.

energies, data on semi-log plots appear to have much lower thresholds than on linear plots.

$$PCl_4^- \rightarrow PCl_3 + Cl^- \qquad (8a)$$

$$\rightarrow PCl_3^- + Cl \qquad (8b)$$

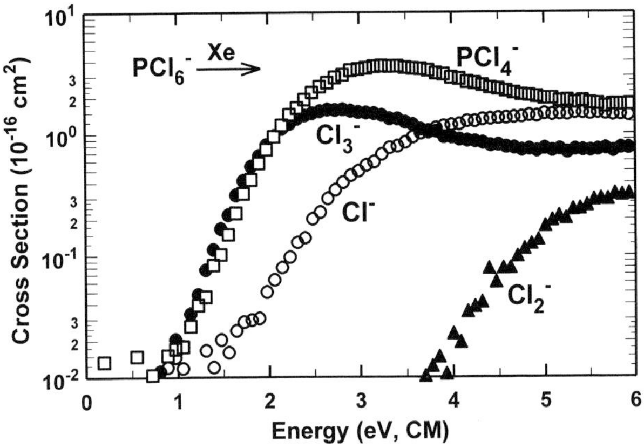

Figure 14. Product ion appearance curves for collision-induced dissociation of PCl_6^- as a function of translational energy in the center-of-mass frame. The (PCl_3 + Cl_3^-) and (PCl_4^- + Cl_2) channels are nearly equal in intensity near threshold.

$$\rightarrow PCl_2 + Cl_2^- \quad (8c)$$

$$PCl_5^- \rightarrow PCl_4^- + Cl \quad (9a)$$

$$\rightarrow PCl_3 + Cl_2^- \quad (9b)$$

$$\rightarrow [PCl_4^-] + Cl^- \quad (9c)$$

$$PCl_6^- \rightarrow PCl_4^- + Cl_2 \quad (10a)$$

$$\rightarrow PCl_3 + Cl_3 \quad (10b)$$

$$\rightarrow [PCl_4] + Cl_2^- \quad (10c)$$

$$\rightarrow PCl_5 + Cl^- \quad (10d)$$

A. PCl_4^-

As is the case with the trihalides, loss of a halogen ion is the dominant reaction channel and loss of a halogen atom is next. A small amount of Cl_2^- loss is also seen. Ligand coupling reactions like reaction 8c will be discussed in the next section.

It would be valuable to obtain the electron affinity of PCl_3 from analysis of the data for reactions 8a and 8b. Unfortunately, the difference in the electron affinities of PCl_3 and Cl atom is sufficiently large that data analysis using the cross-section ratio technique is open to question. How to treat this data will be more clear when higher precision vibrational constants for

PCl_3^- (which has several very low vibrational frequencies) are obtained. However, a preliminary analysis indicates that $EA(Cl) - EA(PCl_3) = 1.0 \pm 0.2$ eV, giving $EA(PCl_3) = 2.6 \pm 0.2$ eV.

B. PCl_5^-

Neutral PCl_5 has a 10-electron trigonal bipyramidal structure around the central phosphorus atom[18] and a single 3c-4e bond. PCl_5^- is an odd-electron anion with the additional electron apparently in an antibonding orbital.[66] Cl and Cl^- loss correspond to homolytic and heterolytic bond cleavage, respectively. Cl_2^- loss must be due to a more complicated mechanism. Dichlorine loss through a ligand coupling reaction is allowed for 10-electron systems such as PCl_4^- and PCl_5,[66,67] and approximate molecular orbital diagrams suggest that the orbital occupied by the extra electron in PCl_5^- is not significantly affected by ligand coupling.[66] Thus, ligand coupling to give $PCl_3^{-\bullet}Cl_2$ followed by electron transfer is a feasible mechanism for this reaction. An alternative pathway involves bond cleavage to form a loosely bound complex, either $PCl_4^{-\bullet}Cl$ or $PCl_4^{\bullet}Cl^-$, followed by chloride or chlorine transfer.[67,68,69]

C. PCl_6^-

While loss of Cl^- is the dominant CID reaction for PCl_4^-, it is a minor product in the CID of PCl_6^-. Instead, two other reactions leading to closed-shell products predominate. Although concerted loss of Cl_2 is allowed, moderately high-level density functional calculations indicate that the lowest-energy reaction path involves almost complete loss of chloride to form an essentially orbiting $PCl_5^{\bullet}Cl^-$ transition state.[65] This species can either lose Cl^- (reaction 11a) or rearrange through chloronium abstraction to $PCl_4^{-\bullet}Cl_2$. This latter species can in turn either dissociate (reaction 11b) or rearrange through chloride abstraction to form $PCl_3^{\bullet}Cl_3^-$. This final intermediate can then dissociate (reaction 11c).

$$PCl_6^- \rightarrow PCl_5 \bullet Cl^- \rightarrow PCl_5 + Cl^- \quad (11a)$$

$$\rightarrow PCl_4^- \bullet Cl_2 \rightarrow PCl_4^- + Cl_2 \quad (11b)$$

$$\rightarrow PCl_3 \bullet Cl_3^- \rightarrow PCl_3 + Cl_3^- \quad (11c)$$

The calculations indicate that $PCl_5 \bullet Cl$ is substantially higher in energy than other stationary points on the potential energy surface.

The very similar onsets for reactions 11b and 11c reflect the fact that they have the same barrier at the $PCl_5 \bullet Cl^-$ transition state. After the transition state, there is a competition between PCl_3 and Cl_2 for the chloride ion. Reaction 11c is less endothermic than reaction 11b by the difference between $D(Cl_2\text{–}Cl^-) = 99 \pm 5$ kJ/mol and $D(PCl_3\text{–}Cl^-) = 91 \pm 7$ kJ/mol, or 8 ± 9 kJ/mol. Fits to the data, which show slightly more Cl_3^- than PCl_4^- near threshold, are consistent with this small energy difference.

The overall threshold for reaction 11 is 150 ± 10 kJ/mol, which corresponds to the barrier height rather than a bond strength. The calculations indicate that the barrier for reaction is only slightly below the endothermicity of Cl^- loss. Thus, $D(PCl_5\text{–}Cl^-)$ can be estimated as 160 ± 20 kJ/mol. More accurate characterization of the transition states involved in the above rearrangement reactions, which will allow more definitive determinations of the reaction thermochemistry, is in progress.

These systems indicate that even relatively small hypervalent molecules can have reaction dynamics far more complicated than simply loss of a ligand. They also exemplify the usefulness of computational chemistry as an adjunct to experimental measurements.

V. PERIODIC TRENDS

In a hypervalent anion AX_n^-, there are three variables that can affect the bond strength: the identity of the central atom A, the identity of the terminal atoms X, and the number of terminal atoms *n*. Although the four homonuclear trihalide anions have similar bond strengths, this comparison involves changing two variables at once. An examination of heteronuclear systems can give a clearer picture of the factors affecting the bond strengths in these systems. These three variables will be discussed below. A summary of the available bond dissociation energies is given in Table 1.

A. The Central Atom

The *s*-block elements (with the exceptions of H and He) do not have sufficient electrons for hypervalent bonding. Transition metals *can* use *d* orbitals in their bonding, and thus do not follow the octet rule. Therefore, with the exception of hydrogen bonding, the central atom in a hypervalent system must be a *p*-block element. Within the *p* block, two obvious issues are how the bond strengths change as the central atom A is varied across or down the periodic table. Sufficient data is currently available to begin to define these periodic trends.

Table 1. Selected Gas-Phase Hypervalent Bond Strengths (kJ/mol)

System	*Bond energy*	*Reference*
F_2–F^-	101 ± 11	60
Cl_2–Cl^-	99 ± 5	58
Br_2–Br^-	127 ± 7	58
I_2–I^-	126 ± 6	37
FCl–F^-	ca. 200	70
Br_2–Br_3^-	40 ± 7	58
I_2–I_3^-	49 ± 6	37
SCl_2–Cl^-	85 ± 8	36
PF_3–F^-	168 ± 8	40
PF_3–Cl^-	65 ± 8	40
PCl_3–Cl^-	91 ± 7	72
PBr_3–Br^-	64 ± 7	72
PI_3–I^-	≤ 50	72
AsF_3–F^-	202 ± 8	40
$AsCl_3$–Cl^-	115 ± 7	72
$SbCl_3$–Cl^-	161 ± 10	72
$BiCl_3$–Cl^-	154 ± 16	72
PF_5–F^-	360 ± 42	40
	329 ± 14	75
PCl_5–Cl^-	160 ± 20	65
$SiCl_4$–Cl^-	101 ± 8	40

Moving Across the Periodic Table

Figure 1 shows that one of the two occupied orbitals in a 3c-4e bond has no population on the central atom. Furthermore, the formal charge on the central atom is zero. This suggests that the nature of the central atom has less effect on the strength of 3c-4e bonding. Indeed, Figure 15 shows that there is little effect on the bond strength as A is varied across the periodic table, given the same 10-electron count on the central atom.

Moving Down the Periodic Table

The same arguments suggest that bonding is not affected by the row of the periodic table in which A is located. However, another factor influencing the bond strength is the spatial extent of the *p* orbitals on A. These orbitals are larger for elements lower in the periodic table, which affects the overlap with the *p* orbitals on the terminal atoms. Because the terminal atoms are

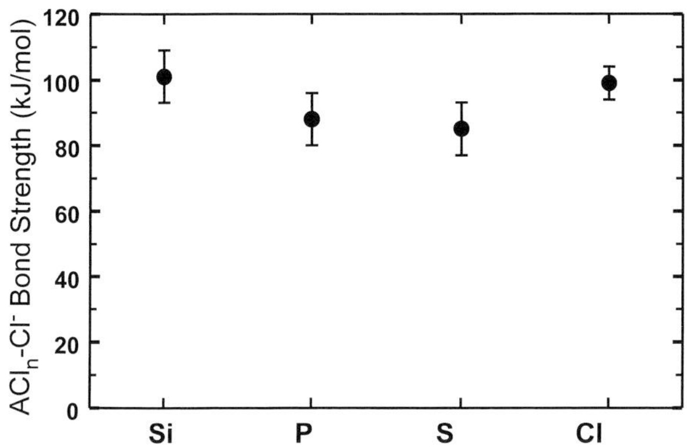

Figure 15. Chloride binding energies in the $SiCl_5^-$, PCl_4^-, SCl_3^-, and $ClCl_2^-$ systems, showing nearly equal bond energies.

Table 2. Anion and Atomic Radii[a]

Anion	*Crystal radius (pm)*	*Atom*	*Covalent radius (pm)*
F^-	119	N	75
Cl^-	167	P	110
Br^-	182	As	121
I^-	206	Sb	139
		Bi	151

[a]Data from Ref. [28].

negatively charged (and the central atom is typically positively charged), it is roughly appropriate to compare the crystal radii of the halogen anions to the covalent radii of the neutral atoms. These data are given in Table 2. It suggests that chloride orbitals should have a better size match and improved overlap with orbitals of elements lower in the periodic table. Figure 16 shows the effect of varying A down the periodic table on the bond strength in group 15 tetrachloride and tetrafluoride anions. The bond energies increase for central atoms lower on the periodic table, in qualitative agreement with the above rationale. $D(AsF_5–F^-)$ is also greater than $D(PF_5–F^-)$,[49] in agreement with the plotted trends.

Preliminary measurements on ClF_2^- give a bond energy of ca. 200 kJ/mol.[70] This is about twice the bond energy in the F_3^- system, 101 ± 11 kJ/mol.[60] It is tempting to interpret these results as indicating that 3c-4e interactions account for about half of the binding energy in ClF_2^-, with expansion of

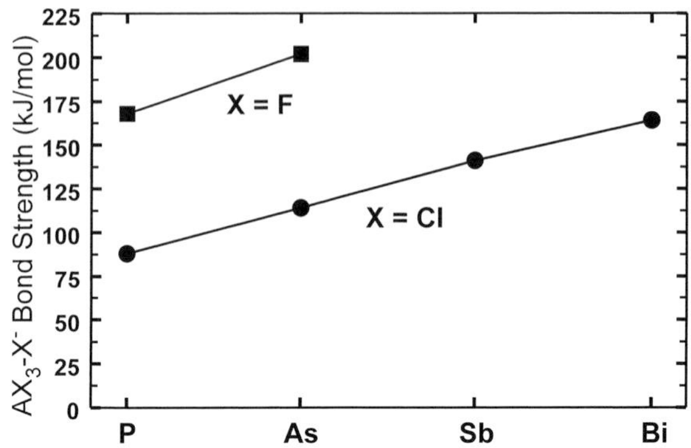

Figure 16. Halide binding energies in the Group 15 tetrachloride and tetrafluoride anions. The lines are aids to visualization only.

the octet responsible for the other half. This is not consistent with the computational results discussed above; instead, weak hypervalent bonds with central atoms from the second row can be attributed largely to electron pair–electron pair repulsion and electronegativity effects.[12–14,27,71] The electronic repulsion is greater for smaller atoms, and also manifests itself in nonhypervalent systems. For example, F has a substantially lower diatom bond energy than Cl: D(F–F) = 155 kJ/mol, D(Cl–Cl) = 239 kJ/mol).[17] The electronegativity effect will be discussed in the next section. Overall, these results are a strong argument in favor of the 3c-4e model being more accurate than the expanded octet model.

B. The Terminal Atoms

The most complete set of data regarding the effect of the terminal atom involves PX_4^-, X = F, Cl, Br, and I.[72] The upper bound on the iodide system is estimated on the grounds that attempts to make this ion in the flowing afterglow have not been successful. Figure 17 shows the PX_4^- bond energies as a function of Pauling electronegativity.[27] Clearly, bond strengths are weaker when X is less electronegative (lower on the periodic table). While this correlation should not be taken too literally, more electronegative terminal atoms stabilize hypervalent bonding.[18,54,58,73] A similar effect is seen in ClX_2^-, X = Cl and F. The bond energy in these systems are 99 and ca. 200 kJ/mol,[70] respectively. The data are consistent with negative charge accumulation on the terminal atoms. The bond strengths also correlate with an increasing size mismatch between terminal atom anionic radii and central atom atomic radii, as discussed above.

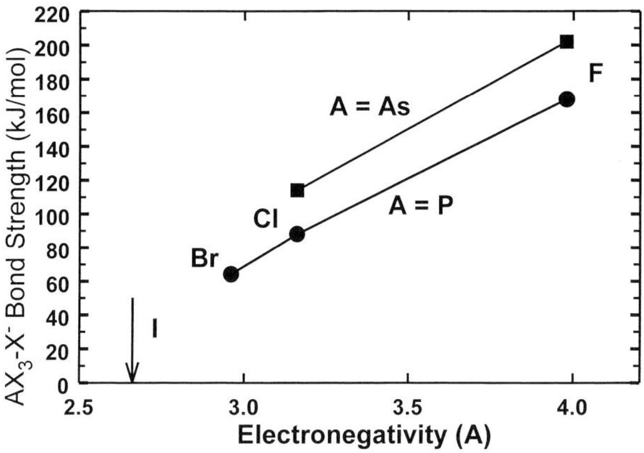

Figure 17. Bond energies in the phosphorus and arsenic tetrahalide anions plotted as a function of halogen electronegativity. The arrow is an upper limit to the PI_3–I^- bond energy. The lines are aids to visualization only.

C. The Number of Ligands

In solution, AX_5 species are known to be much stronger Lewis acids (electron pair acceptors) than AX_3 species.[18] The AX_6^- anions, particularly the fluorides, are noted for their stability, being commonly used as noncoordinating ligands with highly reactive cations.[18,74]

Relatively little gas phase data are available on the effect of varying the number of ligands n. $D(PF_3–F^-) = 168 \pm 8.4$ kJ/mol[40] while $D(PF_5–F^-) = 360 \pm 42$[40] or 329 ± 14 kJ/mol,[75] a factor of two larger. Our data show that $D(PCl_3–Cl^-) = 91 \pm 7$ kJ/mol and $D(PCl_5–Cl^-) = 160 \pm 20$ kJ/mol, again a factor of two larger. The same ordering occurs for arsenic trifluoride and pentafluoride.[49] This is a surprisingly large effect, for which two possible explanations follow.

Sigma-Acceptor Effects

The bonding in these systems can be considered in terms of donor–acceptor interactions. The oxidation states of phosphorus in PX_3 and PX_5 are three and five, respectively. Thus, the phosphorus atom is more positively charged in PX_5. For example, AIM calculations at the B3LYP/6–31+G(d) level give charges on phosphorus of +1.70 in PCl_5, +1.21 in PCl_3, +3.83 in PF_5, and +1.56 in PF_3. Higher charge

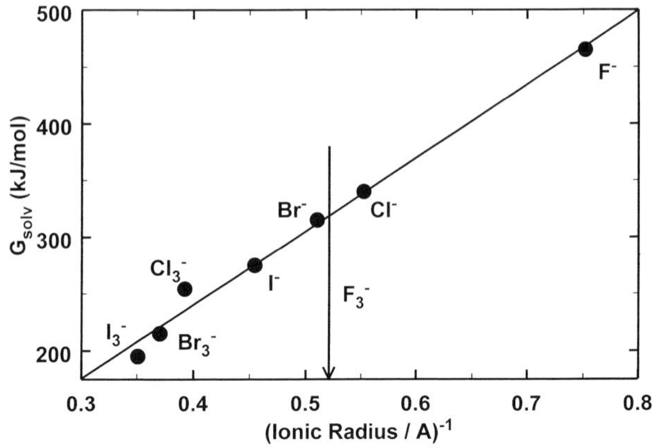

Figure 19. Free energies of solvation of halide and trihalide anions as a function of inverse ionic radius, in accordance with the Born Model. The arrow indicates an upper limit on the trifluoride solvation energy.

enthalpies of the dihalogen molecules are relatively small.[58,60] Therefore, the primary difference between the gas-phase and solution-phase bond strengths is the difference between the solvation enthalpies of X^- and X_3^-. The solvation energies of the monohalide anions were in fact one of the foundations of the Born Model for solvation energetics,[84] which states that the free energy of solvation should be proportional to $1/r_{ion}$. Figure 19 shows experimental free energies of solvation plotted against the inverse ionic radii.[58,60] The arrow for F_3^- represents an upper bound on the free energy of solvation derived from the lack of stability of this ion. This plot illustrates the importance of solvent effects on these bond energies; in particular, the loss of much of the huge solvation energy of F^- explains why F_3^- has not been seen in solution.

Although polyhalide anions are most frequently studied in aqueous solution, they are in fact more stable in less polar solutions. For example, $D(I_2–I^-) = 17$ kJ/mol in aqueous solution, and 47 kJ/mol in acetone.[85] While halide anions are spherical, the trihalide anions are oblate. The Born model is still reasonably accurate for these systems if a spherical ion approximation is used with either an experimental value of the radius[82] or with the assumption that the volume of X_3^- is simply three times the volume of X^-. However, the model does not work for the pentahalide ions I_5^- and Br_5^-.[58] Thus, application of the Born model to larger anions is an oversimplification. As more data become available, it may be possible to use more sophisticated models of solvation to correlate gas- and solution-phase data.

VI. CONCLUSIONS

Although only a small fraction of the accessible gas-phase hypervalent ions have been studied, sufficient data are available to reach several conclusions. The gas-phase and solution-phase thermodynamics of hypervalent bonds are substantially different, and the differences are explained by the effects of solvation. Bond strengths in AX_n^- species tend to be stronger for A lower in the periodic table, but strong hypervalent bonds with second-row elements do exist. Bond strengths are also stronger for X higher in the periodic table, and *n* larger. Finally, long-held explanations that appear to be in complete agreement with experiment can in fact be supplanted by better models as new data become available. Future studies will explore the diversity of hypervalent bonding in more detail.

ACKNOWLEDGMENTS

I thank my NIU coworkers Steven Bachrach, Catharine Check, Khanh Do, BettyCep Gailbreath, Tom Gilbert, Terry Heil, Pamela Keating, Timothy Klein, Katrina Nizzi, Cynthia Pommerening, Aleksandra Strukowska, John Torchia, Barry Walker, and Erica White, and Purdue collaborators Alex Artau, Brian Hill, and Paul Wenthold for their invaluable contributions to this work. Peter Armentrout, Kent Ervin, and Mary Rodgers are thanked for developing the CRUNCH data analysis software and for extensive advice in its use. Greg Gellene, W. Roy Mason, and the late Robert R. Squires made many contributions to my understanding of hypervalent bonding; any remaining inaccuracies are the sole fault of the author. Acknowledgment is made to the National Science Foundation, the Petroleum Research Fund, administered by the American Chemical Society, and the American Society for Mass Spectrometry for the partial support of this research. The College of Liberal Arts and Sciences and the Graduate School at NIU provided student support for this project. Special thanks go to Bob Squires and Peter Armentrout for their development of flowing afterglow-tandem mass spectrometry and their mentoring and support over the years.

REFERENCES

[1] Castleman, A. W., Jr.; Bowen, K. H., Jr.; *J. Phys. Chem.* **1996**, *100*, 12911–12944. Takashima, K.; Riveros, J. M. *Mass Spectrom. Rev.* **1998**, *17*, 409–430.

[2] Chabinyc, M. L.; Craig, S. L.; Regan, C. K.; Brauman, J. I. *Science* **1998**, *279*, 1882–1886.

[3] Lewis, G. N. *J. Am. Chem. Soc.* **1916**, *38*, 762.

[4] Lewis, G. N. *Valence and the Structure of Atoms and Molecules*; Dover: New York, 1966 (reprint of Chemical Catalog Company: New York, 1923 edition).

[5] Musher, J. I. *Angew. Chem. Int. Ed. Engl.* **1969**, *8*, 54–68.

[6] Hach, R. J.; Rundle, R. E. *J. Am. Chem. Soc.* **1951**, *73*, 4321–4324.

[7] Pimentel, G. C. *J. Chem. Phys.* **1951**, *19*, 446–448.

[8] Ramsden, C. A. *Chem. Soc. Rev.* **1994**, 111–118.

[9] Cahill, P. A.; Dykstra, C. E.; Martin, J. E. *J. Am. Chem. Soc.* **1985**, *107*, 6359–6362.
[10] Gillespie, R. J.; Robinson, E. A. *Inorg. Chem.* **1995**, *34*, 978–979.
[11] Mingos, D. M. P. *Essential Trends in Inorganic Chemistry*; Oxford University Press: Oxford, 1998.
[12] Atkins, P. W. *Quanta*. 2nd ed.; Oxford University Press: Oxford, 1991.
[13] Kutzelnigg, W. *Angew. Chem. Int. Ed. Engl.* **1984**, *23*, 272–295.
[14] Cioslowski, J.; Mixon, S. T. *Inorg. Chem.* **1993**, *32*, 3209–3216.
[15] Pauling, L. *The Nature of the Chemical Bond,* 2nd ed.; Cornell University Press: Ithaca, 1940.
[16] Shurki, A.; Hiberty, P. C.; Shaik, S. *J. Am. Chem. Soc.* **1999**, *121*, 822–834.
[17] Afeefy, H. Y.; Liebman, J. F.; Stein, S. E. "Neutral Thermochemical Data," and Bartmess, J. E., "Negative Ion Energetics Data". In *NIST Chemistry WebBook*, NIST Standard Reference Database Number 69, Mallard, W. G.; Linstrom, P. J., Eds.; National Institute of Standards and Technology: Gaithersburg MD, 20899, February 2000. (http://webbook.nist.gov).
[18] Cotton, F. A.; Wilkinson, G.; Murillo, C. A.; Bochmann, M. *Advanced Inorganic Chemistry*, 6th ed.; Wiley: New York, 1999.
[19] Magnusson, E. *J. Am. Chem. Soc.* **1990**, *112*, 7940–7951. Magnusson, E. *J. Am. Chem. Soc.* **1993**, *115*, 1051–1061.
[20] Dobado, J. A.; Martìnez-Garcìa, H.; Molina, J. M.; Sundberg, M. R. *J. Am. Chem. Soc.* **2000**, *122*, 1144–1149.
[21] Reed, A. E.; Schleyer, P. v. R. *J. Am. Chem. Soc.* **1990**, *112*, 1434–1445.
[22] Cleveland, T.; Landis, C. R. *J. Am. Chem. Soc.* **1996**, *118*, 6020–6030.
[23] Novoa, J. J.; Mota, F.; Alvarez, S. *J. Phys. Chem.* **1988**, *92*, 6561–6566.
[24] Babcock, L. M.; Streit, G. E. *J. Chem. Phys.* **1981**, *75*, 3864–3870.
[25] Bader, R. F. W. *Atoms in Molecules: A Quantum Theory*; Clarendon Press: Oxford, 1990.
[26] Cioslowski, J.; Nanayakkara, A.; Challacombe, M. *Chem. Phys. Lett.* **1993**, *203*, 137. Cioslowski, J.; Surjan, P. R. *J. Mol. Struc.* **1992**, *255*, 9. Cioslowski, J.; Stefanov, B. B. *Mol. Phys.* **1995**, *84*, 707. Stefanov, B. B.; Cioslowski, J. R. *J. Comp. Chem.* **1995**, *16*, 1394. Cioslowski, J. *Int. J. Quant. Chem. Quant. Chem. Symp.* **1990**, *24*, 15. Cioslowski, J.; Mixon, S. T. *J. Am. Chem. Soc.* **1991**, *113*, 4142. Cioslowski, J. *Chem. Phys. Lett.* **1992**, *194*, 73. Cioslowski, J. *Chem. Phys. Lett.* **1994**, *219*, 151.
[27] Burdett, J. *Chemical Bonds: A Dialog*; Wiley: New York, 1997.
[28] Wulfsberg, G. *Inorganic Chemistry*; University Science Books: Sausalito, CA, 2000.
[29] See, for example: Atkins, P. W.; Jones, L. L. *Chemistry: Molecules, Matter and Change*, 3rd ed.; W. H. Freeman: New York, 1997. Chang, R. *Chemistry*, 6th ed.; McGraw-Hill: Boston, 1998. McMurry, J.; Fay, R. C. *Chemistry*, 2nd ed.; Prentice-Hall: Upper Saddle River, NJ, 1998. Ebbing, D. D.; Gammon, S. D. *General Chemistry*, 6th ed.; Houghton Mifflin: Boston, 1999. Silverberg, M. *Chemistry*, 2nd ed.; McGraw-Hill: Boston, 2000. Brown, T. L.; LeMay, H. E.; Bursten, B. E. *Chemistry: The Central Science*, 8th ed.; Prentice Hall: Upper Saddle River, NJ, 2000. Brady, J. E.; Russell, J. W.; Holum, J. R. *Chemistry: Matter and Its Changes*, 3rd ed.; Wiley: New York, 2000. Zumdahl, S. S.; Zumdahl, S. A. *Chemistry*, 5th ed.; Houghton Mifflin: Boston, 2000 (this book at least mentions the existence of other models in a footnote).
[30] Akiba, K.-Y. *Chemistry of Hypervalent Compounds*; Wiley-VCH: New York, 1999.
[31] Damrauer, R.; Hankin, J. A. *Chem. Rev.* **1995**, *95*, 1137–1160.
[32] Hajdasz, D. J.; Ho, Y.; Squires, R. R. *J. Am. Chem. Soc.* **1995**, *116*, 10751–10760.
[33] Akiba, K.; Yamashita, M.; Yamamoto, Y.; Nagase, S. *J. Am. Chem. Soc.* **1999**, *121*, 10644–10645.
[34] Wong, M. W. *J. Am. Chem. Soc.* **2000**, *122*, 6289–6290.

[35] Landrum, G. A.; Goldberg, N.; Hoffmann, R. *J. Chem. Soc. Dalton Trans.* **1995**, 3605–3613.
[36] Gailbreath, B. D.; Pommerening, C. A.; Bachrach, S. M.; Sunderlin, L. S. *J. Phys. Chem. A* **2000**, *104*, 2958–2601.
[37] Do, K.; Klein, T. P.; Pommerening, C. A.; Sunderlin, L. S. *J. Am. Soc. Mass Spectrom.* **1997**, *8*, 688–696.
[38] Armentrout, P. B. *Advances in Gas Phase Ion Chemistry, Vol. I*; Adams, N. G.; Babcock, L. M.; Eds., JAI Press: New York, 1992, 83–119.
[39] Hogness, T. R.; Harkness, R. W. *Phys. Rev.* **1928**, *32*, 784–790.
[40] Larson, J. W.; McMahon, T. B. *J. Am. Chem. Soc.* **1985**, *107*, 766–773.
[41] Smith, D.; Adams, N. G. *Adv. At. Mol. Phys.* **1988**, *24*, 1–49, and refs. therein.
[42] Gerlich, D. *Adv. Chem. Phys.* **1992**, *82*, 1–176.
[43] Ervin, K. M.; Loh, S. K.; Aristov, N.; Armentrout, P. B. *J. Phys. Chem.* **1983**, *87*, 3593. Ervin, K. M.; Armentrout, P. B. *J. Chem. Phys.* **1985**, *83*, 166–189.
[44] Aristov, N.; Armentrout, P. B. *J. Am. Chem. Soc.* **1986**, *108*, 1806–1819.
[45] Gaussian 98, Revision A.9: Frisch, M. J.; Trucks, G. W.; Schlegel, H. B.; Scuseria, G. E.; Robb, M. A.; Cheeseman, J. R.; Zakrzewski, V. G.; Montgomery, J. A., Jr.; Stratmann, R. E.; Burant, J. C.; Dapprich, S.; Millam, J. M.; Daniels, A. D.; Kudin, K. N.; Strain, M. C.; Farkas, O.; Tomasi, J.; Barone, V.; Cossi, M.; Cammi, R.; Mennucci, B.; Pomelli, C.; Adamo, C.; Clifford, S.; Ochterski, J.; Petersson, G. A.; Ayala, P. Y.; Cui, Q.; Morokuma, K.; Malick, A D.; Rabuck, K. D.; Raghavachari, K.; Foresman, J. B.; Cioslowski, J.; Ortiz, J. V.; Baboul, A. G.; Stefanov, B. B.; Liu, G.; Liashenko, A.; Piskorz, P.; Komaromi, I.; Gomperts, R.; Martin, R. L.; Fox, D. J.; Keith, T.; Al-Laham, M. A.; Peng, C. Y.; Nanayakkara, A.; Challacombe, M.; Gill, P. M. W.; Johnson, B.; Chen, W., Wong, M. W.; Andres, J. L.; Gonzalez, C.; Head-Gordon, M.; Replogle, E. S.; Pople. J. A. Gaussian, Inc.: Pittsburgh PA, 1998.
[46] PC Spartan Pro, v. 1.01: Wavefunction, Inc.: Irvine, CA, 1999.
[47] Rodgers, M. T.; Ervin, K. M.; Armentrout, P. B. *J. Chem. Phys.* **1997**, *106*, 4499–4508. Rodgers, M. T.; Armentrout, P. B. *J. Chem. Phys.* **1998**, *109*, 1787–1800.
[48] Loh, S. K.; Lian, L.; Hales, D. A.; Armentrout, P. B. *J. Chem. Phys.* **1989**, *90*, 5466–5485. Schultz, R. H.; Crellin, K. C.; Armentrout, P. B. *J. Am. Chem. Soc.* **1991**, *113*, 8590–8601.
[49] Haartz, J. C.; McDaniel, D. H. *J. Am. Chem. Soc.* **1973**, *95*, 8562–8565.
[50] Mathur, B. P.; Rothe, E. W.; Tang, S. Y.; Reck, G. P. *J. Chem. Phys.* **1976**, *65*, 565–569.
[51] Bowen, K. H.; Liesegang, G. W.; Sanders, R. A.; Hershbach, D. R. *J. Phys. Chem.* **1983**, *87*, 557–565.
[52] Taylor, T. R.; Asmis, K. R.; Zanni, M. T.; Neumark, D. M. *J. Chem. Phys.* **1999**, *110*, 7607–7609.
[53] Kaledin, A. L.; Heaven, M. C.; Morokuma, K.; Neumark, D. M. *Chem. Phys. Lett.* **1999**, *306*, 48–52.
[54] Jenkins, H. D. B.; Sharman, L.; Finch, A.; Gates, P. N. *Inorg. Chem.* **1996**, *35*, 6316–6326.
[55] Topol, L. E. *Inorg. Chem.* **1971**, *10*, 736–740.
[56] Finch, A.; Gates, P. N.; Peake, S. J. *J. Inorg. Nucl. Chem.* **1977**, *39*, 2135–2138.
[57] Zanni, M. T.; Taylor, T. R.; Greenblatt, B. J.; Soep, B.; Neumark, D. M. *J. Chem. Phys.* **1997**, *107*, 7613–7619.
[58] Nizzi, K. E.; Pommerening, C. A.; Sunderlin, L. S. *J. Phys. Chem.* **1998**, *102*, 7674–7679.
[59] Tuinman, A. A.; Gakh, A. A.; Hinde, R. J.; Compton, R. N. *J. Am. Chem. Soc.* **1999**, *121*, 8397–8398.
[60] Artau, A.; Nizzi, K. E.; Hill, B. T.; Sunderlin, L. S.; Wenthold, P. G. *J. Am. Chem. Soc.* **2000**, *122*, 10667–10670.
[61] Wenthold, P. G.; Squires, R. R. *J. Phys. Chem.* **1995**, *99*, 2002–2005.
[62] Ayala, J. A.; Wentworth, W. E.; Chen, E. C. M. *J. Phys. Chem.* **1981**, *85*, 768–777.

[63] Nizzi, K. E.; Sunderlin, L. S. manuscript in preparation.
[64] Armentrout, P. B.; Simons, J. *J. Am. Chem. Soc.* **1992**, *114*, 8627–8633.
[65] Check, C.; Strukowska, A.; Walker, B. W.; Gilbert, T. M.; Sunderlin, L. S. work in progress.
[66] Hoffmann, R.; Howell, J. M.; Muetterties, E. L. *J. Am. Chem. Soc.* **1972**, *94*, 3047–3058.
[67] Moc, J.; Morokuma, K. *J. Am. Chem. Soc.* **1995**, *117*, 11790–11797.
[68] Kutzelnigg, W.; Wasilewski, J. *J. Am. Chem. Soc.* **1982**, *104*, 953–960.
[69] Reed, A. E.; Schleyer, P. v. R. *Chem. Phys. Lett.* **1987**, *133*, 553–561.
[70] White, E.; Nizzi, K. E.; Sunderlin, L. S. work in progress.
[71] Gerratt, J.; Cooper, D. L.; Karadakov, P. B.; Raimondi, M. *Chem. Soc. Rev.* **1997**, 87–100.
[72] Heil, T. E.; Walker, B. W.; Sunderlin, L. S. Manuscript in preparation. Walker, B. W. M. S. Thesis, Northern Illinois University, 1999.
[73] Ogawa, Y.; Takahashi, O.; Kikuchi, O. *J. Mol. Struct.* **1998**, *424*, 285–292. Ogawa, Y.; Takahashi, O.; Kikuchi, O. *J. Mol. Struct.* **1998**, *429*, 187–196.
[74] Drews, T.; Koch, W.; Seppelt, K. *J. Am. Chem. Soc.* **1999**, *121*, 4379–4384.
[75] Aleshina, V. E.; Borshchevskii, A.Y.; Korobov, M. V.; Sidorov, L. N. *Russ. J. Phys. Chem.* **1996**, *70*, 1085–1089.
[76] Huheey, J. E.; Keiter, E. A.; Keiter, R. L. *Inorganic Chemistry*, 4th ed.; Harper Collins: New York, 1993.
[77] Most data taken from sources given in Table 2 in Ref. 58. Further data from Ref. 73 and: Malcolm, N. O. J.; McDouall, J. J. W. *J. Chem. Phys.* **1996**, *100*, 10131–10134. Mota, F. Novoa, J. J. *J. Chem. Phys.* **1996**, *105*, 8777–8784. Schuster, P.; Mikosch, H.; Bauer, G. *J. Chem. Phys.* **1998**, *109*, 1833–1844.
[78] King, R. A.; Gailbraith, J. M.; Schaefer, H. F. *J. Phys. Chem.* **1996**, *100*, 6061–6068.
[79] Heard, G. L.; Marsden, C. J.; Scuseria, G. E. *J. Phys. Chem.* **1992**, *96*, 4359–4366.
[80] Sharp, S. B.; Gellene, G. I. *J. Phys. Chem. A* **1997**, *101*, 2192–2197.
[81] Galbraith, J. M.; Schaefer, H. F., III. *J. Chem. Phys.* **1996**, *105*, 862–864. Van Huis, T. J.; Galbraith, J. M.; Scheafer, H. F., III. *Mol. Phys.* **1996**, *89*, 607–631. King, R. A.; Galbraith, J. M.; Schaefer, H. F., III. *J. Phys. Chem.* **1996**, *100*, 6061–6068. Tschumper, G. S.; Fermann, J. T.; Schaefer, H. F., III. *J. Chem. Phys.* **1996**, *104*, 3676–3683. Pak, C.; Van Huis, T. J.; Schaefer, H. F., III. *J. Am. Chem. Soc.* **1998**, *120*, 11115–11121.
[82] Marcus, Y. *Ion Properties*; Marcel Dekker: New York, 1997.
[83] Sato, H.; Hirata, F.; Myers, A. B. *J. Phys. Chem. A* **1998**, *102*, 2065–2071. Lynden-Bell, R. M.; Kosloff, R.; Ruhman, S.; Danovich, D.; Vala, J. *J. Chem. Phys.* **1998**, *109*, 9928–9937.
[84] Atkins, P. *Physical Chemistry*, 6th ed.; Freeman: New York, 1998.
[85] Nelson, I. V.; Iwamoto, R. T. *J. Electroanal. Chem.* **1964**, *7*, 218–221.

ION-MOLECULE KINETICS AT HIGH TEMPERATURES (300–1800 K):

DERIVATION OF INTERNAL ENERGY DEPENDENCIES

A. A. Viggiano and Skip Williams

OUTLINE

Advances in Gas-Phase Ion Chemistry
Volume 4, pages 85–136.

ABSTRACT

This chapter describes the Air Force Research Laboratory's high temperature flowing afterglow (HTFA) apparatus and summarizes the ion-molecule kinetics measured in the HTFA to date. The HTFA apparatus is capable of operating at temperatures as high as 1800 K, and measurements in the 300–1400 K temperature range are made routinely. While numerous motivations for obtaining kinetics information at high temperatures and the practical applications of such data are discussed, the chapter focuses on how the high temperature data are used to derive internal energy dependencies of ion-molecule reactions. Internal energy dependencies are derived from comparisons of the true temperature data measured in the HTFA, for which all degrees of freedom are thermally excited, to data for which only translational energy is increased, i.e. drift tube or beam data. For selected charge transfer reactions, the recombination energy of the reactant ion is varied, and the results are compared to the HTFA data to obtain information regarding the role of electronic energy. For atomic ion–diatomic molecule reactions, separate rotational and vibrational dependencies are derived. However, for reactions involving polyatomic ions and molecules, only dependencies with respect to total internal energy can be obtained from the HTFA data. The chapter concludes with a discussion of the apparent trends that have been observed. In general, rotational and translational energy have been found to be equivalent in controlling total reactivity. Recombination energy and internal energy have also been found to be equivalent in controlling dissociation after charge transfer for molecules with long dissocation lifetimes. Vibrational energy dependencies are complex and highly dependent on the reaction mechanism.

I. INTRODUCTION

Ion chemistry is a mature although continually evolving field, with a wide variety of techniques being exploited to measure ion reactivity over a large range of conditions.[1] In the latest complete compilation of ion-molecule kinetics, now 14 years old, there are over 10,000 separate entries[2] and the number of reactions studied has continued to be impressive. This large body of work has led to many insights into reactivity and numerous generalities have emerged. However, in spite of the large number of studies, there are still several areas of ion kinetics that are largely unexplored. One of these is the subject of this review, namely the study of ion-molecule reactions at high temperatures.

The vast majority of the work on ion-molecule kinetics has been performed at room temperature.[2] A fraction of these room temperature kinetic studies have been made as a function of ion translational energy. However, temperature dependent studies have been limited mostly to the 77 K (liquid nitrogen) to 600 K range because of the significant technical difficulties encountered. Outside of this temperature range, special considerations apply; e.g. the stability of construction materials and reactants at high temperature or condensation of the reactant species at low temperature. Most of the efforts to extend the temperature range have focussed on low temperatures,[3] due to the fact that many of the molecular species made in interstellar clouds are synthesized by ion-molecule reactions at extremely low temperatures.[4] The techniques used to study low-temperature chemistry have been quite successful and have provided good tests of theory, especially with regard to ion-molecule collision rates.[5–7]

In contrast, the number of studies made at high temperature (>600 K) is very limited, particularly before the instrument that is the subject of this review was developed. Previous work on ion-molecule reactivity at temperatures over 600 K was performed in the early 1970s and was limited to temperatures of 900 K and below.[8,9] The impetus for those studies focussed on measurements relevant to ionospheric chemistry. The kinetic temperature of the ionosphere reaches 1800–2000 K at several hundred kilometers under certain conditions.[10] Thus, the 900 K maximum covered only half the temperature range of interest. A further limitation was that product ion branching fractions could not be measured. Nevertheless, the technically challenging measurements provided useful and interesting data on how temperature affected rate constants. However, the generality of the conclusions that could be made was limited because only 10 reactions were studied in total at the highest temperatures.

The approximately 1000 K gap between the previous maximum laboratory operating temperature and the upper ionospheric temperature was covered in other ways. In particular, the reactivity was studied as a function of ion translational energy in drift tubes and beam apparatuses.[1] A limited amount of information on vibrational excitation of the primary reactant ion only could also be derived in the drift tube experiments. This allowed effective temperature dependencies to be calculated assuming kinetic temperature was the main controlling force in driving reactivity, i.e. $KE=(3/2)k_BT_{eff}$.[11–13] As will be shown later, this approach can lead to large errors, although it was the only reasonable way to extrapolate to higher temperature conditions at the time.

Previously, the conventional wisdom was that vibrational energy played a major role in reactivity and that rotational energy could largely be ignored. In the ionosphere, most of the chemistry involves only monatomic and diatomic ions and neutrals, and therefore very little vibrational excitation is present at temperatures below 900 K. Thus, the impact of both rotational and vibrational energy was not seriously considered in the high temperature regions of the ionosphere. A notable exception was for the most important ionospheric reaction,

$$O^+ + N_2 \rightarrow NO^+ + N \tag{1}$$

For this reaction, a separate study of the rate constant dependence on the vibrational temperature of the N_2 reactant was made.[14,15] However, in that study both the ion center of mass (CM) translational energy and the rotational temperature were 300 K. While this was an obviously important step, no true temperature dependent study was made over 900 K. True temperature here refers to the case where the translational, rotational, and vibrational degrees of freedom of the reactants are all in equilibrium and can be represented by a single temperature.

The lack of measurements over the complete temperature range of the ionosphere was one of several driving forces that led us to develop a flowing afterglow capable of reaching temperatures of 1800 K. This apparatus will be hereafter referred to as the High Temperature Flowing Afterglow or HTFA. While the high temperature ionospheric plasma chemistry was an important driver for the development of the HTFA, there are other plasmas that require accurate ion-molecule kinetic measurements at high temperature. Examples include plasma sheathing around high speed vehicles during reentry or hypersonic flight that affect on-board rf-instrumentation, plasma processors in industry, and plasma igniters and pilots for internal combustion engines.

This paper reviews the systems studied to date with the HTFA and compares the HTFA results to those of other experiments when available. Initially, we limited studies to simple test systems. For a number of years only rate constants could be measured, in large part because the impurity ions from outgassing materials prevented branching fraction determinations. Fortunately, many of the reactions produced only one product (at least at lower temperatures) and this limitation did not often result in much information being missed. The early studies included most of the important ionospheric reactions with the exception of the reactions involving O atoms. The upper temperature limit for these studies was at least 1400 K and in some cases 1800 K.

The efficacy of certain types of energy in promoting a particular reaction can be evaluated by comparing the results of different experiments. For instance, molecules are capable of "storing" significant amounts of rovibrational energy at high temperatures. Thus, varying the temperature of the experiment increases the rovibrational energy available for reaction. Translational energy dependencies of ion-molecule reactions are studied by accelerating the primary reactant ion in an electric field as is done in drift tube and beam studies. In charge transfer reactions, varying the recombination energy of the primary ion can yield information regarding the role of electronic energy.[16,17] This statement follows from charge-transfer energy resonance criteria where the charge transfer probability can be described as the product between the pure electronic coupling and the Franck-Condon overlap between reactant and product wave functions.[18]

Thus, the HTFA data are also of fundamental interest because they contain information revealing how internal excitation effects reactivity. For instance, vibrational and rotational information can be obtained by comparing the HTFA data to drift tube and ion beam studies.[19] Note that, while drift tube and ion beam data can cause internal excitation of ions, the distribution will in general be much colder than true temperature data. In particular, the high temperature capability extends our previous work in this area by allowing high frequency vibrations to be excited and substantially increasing the amount of rotational energy that can be added. We have even been able to derive rates for vibrationally excited H_2 and D_2. The limitation of not being able to excite diatomic vibrations in the work done in the 1970s is clearly evident.

Recently, more diverse types of systems have been studied involving large molecules, such as aromatic hydrocarbons containing a large number of low to moderate frequency vibrational modes which become significantly populated at high temperatures. Table 1 shows the average amount of internal energy (vibrational and rotational) available for reaction of benzene, naphthalene, and ethylbenzene at temperatures ranging from 300–1400 K. Over 3 eV of internal energy is available for reactions of naphthalene at 1400 K. This amount of internal energy is significant compared to the differences in energetics for the large number of possible reaction channels. The ability to measure product branching fractions is the key to studying these reactions, and such measurements are now done routinely in the HTFA for ions that can be generated from parent gases, i.e., O_2^+ from O_2. The initial studies in this area have shown that vibrational energy can have a large effect on the product branching fractions as well as the reaction rate constants.

Table 1. Rovibrational Energy (eV) at Selected Temperatures Ranging from 300 to 1400 K for Benzene, Naphthalene, and Ethylbenzene Molecules

T(*K*)	*Benzene*	*Naphthalene*	*Ethylbenzene*
300	0.08	0.15	0.10
500	0.27	0.47	0.36
800	0.72	1.23	1.03
1100	1.29	2.16	1.90
1400	1.93	3.19	2.89

The HTFA provides data which is of fundamental and practical interest to many research areas. However, the emphasis of this review will be on deriving internal energy dependencies by comparing the HTFA data to data taken as a function of translational energy and, for the aromatic hydrocarbon systems, as a function of ion recombination energy. Most often, the comparison is to data taken with drift tubes, but in certain cases comparisons to data taken with ion-beam experiments are also made. In the following experimental section, there is a moderately detailed description of the apparatus and discussion of the derivation of internal energy dependencies. All of the reactions studied to date are then discussed, starting with the simplest systems. The first reactions discussed are covered in detail to illustrate the type of data analysis that is used. Later examples are more focussed on the results. The aromatic hydrocarbon experiments are examined in some detail since they represent a new class of molecules with different experimental considerations. Finally, the trends in the data, and major conclusions that have been found to date, are summarized.

II. EXPERIMENTAL METHODS

The HTFA is in practice operated like a typical flowing afterglow of which the principles have been reported in detail elsewhere.[20,21] A detailed description of the Air Force Research Laboratory flowing afterglow apparatus is found in Ref. 22, so only the key aspects and recent modifications will be discussed here with emphasis on the particular problems associated with high temperature measurements. The remainder of the section outlines how the HTFA data are combined with other measurements to derive internal energy dependencies.

A. Apparatus

Our first attempt at making measurements at very high temperature involved a stainless steel flow tube wrapped with heating tape. This apparatus had a maximum temperature of 1300 K,[22,23] and has been out of service for a number of years. This version of the HTFA will not be discussed further. Figure 1 shows a schematic of the second generation HTFA currently in use. In this version, either an alumina or an industrial strength quartz flow tube is inserted into the center of a three-zone commercial furnace. The inner diameter of the flow tubes is 7.6 cm and the tubes are 106 cm long. Ions are created upstream in a cooled section of the

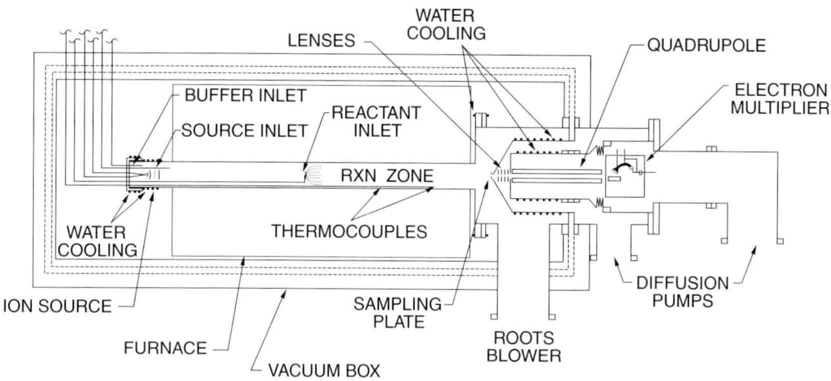

Figure 1. Schematic drawing of the high temperature flowing afterglow (HTFA).

flow tube by electron impact. A buffer carrier gas, usually helium, carries the ions downstream. The reactant neutral is added 50 cm from the end of the flow tube and kinetic measurements are taken as a function of the reactant neutral flow rate. A small amount of the gas mixture is sampled by a 400 micron hole in a truncated nose cone, which terminates the flow tube. Ions are separated by a quadrupole mass filter and detected by a discrete dynode electron multiplier. The bulk of the carrier gas is pumped away by a 1449 CFM Roots type blower.

Reaction rate constants are measured by following the change in primary ion signal as a function of added neutral reactant flow rate. Concentrations are such that [buffer] $\gg$ [reactant neutral] $\gg$ [ions]. Under these conditions, pseudo-first order kinetics apply, and the rate constant is given by

$$k = \frac{1}{[B]\tau} \ln\left\{\frac{[A_o^{\pm}]}{[A^{\pm}]}\right\} \tag{2}$$

where k is the rate constant, τ is the reaction time, $[B]$ is reactant neutral concentration, $[A_0^{\pm}]$ and $[A^{\pm}]$ are the primary ion concentrations in the absence and presence of reactant neutral. The reaction time, τ, is the reaction distance divided by the buffer velocity and multiplied by a correction factor that accounts for the fact that both the ion velocity and ion concentration are at a maximum along the axis of the flow tube. A typical value for the correction factor is 1.55, which will be discussed later. The buffer gas velocity is obtained from the mass flow rate of the buffer, the flow tube cross section, temperature and pressure in the normal manner.[21] Up to now, we have described a typical flowing afterglow. The numerous details pertaining to high temperature operation are discussed below.

The furnace is 200 kg, has a maximum operating temperature of 1875 K, and at full power, consumes 13.5 kW. The heating elements are silicon carbide and the heated region of the furnace is divided into three zones. The temperature is controlled in each zone to ± 2 K as measured by thermocouples on the outside of the flow tube. Measurements on a movable injector have confirmed that the temperature inside the flow tube is the same as the furnace temperature. The downstream sampling region is sealed by abutting the furnace to the sampling structure, which is water-cooled. The entire furnace is in a vacuum box to reduce gas leaks into the flow tube.

The alumina flow tube is used to run at temperatures up to 1800 K since it has a higher thermal stability than the quartz one. While the furnace can reach a temperature of 1875 K, the maximum operating temperature has been limited to 1800 K because lifetime of the heating elements decreases substantially as the maximum temperature is approached. Hot alumina reacts with several of the neutrals, including NO. The industrial grade quartz tube, on the other hand, has been found to be less reactive but has an upper temperature limit of 1400 K. At present, we routinely use the quartz flow tube since most molecules of interest have appreciable levels of rovibrational excitation at 1400 K. If higher temperatures are required to elucidate the chemistry for a particular reaction system, it takes about 1–2 days to switch over to the alumina flow tube.

The upstream end of the flow tube protrudes from the furnace approximately 11 cm. All the electrical and gas feedthroughs are in a water-cooled cap that is sealed to the flow tube via a Teflon washer. The ion source consists of a filament biased with respect to a nearby grid. The source is baffled to prevent UV light from creating ions in the reaction region. The source region is presently being replaced with the ion source in a "T" outside the diameter of the flow tube. This will ensure better baffling, allow the ion source to run at higher pressure so that lower source gas flows can be used, and cut down on the physical congestion in the source region. The source contains two interchangeable filaments to minimize system maintenance requirements, since several days of baking are required after venting to reduce background impurities to acceptably low levels.

Several gas inlets are fed through the endcap. The helium inlet and an upstream ion source inlet are located in the cap, i.e. the furthest upstream position possible. A downstream source gas inlet is located about 10 cm downstream from the ion source. The reactant inlet is located 50 cm from the nose cone. These latter two inlets are either ceramic or quartz tubes resting on the bottom of the flow tube with a bend at the end of the tube that allows gas to be added at the center of the flow tube. Pressure is measured by another tube truncated in the reaction zone with the hole perpendicular to the gas flow. The new "T" endcap being developed will allow for more gas feedthroughs.

In this design, the buffer gas is added upstream of the furnace. Therefore, sufficient time must be allowed for the buffer to equilibrate with the flow tube wall. Calculations indicate that the buffer temperature at the centerline is within several percent of the flow tube walls by the reactant inlet position.[24] This equilibration has been confirmed by thermocouple measurements in the gas flow. Furthermore, mixing of the neutral reagent is estimated to result in an end correction of 2 cm or smaller. This end correction was confirmed by measurements at two different inlet distances. This measurement is not made routinely since the pressure inlet had to be eliminated for the measurement. Diffusive loss varies as $T^{+1.5}$, which makes diffusive losses much greater at higher temperatures. Therefore, the helium flow rate and pressure are typically increased with increasing temperature. Another potential temperature problem results from the fact that the flow tube pressure is measured by an external capacitance manometer which is maintained at 318 K and is connected with 0.25 inch diameter tubing. The temperature gradient across this gas-filled tube leads to the development of a pressure gradient due to an effect known as thermal transpiration. The equation developed by Takaishi and Sensui[25] and discussed by Poulter et al.[26] is used to correct the thermal transpiration error recorded by the capacitance manometer. When lower pressures are used, the thermal transpiration correction can be as high as 20% at the highest temperatures studied. Typical corrections are in the 0–10% range.

A key parameter in calculating the rate constants is the ion velocity or the ratio of the ion velocity to the buffer velocity. The ion velocity was measured by replacing the reactant inlet with two separate molybdenum grids. The grids were pulsed individually and the arrival time spectra of the ions at the electron multiplier detector (see Figure 1) were measured. The difference in the arrival times of the individual disturbances from the two grids and the distance between them is used to determine the velocity. This difference method cancels any end corrections associated with the pulsing technique. We found that the ion velocity to the buffer velocity is roughly constant at 1.5–1.6 at temperatures up to 1000 K. Above 1000 K, the pulsing produced a dip in the ion signal too small to measure. Therefore, a value of 1.55 has been adopted for the whole temperature range. This is consistent with the value measured in the selected ion flow drift tube in our laboratory. We estimate that the uncertainty in the ion velocity adds at most a 10% error to the rate constant determination. This is verified by the fact that, for numerous reactions which proceed at the collision rate, the rate constant does not vary up to the maximum temperature studied.

The inability to completely isolate the flow tube from its surroundings initially prevented branching fraction determinations. Firebricks constitute much of the furnace material and emit a lot of gas. Initially, the vacuum box had insufficient pumping (a 300 CFM Roots pump) to reduce the pressure in the vacuum box much below that of the flow tube. This resulted in the emitted gases entering the flow tube through leaky seals. The addition of a 1300 CFM Roots pump to the vacuum box reduced the pressure sufficiently to ensure that "all" leaks were from the flow tube to the vacuum box. This resulted in much cleaner mass spectra. For instance O_2^+ could be made with greater than 99% purity and N_2^+ with greater than 95% purity. Thus, branching fractions can now be determined for selected ions that can be made cleanly. A potential complication is due to the source gas being in the flow tube. This could complicate the branching ratio determinations through reaction with the product ions. With the new ion source configuration we expect that less source gas will be needed, reducing this potential source of error.

At temperatures above 800 K, emission of alkali ions from the hot surfaces becomes comparable to the primary reactant ion intensity under standard operating conditions. The main alkali is K^+, followed by Na^+ and Rb^+. Months of running at elevated temperatures have resulted in only a slight reduction in these signals. Fortunately, these low ionization energy ions are unreactive and only serve to congest the mass spectra. For instance, a large (off-scale at high temperatures) K^+ peak makes Ar^+ measurements difficult at high temperature and prevents ions such as $C_3H_3^+$ from being detected with any reasonable sensitivity. An alkali ion signal independent of the neutral flow is used to confirm that these ions are not participating in the ion-molecule reactions being studied.

Error limits on the rate constants are estimated as follows. The reproducibility of the rate constants is typically $<10\%$. Sources of error include the above reproducibility ($\pm 10\%$), measurement of temperature ($\pm 2\%$), pressure ($\pm 1\%$), ion flight time ($\pm 5\%$), an end correction related mainly to reactant gas mixing (± 2 cm or $\pm 4\%$), buffer gas flow rate (relative, $\pm 1\%$, or absolute, $\pm 2\%$), and neutral reactant gas flow rate (relative, $\pm 3\%$, or absolute, $\pm 15\%$). Propagation of errors leads to a net relative uncertainty of $\pm 13\%$ and a net absolute uncertainty of $\pm 19\%$. The reaction rate constants measured with the HTFA apparatus are quoted with relative and absolute uncertainties of $\pm 15\%$ and $\pm 25\%$, respectively, to account for the possibility of systematic errors not treated in the above analysis.

B. Internal Energy Dependencies

The principle used to derive internal energy dependencies is similar to that used at lower temperatures[19] and is described below. The various forms of energy probed by the different experiments are defined as follows. The average reactant rotational energy, $\langle E_{rot} \rangle$, is 1/2 k_BT for each rotational degree of freedom, and the average reactant vibrational energy, $\langle E_{vib} \rangle$, is an ensemble average over a Boltzmann distribution of vibrational energy levels. The average translational energy, $\langle E_{trans} \rangle$, is 3/2 k_BT in flow tube experiments and is the nominal CM collision energy in drift tube and ion beam experiments. RE_{ion} is the recombination energy of the reactant ion, which for most ions is simply the ionization energy of the neutral precursor. However, if the ground electronic state of the neutral precursor is repulsive, e.g. Xe_2^+, the recombination energy corresponds to the vertical recombination energy to the repulsive region of the neutral potential.

In the HTFA, all degrees of freedom are thermally excited by heating the apparatus, i.e. the rotational, translational, and vibrational temperatures are the same. In a drift tube or beam apparatus, the translational energy of the ion is increased by the use of electric fields. Fortunately, the translational energy distribution in a drift tube with a helium buffer can be approximated by a Maxwellian distribution.[27–30] The average translational energy can be converted to an effective translational temperature by $KE = (3/2)k_B T_{eff}$ and can be directly compared to the HTFA data since the translational energy distributions are similar. In beams, the internal temperature of the ion depends on the source conditions. The internal energy dependence is derived by comparing data taken at the same translational temperature or average energy, but with the neutrals at different temperatures. The internal energy dependence is most easily observed by plotting the rate constant data as a function of translational energy or temperature. In this type of plot, differences along the vertical axis reflect the effect of internal energy on reactivity. Comparison to beam data is done the same way, but allowances for different translational energy distributions increases the uncertainty of the analysis.

The analysis of atomic ions reacting with diatomic neutrals is relatively straightforward. For most diatomics, little or no vibrational excitation occurs below ca. 1000 K. Therefore, at lower temperatures, any internal energy dependence is due solely to the rotational excitation of the reactant neutral. Vibrational excitation may contribute at higher temperatures. To elucidate the energy effects further, it is useful to plot the data as a function of average translational plus rotational energy. For drift tube data at 300 K, a constant value of $k_B T = 0.026$ eV is added to the translational energy, and the average translational energy in the HTFA is multiplied by 5/3. As will be shown in the results section, plots of this type often have the drift tube and HTFA data overlapping below 1000 K or 0.2 eV. This agreement suggests that rotational and translational energy control the reactivity equally, at least in an average sense.

If rotational and translational energy are found to be equivalent at lower temperatures, it is assumed that they are equivalent at higher temperatures and that any differences observed at high temperatures are due to vibrational excitation. In this case, the HTFA rate constants can be written as

$$k = \sum_i \mathrm{pop}(i) \times k_i \tag{3}$$

where i represents the vibrational level, pop(i) is the fraction of the molecules in the ith state, and k_i is the rate constant for the ith state. The populations of the various states can be calculated assuming a Boltzmann distribution. At the temperatures of the drift tube measurements, the diatomic molecules studied typically contain no vibrational excitation. Therefore, assuming all excited states react at the same rate, the $v \geq 1$ rate constant can be extracted with the aid of equation 3. All excited states are included in one group for practical reasons. In most cases, the derived $v \geq 1$ rate constant represents the $v = 1$ rate constant, because even at the temperatures achieved in the HTFA, most of the vibrational excitation is limited to $v = 1$. For some systems, either the HTFA or drift tube data are multiplied by a constant near unity to account for systematic errors in results taken in two different apparatuses.

As the reactants become more complex, the analysis becomes more difficult. However, if both reactants are diatomic, the analysis is almost as straightforward. The low temperature data contain only rotational excitation, however, both reactants are rotationally excited. The rotational temperature of the ionic reactant in a drift tube is calculated from the CM energy with the buffer mass substituted for the reactant mass by assuming $\mathrm{KE}_{\mathrm{buf}}^{\mathrm{cm}} = 3/2\mathrm{kT}_{\mathrm{rot}}$.[31,32] The rotational distribution is expected to be close to Boltzmann. Vibrational excitation also occurs in both reactants. The ionic vibrational temperature is the same as that for rotations for a steady state situation and this assumption is used in all the work described here. The contributions from ionic and neutral vibrations can only be separated if independent information exists regarding how vibrational excitation of one of the reactants affects the reactivity. In practice, if any such information is available, it is likely to be the vibrational dependence of the primary reactant ion.

For atomic and polyatomic ions reacting with polyatomic molecules, it is often useful to plot the data as a function of total energy, i.e. the sum of vibrational, rotational, and translational energy. This type of analysis does not allow for separation of the effects resulting from the various types of energy, but it does provide a test to determine if all types of energy control the reactivity similarly. For molecules like CH_4, where little vibrational excitation occurs below 500 K, it is still possible to obtain a limited amount of energy-specific information. For the charge-transfer reactions involving aromatic reactants, plotting the data as a function of total energy plus the primary ion recombination energy has proved to be a valuable method for characterizing the reactivity.[17,33,34] The comparison is particularly relevant because the amount of internal energy of the aromatic reactants at high temperatures (see Table 1) is significant compared to the differences in the recombination energies of the ions involved in the study which have recombination energies as follows: NO^+, 9.26 eV; Xe_2^+, 10.4 eV; O_2^+, 12.07 eV; Xe^+, 12.13 eV; N_4^+, 12.9 eV; O^+, 13.62 eV; $Kr^+(^2P_{3/2})$, 14.00 eV; N^+, 14.53 eV; $Kr^+(^2P_{1/2})$, 14.66 eV; N_2^+, 15.58 eV; Ar^+, 15.76 eV, F^+, 17.42 eV; and Ne^+, 21.56 eV.

Thus, there are four types of plots used to facilitate the discussion: reactivity versus (1) translational energy or temperature, (2) rotational plus translational energy, (3) total energy, and (4) total energy plus recombination energy. Each plot type yields useful information and examples of each type are given in the next section. For a few of the reactions, the previously published analysis did not include plots as a function of translational plus rotational energy. In this review, these data have been reanalyzed and slightly different conclusions are sometimes reached and these are noted.

III. RESULTS

A. Three Atom Systems

As mentioned in the Introduction, much of the impetus for building the HTFA came from the desire to measure relevant ion-molecule reactions over the complete temperature range of the Earth's ionosphere. Figure 2 shows a simplified schematic of the ion chemistry of the upper ionosphere.[35–38] While several ions are produced initially, the diatomic ions recombine with electrons rapidly and the slowly recombining O^+ is left as the dominant ionic species. Reactions that convert O^+ to diatomic ions, speed up recombination, and are therefore extremely important in controlling the plasma density of the ionosphere. In order of importance, the three most important neutrals are N_2, O_2, and NO and these are discussed first. The N^+ atomic ion, not shown in Figure 2, is present in small concentrations. The reaction of N^+ with O_2 has been studied as a function of temperature as well as several reactions of diatomic ions of ionospheric importance and these are also discussed below.

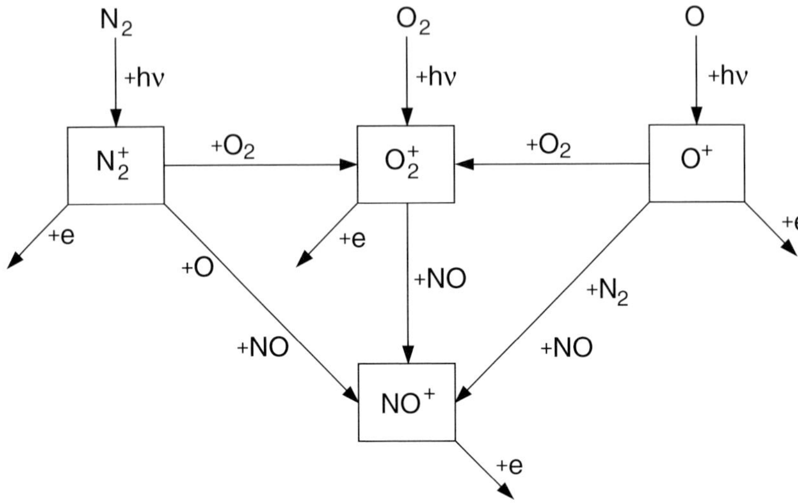

Figure 2. Simplified ion chemistry scheme for the E-region of Earth's ionosphere.

The reaction of O^+ with N_2,

$$O^+ + N_2 \rightarrow NO^+ + N \tag{4}$$

was thoroughly studied in the 1960s and 1970s.[8,12,14,28,39–42] During that time period, the temperature dependence of this reaction was measured up to 900 K.[8,9] However, at 900 K only 2% of the N_2 molecules are vibrationally excited. To overcome this shortcoming, both the translational energy dependence and the dependence on the N_2 vibrational temperature were measured independently.[14,15] Figure 3 shows our measurements[43] up to 1600 K along with the ones of the previous temperature dependent studies[9] and a drift tube study of the energy dependence.[28] The data from the drift tube study is converted to an effective temperature by assuming that the average translational energy equals $(3/2)k_BT_{\text{eff}}$. The two thermal experiments agree very well, and the other temperature dependent study[8] (not shown) is similar and shows the rate constants decreasing up to temperatures of 900 K. The drift tube study also shows good agreement in this range, although the values are slightly below the thermal rate constants. This may be due in part to the difficulty of measuring such slow rate constants, which are approaching the lower limit that can be measured accurately in low-pressure flow tubes. The agreement between the drift tube data and the thermal data shows that rotational energy does not have a big effect on the reactivity. Above 1200 K, the HTFA and drift tube

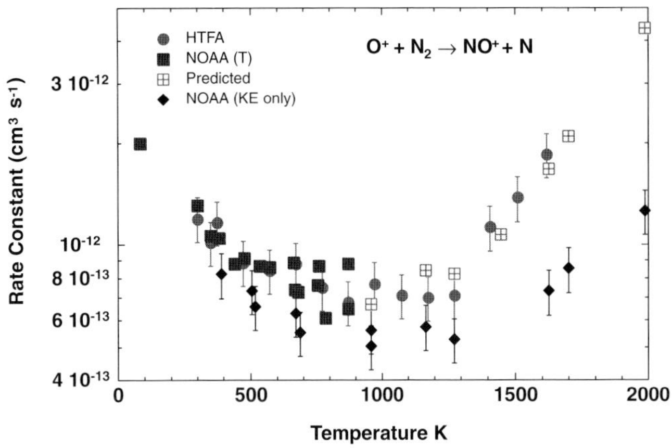

Figure 3. Rate constants for the reaction of O^+ with N_2 as a function of temperature. The HTFA,[43] the NOAA (T),[9] and NOAA (KE)[28] data are shown as circles, squares and diamonds, respectively. See the text for a description of the predicted values.

data start to increase with increasing temperature although the thermal data increase at a lower temperature and increase more rapidly. This shows that vibrational excitation increases the rate constants substantially and why true temperature data is needed.

A previous study measured the effect of the vibrational temperature of N_2 on the rate constant.[14,15] The combination of the translational energy dependence of the drift tube data with the vibrationally excited N_2 data provides an interesting comparison to the present data. The vibrational temperature data were reported relative to the 300 K rate constant. Scaling these data to the drift tube translational temperature ($T_{vib} = T_{trans}$), however, allows a thermal rate constant to be predicted with both vibrational and translational effects included, i.e., each drift tube translational energy data point is scaled according to the vibrational energy dependence at the corresponding effective temperature. This procedure ignores the effects of rotational excitation which was shown above to be small at temperatures below 900 K. This also assumes that the translational energy dependence of the vibrationally excited species is similar to that for $v = 0$. The results of this prediction are shown in Figure 3. Very good agreement is found with the thermal rate constants. The agreement between the different sets of data indicates that the above assumptions are justified. Not surprisingly, unsatisfactory agreement is obtained (not shown) if the vibrational temperature data are plotted relative to the 300 K rate constant.

The large upturn in the rate constant above 1200 K is due to vibrational excitation. At first glance one would assume that it was due to N_2 ($v=1$). However, the NOAA group has shown that $v=1$ reacts at almost the same rate as $v=0$ and that it is $v=2$ and higher vibrational levels that react a factor of 40 faster than the lower energy states.[14,15] Thus, the rather large difference between the HTFA and drift tube data is due to the less than 2% of the N_2 molecules that are excited to $v=2$ or higher in the HTFA experiments.

The rate constants for the reaction of O^+ with O_2 are shown in Figure 4 as a function of temperature.[43] This is one of only two reactions which was studied up to the full temperature range of 1800 K. The temperature data decrease with temperature up to about 800 K, go through a minimum about 300 K wide and increase dramatically above that point. Two other data sets are shown for comparison.[9,12,44] The previous temperature-dependent data taken up to 900 K are in good agreement with the present data except for the 900 K point which still agrees within the combined error limits. Only the NOAA drift tube data are shown and are slightly higher than the present values at low temperature, with the difference increasing with higher temperatures. The drift tube study also has a much wider minimum and increase more slowly. Another drift tube study found values somewhat higher but with similar trends.[39]

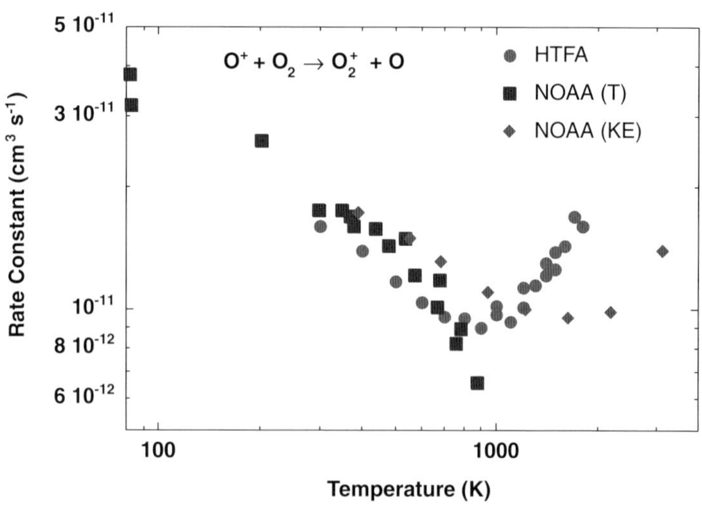

Figure 4. Rate constants for the reactions of O^+ with O_2 as a function of temperature. The HTFA,[43] the NOAA (T),[9] and NOAA (KE)[12] data are shown as circles, squares and diamonds, respectively.

Figure 5 shows a plot for the HTFA and NOAA drift tube data vs. rotational plus translational energy for the reaction of $O^+ + O_2$, the NOAA data has been scaled by 0.88 to better match the lowest energy HTFA points. This is a small correction, considerably less than the error limits, which accounts for a small systematic difference between the data sets. The data agree almost perfectly up to almost 0.2 eV. In this range very little of the O_2 is vibrationally excited. Since the two data sets have considerably different contributions from the two types of energy, the agreement indicates that rotational and translational energy affect reactivity similarly, at least in an average sense. At higher energies, the HTFA rate constant is significantly greater than the drift tube data. The separation between the two curves occurs at the temperature where an appreciable fraction of O_2 is vibrationally excited.

For most of the high temperature range, only $v = 0$ and 1 of O_2 are significantly populated.[45] This allows determination of the rate constant for O_2 in the $v = 1$ state. To facilitate the derivation, the two data sets are fit to a power law plus Arrhenius type exponential. The results of the fits are shown in Figure 5 and are excellent representations of the data. The rate constants for vibrational excited O_2 can then be derived, by assuming that all excited vibrational states of O_2 react at the same rate. Since most of the

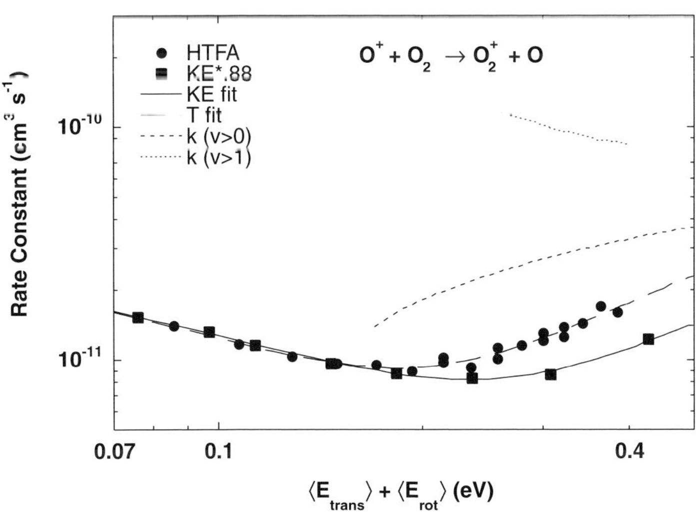

Figure 5. Rate constants for the reaction of O^+ with O_2 as a function of average translational plus rotational energy. The HTFA[43] and the NOAA (KE)[9,44] data are shown as circles and squares, respectively. See the text for a description of the fits and predicted rate constants.

excited population is in $v = 1$, this appears to be a reasonable assumption. The populations of $v = 0$ and $v > 0$ are calculated using the harmonic oscillator approximation and the rate constant for $v = 0$ is taken as the drift tube rate constant. Equation 3 is then solved for k_1. The result is shown in Figure 5 as the dashed line. The vibrationally excited rates are about 2–3 times higher than the ground state rate. Note that this analysis is different from our original paper[43] where we assumed rotational energy did not influence the rate constant. The increase in rate constant may be attributed to changes in Franck-Condon factors. For near resonant states, the Franck-Condon factors are larger for the $v = 1$ state than the $v = 0$ state.[46,47] As an alternative, rate constants were also derived for the assumption that $v = 1$ reacts similar to $v = 0$. This is shown in Figure 5 as k ($v > 1$).

The last of the O^+ reactions to be discussed, which is of ionospheric importance, is the charge transfer reaction of O^+ with NO.[48] Original attempts to study this reaction at high temperature failed because the NO reacted on the ceramic flow tube walls at temperatures over 1000 K. For that reason, we replaced the ceramic flow tube with a industrial grade quartz tube and have used it ever since. Figure 6 shows the rate constants for this reaction plotted as a function of average rotational and translational energy, as well as previous drift tube[28,49] and ultralow temperature data[50] corrected as described in our original paper. The combined data sets fit on one curve, showing the equivalence of rotational

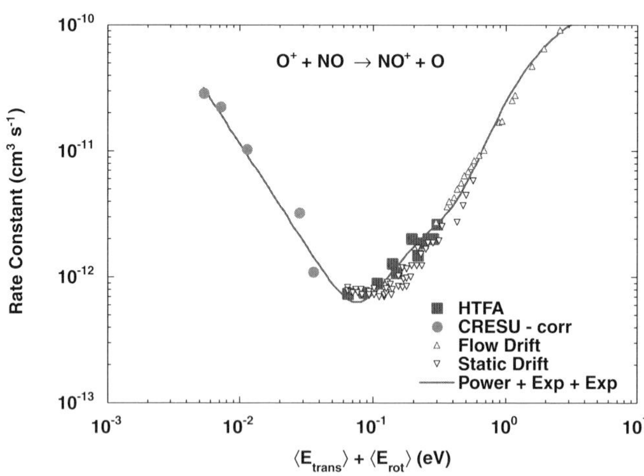

Figure 6. Rate constants for the reaction of O^+ with NO as a function of average translational plus rotational energy. The HTFA,[48] CRESU,[50] flow drift tube,[28] and static drift tube data[49] are shown as squares, circles, triangles and inverted triangles, respectively.

and translational energy in controlling the reactivity. The agreement between the highest temperature points and the drift tube data indicate that vibrational excitation to $v = 1$ does not substantially increase the rate constant.

Only by combining several data sets can the typical behavior for a slow ion-molecule reaction be observed, i.e. an initial decline in the rate constants followed by an increase at higher temperature/energy. The minimum does not show up clearly in any one data set. The combined data looks like they could be fit to a power law plus exponential, similar to the O_2 reaction. However, this does not fit the data well, but a power law plus two exponentials does. This fit is shown in Figure 6. The small rate constants have been attributed to a spin forbidden process.[51] The lower energy activation energy (0.25 eV) appears well correlated with the $^3A_1$ and $^3B_1$ states of the NO_2^+ intermediate.[52] Production of $NO^+(^3S)$ is endothermic by approximately 2 eV, correlating well with the 2.3 eV second activation energy.

The published data on the reaction of N^+ with O_2 is more complicated than the reactions discussed above. Three drift tube studies show flat translational energy dependencies with the rate constant approximately half the collision rate.[12,41,53] In contrast, both HTFA data[54] and a previous NOAA temperature dependence data[9] found the rate constant to increase with increasing temperature until the rate saturates at approximately the collision limit. The results indicated that rotational energy had a large influence on the reactivity and translational energy did not. This is in contrast to the other results summarized in this chapter. Therefore, we have very recently reexamined the rate constants for this reaction in both the HTFA and a selected ion flow tube (SIFT) in our laboratory.[55]

The results are shown in Figure 7. The NOAA (KE) data are plotted and agree with all three drift tube studies with no change in translational energy. The older HTFA data has been corrected for thermal transpiration. The previous NOAA temperature data are in good agreement with the older HTFA data. Both older temperature data sets increased with temperature in contrast to the drift tube data. The SIFT data are flat from 200–530 K. The new HTFA are distinctly lower than the previous HTFA data and show a slight increase with temperature. However, two old HTFA data points, taken at higher pressure (1.5 Torr), agree well with the new data set. The new rate constants were checked as a function of pressure, He flow rate, N_2 source gas flow, electron bias voltage, and emission current. Nothing was found to affect the rate constants within error. The branching ratios in the SIFT were also found to be invariant with temperature in agreement with the drift tube data.[53,55]

The good agreement of the new HTFA data and SIFT data, the simple dynamical conclusions (see below), the checks on the data, and the lack

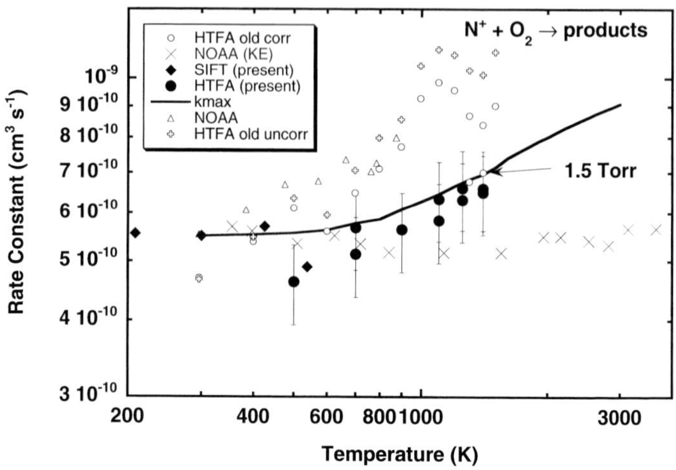

Figure 7. Rate constants for the reaction of N^+ with O_2. The SIFT and HTFA new points are from Viggiano et al.[55] The NOAA kinetic energy (KE) data are from McFarland et al.[12], the HTFA old data are from Dotan et al.[54] corrected for thermal transpiration. k_{max} assumes the $v=0$ rate constant equals the drift tube values and the $v>0$ rate constant equals the collision rate. NOAA (T) are from Lindinger et al.[9]

of temperature or kinetic energy dependencies in the branching ratios, all lead one to conclude that the new data are correct. The fact that higher pressure points agree with the new data may indicate that the source chemistry was not complete, i.e. all the He^+ was not reacted away at the start of the reaction zone. Throttling the pump increases both the reaction time and number density and therefore has a large effect on the distance needed to complete the source chemistry. The new data were taken with the new source configuration, which allows for better control of the source chemistry as well as a cleaner system.

The new data are much easier to explain dynamically. At low temperatures, the drift tube and temperature data are in agreement. This shows that neither translational nor rotational energy influences the rate constants to a large degree. At higher temperatures, the rate constants increase slightly just outside of our relative error. The most likely explanation for the increase is that vibrational excitation increases reactivity. The maximum rate (kmax) constants shown in Figure 7 assume that $v=0$ rate constants are the drift tube values and that the $v>0$ rate constants are equal to the collisional value. The new HTFA data agree very well with this curve, implying that vibrationally excited O_2 does react close to the collisional value. In light of the discrepancy between data sets and the small differences involved, this conclusion needs further verification.

The reactions of C^+ with H_2 and D_2 are other interesting applications of the technique. The rate constants are the smallest we have measured in the HTFA,[56] two products are produced, the temperature dependence is very steep, and there are good beam[57] and drift tube data[58] to compare to. The reactions proceed by two channels for hydrogen,

$$\begin{aligned} C^+(^2P) + H_2 &\rightarrow CH^+(X^1\Sigma^+) + H(^2S) - 0.398\ \text{eV}, \\ &\rightarrow CH_2^+ + 4.3\ \text{eV} \end{aligned} \tag{5a}$$

and for deuterium,

$$\begin{aligned} C^+(^2P) + D_2 &\rightarrow CD^+(X^1\Sigma^+) + D(^2S) - 0.430\ \text{eV} \\ &\rightarrow CD_2^+ + 4.3\ \text{eV}. \end{aligned} \tag{5b}$$

The association process is only detectable below 600 K. For this reaction, branching fractions could not be measured directly in most apparatuses since both products of the reaction react rapidly with H_2 to produce CH_3^+.[2] At the time the measurements were made, we could not study branching fractions directly either. Therefore, the two and three-body processes were separated by studying the reactions as a function of pressure. The rate constants for the H_2 reaction at 400 K are shown in Figure 8 as a function of the helium concentration. The overall measured rate constant is as low

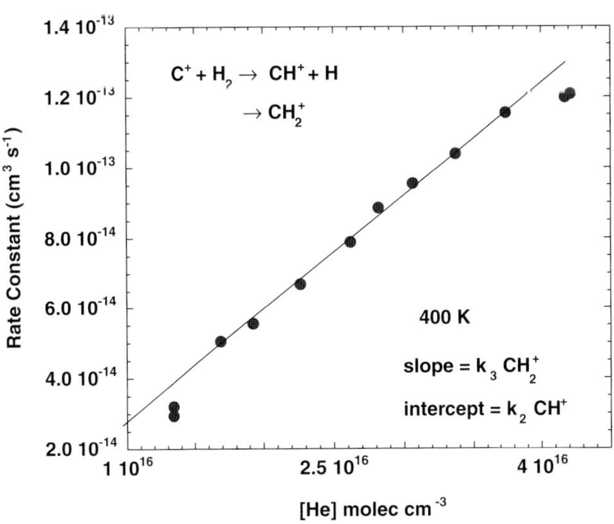

Figure 8. Rate constants as a function of helium concentration for the reaction of C^+ with H_2 at 400 K.[56]

as 3×10^{-14} $cm^3\,s^{-1}$ and varies linearly with pressure. The intercept yields the two-body rate referring to CH^+ production and the slope yields the three-body rate for CH_2^+ production. The three-body rate constants derived in this way agree reasonably well with lower temperature data taken elsewhere.[59–61] The two body rate constant is less than 10^{-15} $cm^3\,s^{-1}$. While this value is very low, it is consistent with higher temperature data points where the extrapolations were less problematic.

The rate constants for the hydrogen transfer channel are shown in Figure 9 as an Arrhenius plot for both H_2 and D_2. The rate constants vary rapidly with temperature and the dependencies are extremely linear. The activation energies of 0.391 ± 0.006, and 0.422 ± 0.013 eV for H_2 and D_2, respectively, compare very favorably to the endothermicities of 0.398 and 0.430 eV. The pre-exponential factors are about 50% of the collisional value. The fact that they are less than collisional has been explained by ab initio calculations which indicates that only two of six possible states couple to products without barriers.[62–65]

Figure 10 shows the same data plotted as a function of rotational plus translational energy. Also shown are values taken from a beam study.[57] Drift tube values are similar to the beam data but start at higher energy.[58] The H_2 HTFA data agree with the translational energy data when rotational energy is added, except at the highest values. The disagreement at the highest energies shows the influence of vibrations. The influence of vibrations stands out more clearly for D_2 than for H_2 due to the lower frequency of D_2, i.e. the translational energy and HTFA data agree at low temperatures but not as well at high temperature.

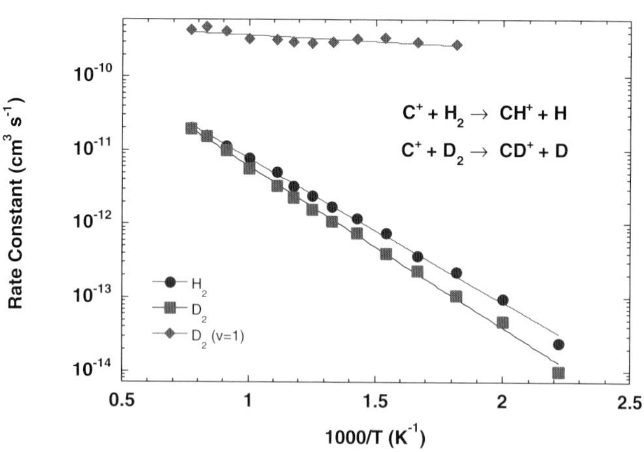

Figure 9. Rate constant for the reactions of C^+ with H_2 and D_2 displayed as an Arrhenius plot.[56]

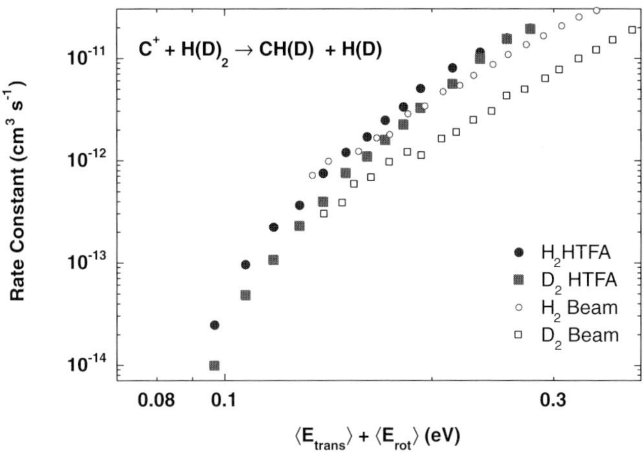

Figure 10. Rate constants for the reactions of C^+ with H_2 and D_2 as a function of average translational and rotational energy. H_2 and D_2 HTFA,[56] and H_2 and D_2 beam data[57] are shown as solid circles and squares and open circles and squares, respectively.

Rates for $v=1$ have been derived. The D_2 ($v=1$) rates are shown in Figure 9. They increase slightly with temperature because the reaction is still endothermic for $v=1$ (0.047 eV). This compares well with the activation energy of 0.034 ± 0.010 eV, given the uncertainties in the derivation. At 500 K, the rate constant for the $v=1$ channel is 6000 times greater than for $v=0$. For H_2, the values are larger than for D_2 and decrease with increasing temperature. The reaction is energetically allowed for $v=1$. The decrease may be an artifact arising from the small difference between the translational energy data and the HTFA data at low temperatures, resulting in a relatively large error. The ratio of the $v=1$ to $v=0$ rate decreases from 1000 to 50 as temperatures increase from 800 to 1300 K.

This system has provided a very good test of the apparatus, in part because changes in the rate constants with temperature are so large. Three body rate constants agree well with other techniques. There is excellent agreement between the HTFA and previous translational energy dependencies at low energies where no vibrational excitation occurs. This shows that rotational energy and translational energy have similar effects on reactivity. The difference in vibrational frequencies between D_2 and H_2 are easily seen. H_2 has a much higher frequency and differences from translational energy data occur only at higher temperatures, not because the rate enhancement differs dramatically, but because the population of the excited state is less for H_2. Finally, there is excellent agreement for the activation energies and endothermicities for reactants in both the $v=0$ and $v=1$ states showing that good rate constants for excited states can be derived.

While nothing inherently prevents negative ion systems from being studied in the HTFA, only three have been studied to date. The reactions of O^- with NO and CO were the first systems studied in the original stainless steel flow tube.[23] Both of these are associative detachment reactions, forming an electron and NO_2 or CO_2. The data are shown in Figure 11. While the trends in the data mimic previous work, the scatter is larger. Relative errors of 30% are probably more appropriate and comparisons of translational and rotational energy are inconclusive. Some of this scatter is the result of unwanted chemistry in the flow tube. O^- is normally made in flowing afterglows from electron attachment to N_2O. At low temperature, N_2O does not attach electrons. However, at high temperatures a distributed source of O^- was found, which was believed to be the result of the electrons from the detachment reactions reattaching to N_2O in the flow tube. To circumvent this problem, CO_2 was used as the source of O^-, and SF_6 was used to scavenge electrons. In retrospect, the scatter in the data probably indicates that small problems remained. In addition, since the time at which these measurements were made, we have realized that NO reacts on hot ceramics and the possibility exists that NO may have also reacted on hot

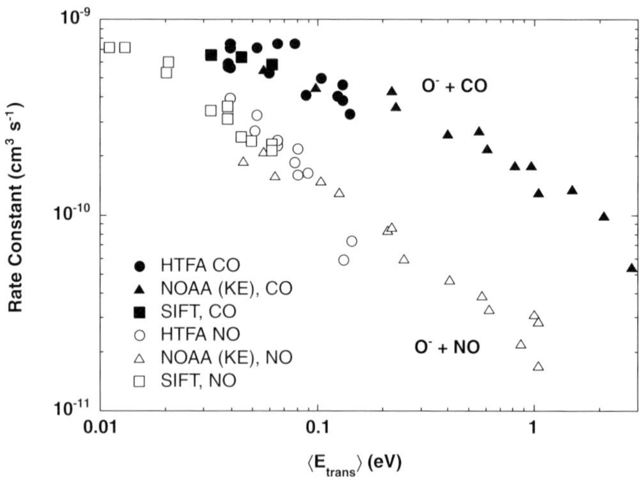

Figure 11. Rate constants for the reactions of O^- with CO and NO as a function of average translational energy. Closed and open circles refer to HTFA data for CO and NO, respectively.[23] Closed and open triangle refer to NOAA drift tube data for CO and NO.[13] Closed and open squares refer to SIFT data for CO and NO.[154,155]

stainless steel. In particular, the highest temperature point is lower than the data trends which indicates that NO was destroyed on the surface. Taken at face value, these data seem to indicate that rotational energy does not change the rate constants.

Other interesting examples of vibrational enhancement are the reactions of Ar^+ with CO and O_2 which are very similar.[66] The rate constants for both reactions are in the 10^{-11} $cm^3 s^{-1}$ range and initially decrease with temperature, have minimums at about 1000 K, and increase at higher temperatures. Comparing rate constants from the HTFA to those from drift tube experiments[67] at the same sum of translational and rotational energy shows good agreement before the minimum, indicating that the two forms of energy control the reactivity in a similar manner.

The higher temperature data for these two reactions not only indicate that vibrational excitation increases the rate constants, but also that vibrational energy changes the rate constants more rapidly than do other forms of energy. In deriving state specific rates from comparisons to translational energy data, it is usually assumed that all vibrationally excited states react at the same rate. However, a couple of observations lead one to believe that $v = 1$ reacts more like $v = 0$ and that $v = 2$ has the larger effect. Little or no enhancement of the rate constants occurs at temperatures where appreciable excitation of the $v = 1$ state occurs. Fits to a power law plus exponential yields activation energies (41.8 and 57.4 kJ/mol for O_2 and CO, respectively) in line with two quanta of vibrational excitation (37.8 and 51.84 kJ/mol for O_2 and CO, respectively).[45] If one assumes that only states in $v \geq 2$ enhance the rate constants, one finds the values about a factor of 100 greater than the $v = 0$ rate constants, very close to the collisional limit, and independent of temperature. When assuming that all states in $v \geq 1$ react at the same rate, one finds about a factor of 5 enhancement and rates that increase with increasing temperature. In either case the enhancement is much greater than can be explained by energy arguments. The production of O_2^+(a) and CO^+(A) states may lead to the observed behavior. The O_2 reaction will be compared to the similar reaction of N_2^+ below.

B. Four and Five Atom Systems

As the molecular complexity increases, the detailed derivation of internal energy effects becomes less clear. Nevertheless, important and useful data can still be derived. We have studied four 4-atom systems, two with both reactants diatomic and two with an atomic ion and a triatomic neutral.

The charge transfer reaction of N_2^+ with O_2 provides another example of the equivalency of translational and rotational energy in controlling the

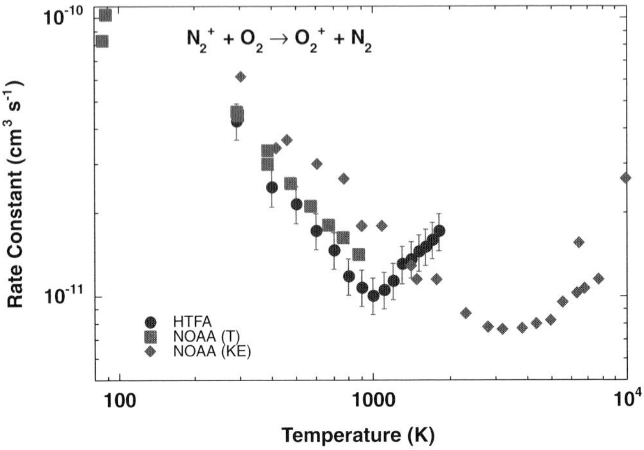

Figure 12. Rate constants for the reaction of N_2^+ with O_2 as a function of temperature. The HTFA,[54] the NOAA (T),[9] and NOAA (KE)[12] data are shown as circles, squares and diamonds, respectively.

reactivity.[54] This reaction occurs in the ionosphere (see Figure 2), but is of minor importance since it converts one diatomic ion to another. Figure 12 shows a plot of the rate constants versus temperature. The rate constants for the thermal data decrease by a factor of 4 from room temperature to the minimum value at 1000 K, and increase by a factor of 2 from 1000 to 1800 K. Excellent agreement is found between the HTFA results and the previous temperature study up to 900 K.[9] The drift tube study is distinctly different.[12] The rate constants decrease with increasing translational energy, but quite a bit more slowly. The minimum is at a distinctly higher energy. At the minimum, the drift tube rate constants are a factor of 2 larger than those measured in the HTFA, i.e., a large difference. A power law plus exponential fits the data well, with all residuals less than 11% of the rate value. The activation energy for the HTFA is 0.29 eV. This form does not fit the drift tube data well.

The data are shown replotted as a function of rotational (ion and neutral) plus translational energy in Figure 13. In this plot, there is excellent agreement between the two data sets up to the minimum in the HTFA rate constants. This shows that rotational energy and translational energy are equivalent in controlling the reactivity. The factor of two differences between the two data sets in Figure 12 disappears. The fact that the rotational effect is so large is due in part because both reactants have rotational energy as opposed to the reactions described above where only one reactant had rotational energy. This is one of the few cases where we

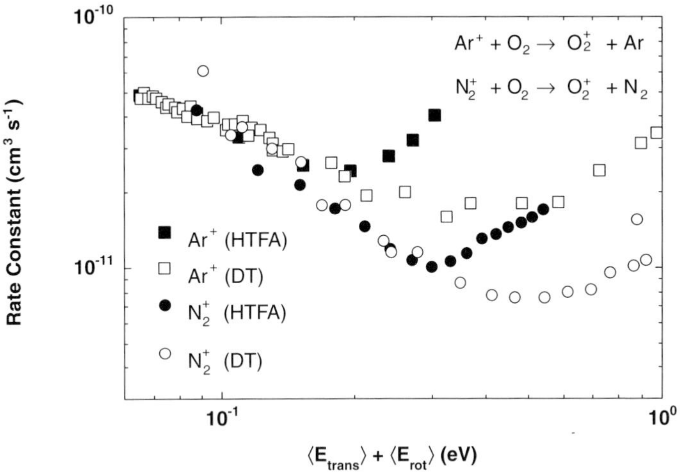

Figure 13. Rate constants for the reactions of Ar^+ and N_2^+ with O_2 as a function of average translational and rotational energy. The HTFA data for Ar^+ and N_2^+ are shown as solid squares[66] and circles,[54] respectively. Drift tube data for Ar^+ and N_2^+ are shown as open squares[67] and circles,[12] respectively.

have been able to make conclusions about the rotational energy of the ion. Above the minimum in the HTFA data, the two data sets diverge due to vibrational excitation in the HTFA experiment. As explained below it is the O_2 vibrations that enhance the rate constant.

Several previous studies have shown that vibrational excitation of N_2^+ does not affect the reactivity.[68–71] This is probably a result of the fact that there is good Franck-Condon overlap between N_2^+ and N_2 in the same vibrational levels. The good overlap indicates that there is little hindrance in this reaction due to N_2 changing its charge state. This is corroborated by the fact that several O_2 reactions behave similarly as shown in the last section of this chapter. This suggests that the differences seen in Figure 13 above 0.3 eV are due mainly to O_2 vibrations. If this assumption is correct, then the reaction of Ar^+ with O_2, which has similar energetics, should behave similarly. A power law plus exponential fit to the HTFA data yields an activation energy between the values for one and two quanta of O_2 vibrations. Therefore, we derive rate constants for two cases assuming: (1) that the rate constant for $v=1$ equals $v=0$; and (2) all vibrationally excited states react at the same rate. The latter assumption yields rate constants a factor of six higher than those for $v=0$ while the former yields rate constants about a factor of 20 higher. In both the Ar^+ and N_2^+ reactions, the upturn has previously been attributed to the production of the $O_2^+(a\Pi_u)$ state,[72] which is endothermic in both reactions. For the reaction of Ar^+ with O_2, it appeared that O_2 ($v \geq 2$) was the most likely explanation for the upturn in the data. However, for the N_2^+ reaction, the activation energy is in between that for the $v = 1$ and 2 states. This also shows up in the minimum between the two data sets. If exactly the same processes are occurring, the minimum between the two curves should shift by the recombination energy difference of 0.178 eV. However, the difference in the minimums is slightly less than this, which is a further indication that O_2 ($v = 1$) must already be enhancing the reactivity for the N_2^+ reaction.

Another reaction of minor importance to ionospheric chemistry is the reaction of O_2^+ with NO. Previous studies have shown that the drift tube dependence and the temperature dependence up to 900 K are flat.[9,73] The measurements up to 1400 K, continue this trend and show that neither translational, rotational, or vibrational energy has a large effect on the reactivity.

Ar^+ and N_2^+ have similar recombination energies and for some reactions have similar reactivity even though one is atomic and the other diatomic. The similarities and differences in the reactions of these two ions with O_2 was described above. The reactions of these ions with CO_2 and SO_2 have also been studied in the HTFA.[74,75] The reactions with CO_2 proceed exclusively by charge transfer and the SO_2 reaction is mainly charge transfer except at high temperature/energy where SO^+ is produced by dissociative charge transfer, a process which is endothermic at room temperature.

Plots of rate constants versus temperature show clear differences between real temperature and kinetic temperature for the reactions of both Ar^+ and N_2^+ with CO_2,[75,76] showing that internal energy has some effect on reactivity. The ability to separate rotational effects diminishes for molecules with three heavy atoms since vibrations are excited at low temperatures. Therefore, the data are replotted as a function of average rotational, translational, and vibrational energy instead of just rotational and translational energy. Such a plot for both reactants with CO_2 is shown in Figure 14. The Ar^+ data fall on the same line up to energies of 0.4 eV after which the temperature data are lower than the drift tube data. To test the high temperature behavior, data were taken in both the ceramic and quartz flow tubes, and similar results were found. In contrast, the N_2^+ temperature data are lower than the drift tube dependencies at all energies. Therefore, in both of these reactions, internal excitation hinders the reactivity more than translational excitation. As shown above, rotational energy only occasionally has a different effect than translational energy and the differences are probably due to CO_2 vibrations since N_2^+ vibrations are mostly unexcited (5% at 1400 K).

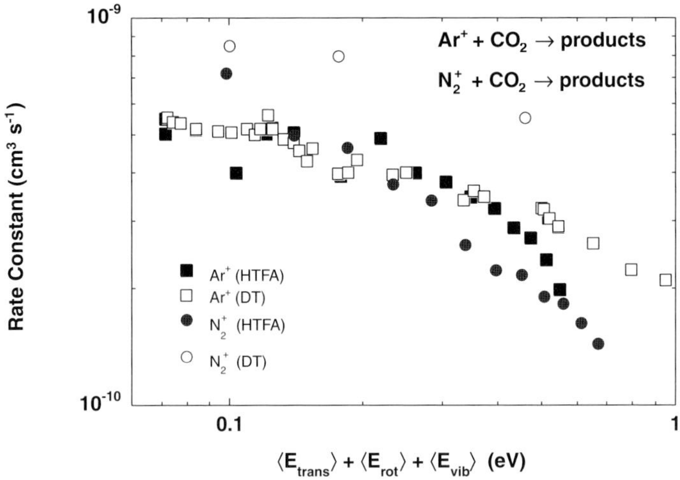

Figure 14. Rate constants for the reactions of Ar^+ and N_2^+ with CO_2 as a function of average translational, rotational, and vibrational energy. The Ar^+ HTFA,[74], N_2^+ HTFA,[75] Ar^+ drift tube,[76] and N_2^+ drift tube[75] data are shown as solid squares, solid circles, open squares and open circles, respectively.

The data show that the rate constant difference is bigger for the N_2^+ reaction.

The reactions of Ar^+ and N_2 with SO_2 also show interesting differences.[74,75,77,78] The temperature plot for the Ar^+ reaction shows good agreement between translational energy and drift tube dependencies at temperatures below 1000 K. Above 1000 K, the temperature data are clearly higher. In contrast, the HTFA rate constants for the N_2^+ reaction are lower than the drift tube data at all temperatures. Figure 15 shows a total energy plot for these reactions; vibrational energy is included. The HTFA data for the Ar^+ reaction is always higher than that from the drift tube showing that internal energy is more effective in driving the reaction than translational energy. For the N_2^+ reaction, the HTFA and drift tube data agree except at the highest temperatures, showing that all forms of energy behave similarly for most of the energy range, in contrast to the Ar^+ data. The disagreement at high temperature may indicate that the lifetime of the complex is becoming short enough so that complete energy randomization cannot occur.

In addition to measuring total rates for the reaction of N_2^+ with SO_2, the branching between SO^+ and SO_2^+ was also measured. Figure 16 shows the rate constants for the individual channels as a function of total energy. For the SO_2^+ channel, the HTFA and drift tube data agree throughout the whole energy range indicating that all forms of energy affect

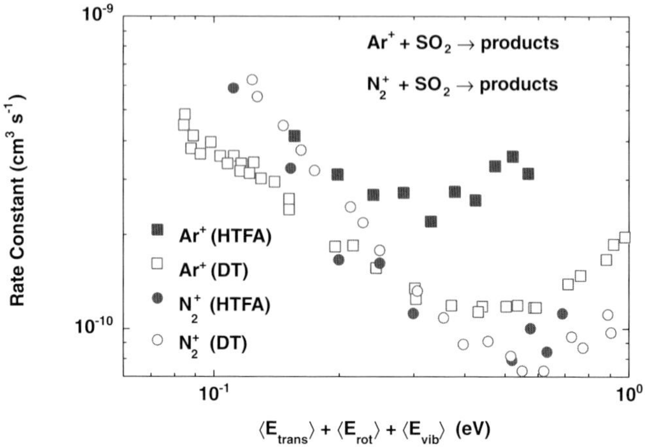

Figure 15. Rate constants for the reactions of Ar^+ and N_2^+ with SO_2 as a function of average translational, rotational, and vibrational energy. The Ar^+ HTFA,[74] N_2^+ HTFA,[75] Ar^+ drift tube,[77] and N_2^+ drift tube[77] data are shown as solid squares, solid circles, open squares and open circles, respectively.

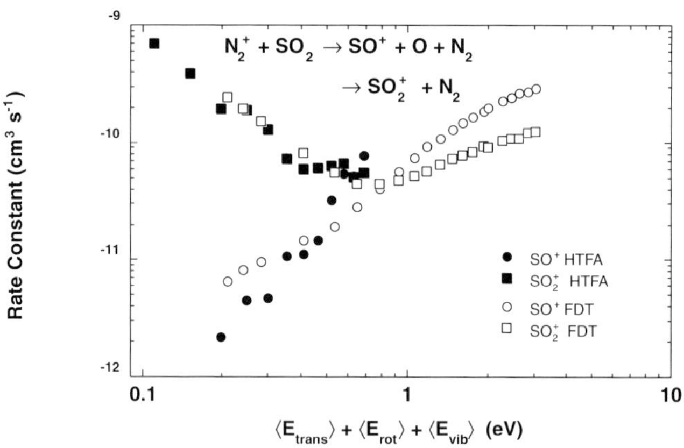

Figure 16. Rate constants for the individual channels in the reaction of N_2^+ with SO_2 as a function of total energy. HTFA data[75] and drift tube data[77] are shown as open and closed points. Circles and squares refer to SO_2^+ and SO^+ channels.

reactivity equally for this channel. This is indicative of a long-lived complex. The two data sets disagree for the SO^+ channel. The disagreement at low energy was attributed to a <1% high energy impurity (N_2^{+*} or He^+) in the drift tube experiment. If this assumption is correct, it appears that all the

SO^+ comes from N_2^+ ($v > 0$) reacting at about 0.5 of the collisional value. This is consistent with the conclusions of Orth et al.[78] who have found that vibrationally excited N_2^+ produces an excited state of SO_2^+ that decomposes into SO^+ if not quenched. Thus, the three systems (O_2, SO_2, and CO_2), for which N_2^+ and Ar^+ reactions can be compared, show contrasting behavior. This calls into question the equivalency of the ions in controlling reactivity.

C. Reactions with CH_4

Methane is a polyatomic molecule for which relatively detailed dependencies on various types of energy can be obtained from comparisons of temperature and translational energy data. Since there is only one heavy atom, little vibrational excitation occurs at 300 K and below. Both, the O^- and O_2^+ reactions with CH_4 have been measured at high temperature.[79,80] The O^- reaction is much simpler and proceeds exclusively by hydrogen abstraction. For this reaction, we have studied drift tube dependencies at multiple temperatures from 93 K to 565 K. At low energy, rate constants obtained at lower temperatures are larger than those at higher temperature and the opposite is true at high energies. Figure 17 shows the data replotted as a function of translational and rotational energy. Data at energies lower than the minimum in the rate constant agree extremely well, showing that in this energy range rotational and translational energy

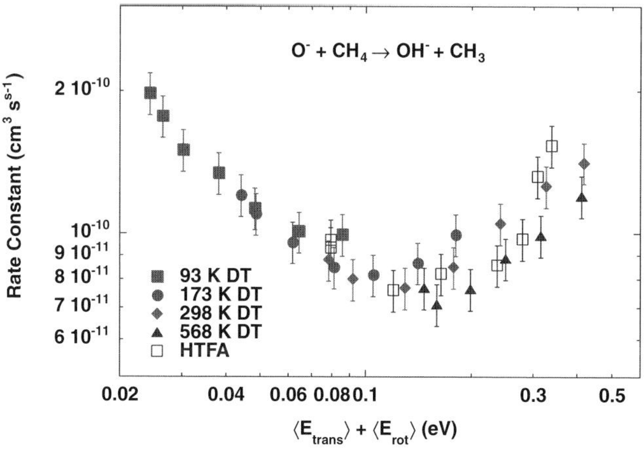

Figure 17. Rate constants for the reaction of O^- with CH_4 as a function of average translational and rotational energy. Solid points refer to SIFDT data and the open squares to HTFA data.[79]

are equivalent in controlling the reactivity. Above the minimum, slight differences in the rate constants are found. The 568 K data seems systematically lower than the 298 K data, which is slightly lower than the 173 K data. The trend may indicate that rotational energy is hindering the reactivity slightly in this energy range but the differences are small. In addition, at 568 K, excitation of the CH_4 bending type modes occurs. The HTFA data fell in between the 298 K and 568 K SIFDT data. While the agreement is good, it is not good enough to corroborate this small trend. In any case, rotational and low frequency vibrations for this system do not have a strong influence on reactivity.

The HTFA data agree well with the drift tube data although the trend at the uppermost temperatures is that the HTFA rate constants increase more rapidly than the drift tube data. The trend changes are outside of the relative error, and these changes most likely indicate that higher order vibrations are starting to increase the reactivity. Since the reaction proceeds via a hydrogen abstraction, it seems plausible that excitation of the C–H stretches could increase the reactivity. The data upturn occurs at the temperatures where such stretching vibrations are becoming excited. The data are, however, too scattered to calculate excited state rates. Note that in the original paper, only total energy plots were made. Rate constants for molecules with stretches excited were found to be five times greater than for molecules with no vibrational excitation or excitation in the bending modes. Based on the present plot, this appears to be too large a factor.

The reaction of O_2^+ with CH_4 is one of the most studied ion-molecule reactions[80–96] and the high temperature behavior has turned out to be very complicated and therefore interesting. Originally, only rate constants were studied at high temperature.[79] The trends in the rate constants agreed well with the findings of a study of translational energy dependencies at various temperatures,[95] i.e. vibrational excitation greatly increased reactivity. This reaction was revisited after the instrument was modified to be able to measure branching fractions. Previously, it had been found that branching into CH_3^+ or CH_4^+ occurred at high translational energy[88] and branching fractions in the HTFA could be compared to that work to see if the enhanced reactivity was due to specific channels.[86,96]

At 1400 K, seven product channels were found.[80] The reactions are

$$O_2^+ + CH_4 \rightarrow CH_2O_2H^+ + H + 22\ \mathrm{kcal/mol} \tag{6a}$$

$$O_2^+ + CH_4 \rightarrow CH_3^+ + HO_2 - 0.4\ \mathrm{kcal/mol} \tag{6b}$$

$$O_2^+ + CH_4 \rightarrow CH_4^+ + O_2 - 12.45\ \mathrm{kcal/mol} \tag{6c}$$

$$O_2^+ + CH_4 \rightarrow H_3O^+ + HCO + 113\ \mathrm{kcal/mol} \tag{6d}$$

$$O_2^+ + CH_4 \rightarrow HCO^+ + H + H_2O + 69\ \text{kcal/mol} \qquad (6e.1)$$

$$\rightarrow HCO^+ + OH + H_2 + 54\ \text{kcal/mol} \qquad (6e.2)$$

$$O_2^+ + CH_4 \rightarrow CH_3O^+ + OH + 78\ \text{kcal/mol} \qquad (6f)$$

$$O_2^+ + CH_4 \rightarrow H_2O^+ + CH_2O + 53\ \text{kcal/mol.} \qquad (6g)$$

Production of $CH_2O_2H^+$ is the primary channel observed at lower temperature and energy. At 1400 K, only 2% of this product remains. Charge transfer (34%) and production of HCO^+ (30%) are the dominant channels at 1400 K. The later channel had not been identified previously, in part because it has the same mass as $C_2H_5^+$ which forms in the secondary reaction of CH_3^+ with CH_4. The branching percentage for the CH_3^+ product was found to be 15%. Two other products were never previously identified, namely H_2O^+ (2%) and CH_3O^+ (5%). The production of H_3O^+ (14%) was much larger than seen at high translational energy and room temperature. Thus, three product channels were found that were never previously observed and another was found to be in much greater abundance than observed previously. Comparison of the rate constants for the individual channels to those found in a beam experiment shows that most of these channels are enhanced by vibrational excitation. Detailed calculations on the potential energy surface are underway to help explain the data and will be reported with the complete temperature dependence of the branching fractions.

D. Aromatics

Recently, numerous ion-molecule reactions involving aromatic hydrocarbons at high temperature have been investigated, and these data are being compiled for use in computational modeling studies regarding the effects of ionization in combustion.[17,33,34,97] High temperature rate constants and branching fractions have been studied in the HTFA for the reactions of O_2^+ and N_2^+ with benzene (C_6H_6) and naphthalene ($C_{10}H_8$) up to 1400 K. Furthermore, reactions of NO^+ and O_2^+ with toluene (C_7H_8), ethylbenzene (C_8H_{10}), and *n*-propylbenzene (C_9H_{12}) have been studied up to temperatures of 1300 K, 1000 K, and 900 K, respectively. The upper temperature limit studied is determined by the thermal stability of the neutral reactant. Overall the reactions show a progression toward greater extent of dissociative charge transfer (versus nondissociative) as both the reactant ion recombination energy increases and the flow tube temperature increases. The total reaction rate constants reported are large and are approximately equal to the thermal energy capture rate constant given by the Su-Chesnavich equation based on average dipole orientation theory.[98,99]

This is approximately 2×10^{-9} $cm^3 s^{-1}$ for all reactions studied. In contrast to the branching fractions, the total reaction rate constants do not vary significantly with temperature over the temperature ranges studied. A complete compilation of the rate constants and branching fractions measured to date can be found in references 17, 33, 34, and 96.

An example of the temperature invariance of the rate constants is shown in Figure 18 which is a plot of the rate constants for the reactions of O_2^+ with C_6 to C_{10} aromatic hydrocarbons as a function of temperature.[17,33,34] The upper and lower dashed lines are the calculated Su-Chesnavich[98,99] capture rate constants as a function of temperature for *n*-propylbenzene ($k_c = 2.0 \times 10^{-9}$ $cm^3 s^{-1}$) and benzene ($k_c = 1.5 \times 10^{-9}$ $cm^3 s^{-1}$), respectively. The calculated capture rate constants for toluene, ethylbenzene, and naphthalene are 1.8, 1.9, and 1.9×10^{-9} $cm^3 s^{-1}$, respectively. The rate constants plotted in Figure 18 are large and well represented by the calculated capture rate constants given the experimental error of ±25%. The scatter in the data increases as the vapor pressure of the aromatic reactant decreases, because low vapor pressure reactant gases are more difficult to deliver to the flow tube. The vapor pressures of *n*-propylbenzene and naphthalene are so low that a bubbler system was used to deliver the gas. Despite the scatter in the data, it is clear that the rate constants do not exhibit a significant temperature dependence, especially compared to the temperature dependence exhibited for the smaller systems discussed earlier. The energy dependence of these reactions appears in the branching fractions.

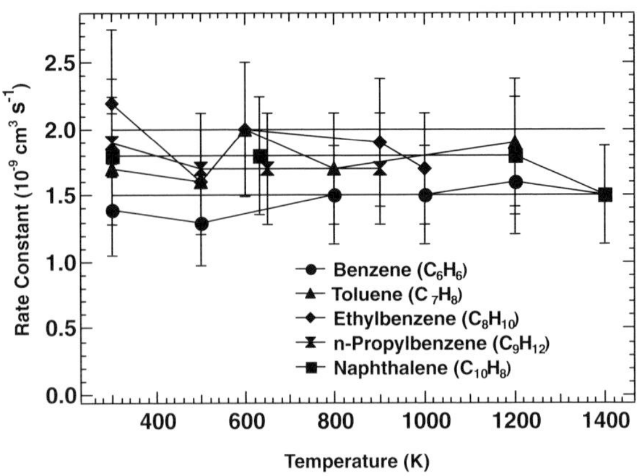

Figure 18. Rate constants for the reactions of O_2^+ with C_6 to C_{10} aromatic hydrocarbons as a function of temperature.[17,33,34] The upper and lower dashed lines are the calculated Su-Chesnavich[98,99] capture rate constants as a function of temperature for *n*-propylbenzene and benzene, respectively.

The construction of product ion breakdown diagrams has proved to be a valuable method for characterizing the observed branching fractions for charge-transfer reactions involving aromatic reactants.[17,33,34] A breakdown diagram for a particular reactant consists of plotting the parent and various fragment ion branching fractions as a function of total energy plus primary reactant ion recombination energy. The energy dependent breakdown curves for benzene are shown in Figure 19. Reactions of benzene with N_2^+ and O_2^+ between 500 and 1400 K in the HTFA are shown along with branching percentages obtained in the SIFT for ions having

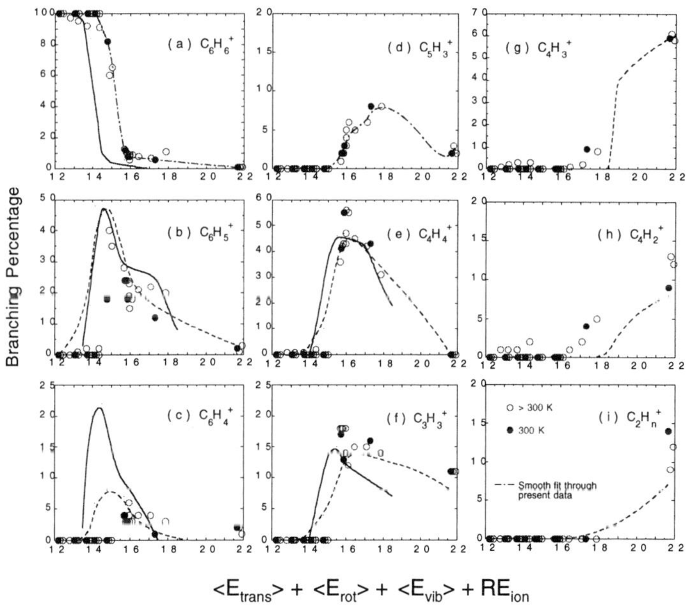

Figure 19. Branching percentage for each product channel observed in the benzene (C_6H_6) charge transfer reactions. Solid circles are 300 K SIFT data. Open circles are all other higher temperature data from the SIFT (330–500 K) and the HTFA (500–1400 K).[33] Solid lines represent breakdown curves obtained previously by the photoelectron–photoion coincidence technique.[100] Dashed lines represent previous charge transfer mass spectral breakdown curves.[101] The reactant internal and translational energies have been added to the energy scale of both the photodissociation and the previous charge transfer mass spectral data. (a) $C_6H_6^+$, dashed–dotted line is an interpolation through the 300 K VT-SIFT data. Because the photodissociation and the previous charge transfer mass spectral data are essentially identical, only one set of previous data is shown for this channel. (b) $C_6H_5^+$, (c) $C_6H_4^+$, (d) $C_5H_3^+$, dashed–dotted line is a smooth fit through all flow tube data. There are no previous data for comparison. (e) $C_4H_4^+$, (f) $C_3H_3^+$, (g) $C_4H_3^+$, (h) $C_4H_2^+$, (i) $C_2H_n^+$.

recombination energies ranging from 9.26 eV (NO^+) to 21.56 eV (Ne^+). In order to distinguish the data, the 300 K data are represented as solid circles, and all other higher temperature data are represented as open circles. For reactions with benzene, the mechanism changes from association to nondissocative charge transfer and then to dissociative charge transfer with an increase in both reactant ion recombination energy and reaction temperature. A total of 10 dissociation products are observed: $C_6H_5^+$, $C_6H_4^+$, $C_5H_3^+$, c-$C_4H_4^+$, linear $C_4H_4^+$ and $C_3H_3^+$, $C_4H_3^+$, $C_4H_2^+$, $C_2H_3^+$, and $C_2H_2^+$. In general, there is very good agreement between the closed and open symbols in the branching percentages shown in Figure 19 suggesting that electronic energy (open symbols) and rovibrational energy (solid symbols) are equally effective in promoting fragmentation.

The current results are similar to data previously reported by Eland et al.[100] and by Jonsson and Lindholm[101] (Figure 19) obtained by photodissociation and charge transfer mass spectral studies, except that there is a shift in the observed threshold energies ranging from ~0.7–2 eV. The difference in threshold energies is due to a kinetic and pressure shift, resulting from slow fragmentation of the energized $C_6H_6^{+*}$ complex combined with collisional stabilization of the complex by the He buffer gas. The slow dissociation rate of the $C_6H_6^{+*}$ energized complex is a likely explanation for the close agreement between the solid (electronic energy) and open (rovibrational energy) symbols in Figure 19, i.e. enough time is available for intramolecular vibrational relaxation to occur.

Reactions of naphthalene with ions having recombination energies ranging from 9.26 eV (NO^+) to 21.56 eV (Ne^+) were studied at 300 K in the SIFT and were compared to reactions of N_2^+ and O_2^+ between 500 and 1400 K in the HTFA apparatus to evaluate the relative effectiveness of rovibrational energy and electronic energy.[34] The branching percentages of the $C_{10}H_m^+$, $C_8H_m^+$, and $C_6H_m^+$ products produced in the naphthalene reactions are plotted versus the average total energy in Figure 20. As shown in the figure, the amount of nondissociative charge transfer products decreases with a concomitant increase in the fraction of dissociative charge transfer products as the total energy increases. The observed product dissociation threshold energies are shifted by as much as 3 eV compared with previous time-dependent photodissociation results of Lifshitz and co-workers.[102,103] The onsets of the dissociative channels exceed the thermodynamic thresholds due to both kinetic shifts and collisional quenching by the helium buffer gas as was observed for the benzene reactions. If the effect of rovibrational energy (open symbols) and electronic energy (solid symbols) is compared, both types of energy appear to be equally effective, however, some differences between the high temperature, N_2^+ data and the 300 K, F^+ data are observed. This difference may be a real effect or may reflect the fact that the 300 K point at 17.42 eV

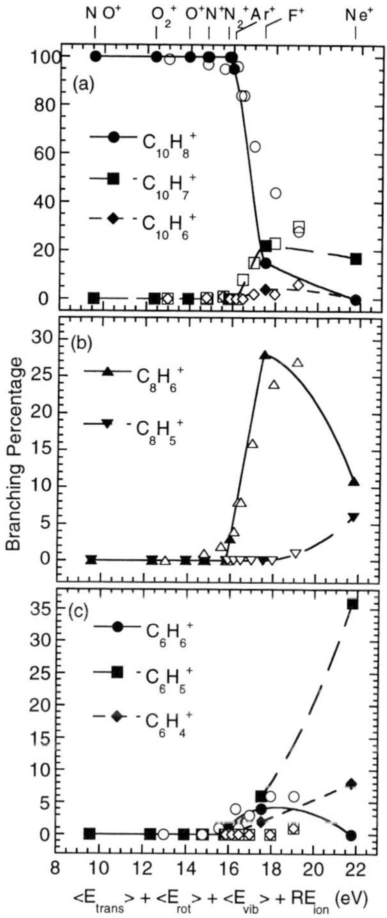

Figure 20. Branching percentages for the reactions of naphthalene (C_8H_{10}) with various ions showing (a) $C_{10}H_m^+$ for m = 6–8, (b) $C_8H_m^+$ for m = 5–6, and (c) $C_6H_m^+$ for m=4–6, plotted versus the sum of the translational, rotational, vibrational, and recombination energy in eV. The solid symbols reflect 300 K SIFT data, and the line is interpolated to fit the points. The open symbols reflect data measured from 500 to 1400 K.[34]

represents the reaction of F^+ with naphthalene for which hydride transfer giving HF is highly exothermic.

Reactions of toluene, ethylbenzene, and *n*-propylbenzene with ions having recombination energies ranging from 9.26 eV (NO^+) to 15.58 eV (N_2^+) were studied at 300 K in a selected ion flow tube (SIFT) yielding information regarding the role of electronic energy in these reactions. These results were compared to reactions of NO^+ and O_2^+ between 500 and 1300 K in the HTFA apparatus to evaluate the relative effectiveness of rovibrational energy and electronic energy.[17] The effect of translational energy in these reactions was also investigated using a guided-ion beam (GIB) apparatus using a high temperature octopole (HT8P) operating under single collision conditions. In the GIB-HT8P experiments, the CM collision energy was varied from ca. 0.1 to 7 eV for reactions of NO^+ and O_2^+, and the target gas temperature was varied between 300 and 800 K.[17]

The $C_7H_7^+$ ion is the dominant fragment ion observed for all three ion-alkylbenzene reaction systems over the energy ranges studied. Breakdown diagrams for the aromatic charge transfer and $C_7H_7^+$ product branching fractions are shown in Figure 21 for toluene as a function of the average total reaction energy for each of the different experiments. The lines are an interpolation through the data and are meant to guide the eye. The minimum energy for each primary reactant ion studied corresponding to the 300 K SIFT data point is marked at the top of each figure. In Figure 21, the data labeled SIFT correspond to measurements taken at 300 K for each of the primary ions listed on the top of the figure. The HTFA data are represented by two sets of open diamonds corresponding to reactions of NO^+ and O_2^+ at temperatures above 300 K. Each HTFA data set originates at either the NO^+ or O_2^+ minimum energy point noted on the top of the figure. The HT8P data for NO^+ or O_2^+ also originate near their respective minimum energy markers shown on the top of the figure but at slightly higher energy.

The energy dependence of the branching fraction of the toluene charge-transfer product ($C_7H_8^+$) is shown in Figure 21a. For each set of data plotted in Figure 21a, the falloff in the $C_7H_8^+$ branching fraction is associated with a concomitant increase of the $C_7H_7^+$ product branching fraction as seen in Figure 21b. The data shown in Figure 21a exhibit a sharp falloff of the $C_7H_8^+$ branching fraction at an average total energy of approximately 12 eV, with the NO^+ HTFA data being the exception. The NO^+ data falls off at approximately the thermodynamic threshold for formation of $C_7H_7^+$ at 10.5 eV. This fragmentation at lower total energies can be explained by the thermal decomposition of the $C_7H_8^+$ ion resulting from collisions with the helium buffer gas, i.e., $C_7H_8^+ + M \longleftrightarrow C_7H_7^+ + H + M$, at temperatures greater than 1000 K. This is not a problem in the O_2^+ reactions since the charge transfer product is already largely dissociated at lower temperatures.

Assuming that the fragmentation of the NO^+ HTFA data below 12 eV is due to thermal decomposition, the remaining data plotted in Figure 21 suggest that rovibrational energy is nearly as effective as electronic energy in promoting fragmentation and that both of these are much more effective than translational energy. The equivalence of electronic and

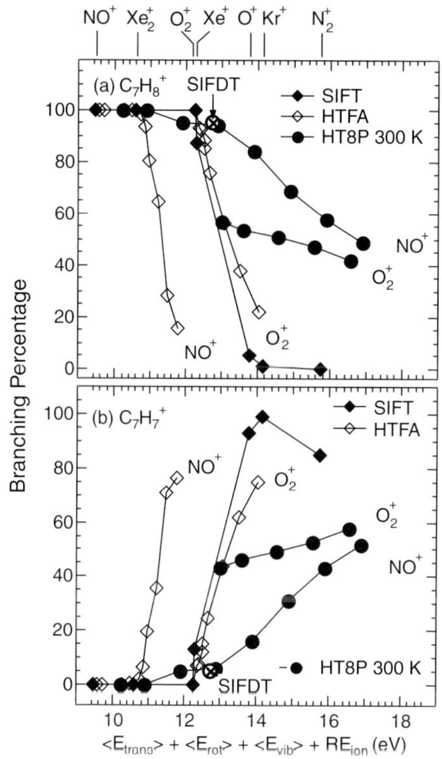

Figure 21. Branching percentages for the reactions of toluene (C_7H_8) with various ions showing (a) $C_7H_8^+$ and (b) $C_7H_7^+$ as a function of average the sum of the translational, rotational, vibrational, and recombination energy. SIFT data (solid diamonds) correspond to 300 K data. HTFA data (open diamonds) for NO^+ and O_2^+ reactions are recorded at temperatures from 500–1300 K. HT8P data (solid circles) for NO^+ and O_2^+ reactions recorded at 300 K. An ⊗ symbol marks an O_2^+ SIFDT result corresponding to 300 K and a CM collision energy of 0.52 eV.[17]

rovibrational energy in promoting fragmentation follows from the fact that the SIFT and O_2^+ HTFA breakdown curves nearly overlap. Recall that at 300 K, very little internal energy is available for reaction (Table 1), so that increasing the recombination energy of the primary reactant ion, as in the SIFT experiments, results in additional electronic energy being available for reaction. In the O_2^+ HTFA experiments, on the other hand, the recombination energy is held constant and the rovibrational energy is increased. Therefore, the nearly identical and rapid decrease in the $C_7H_8^+$ branching fraction with energy is indicative of the effectiveness of these types of energy in promoting dissociation.

Both the recombination energy and rovibrational energy are held constant in the HT8P experiments, and the CM collision energy is increased independently. As shown in Figure 21a, the decrease in the $C_7H_8^+$ branching fraction with increasing average total energy is substantially more gradual for the HT8P data than in the flow tube studies, indicating that translational energy is less effective than electronic and rovibrational energy in promoting fragmentation. Note that translational energies well in excess of the dissociation threshold do not produce 100% fragmentation. In addition, the HT8P NO^+ and O_2^+ data at high average total energy level off at different $C_7H_8^+$ branching fractions, with the O_2^+ primary ion at 650 K producing the most fragmentation. The data suggest that the maximum degree of fragmentation at high translational energy is related to the amount of electronic and rovibrational energy available. To confirm that the differences observed between the HT8P and flow tube measurements are not artifacts of the different methods, an O_2^+ flow drift tube measurement (marked SIFDT) was made at a collision energy of 0.52 eV and is shown in Figure 21. As can be seen in the figure, the SIFDT data point is in good agreement with the translational energy dependence measured with the HT8P.

The charge-transfer product and fragment ion branching fractions for ethylbenzene and *n*-propylbenzene reactions have similar energy dependencies to the toluene reactions. The NO^+ HTFA data falls off at the energetic threshold for formation of $C_7H_7^+$ at 10.0 eV and 9.9 eV for ethylbenzene and *n*-propylbenzene, respectively. This feature is similarly attributed to thermal decomposition of the charge-transfer products. Because the O_2^+ and Xe^+ 300 K data exhibit much more fragmentation at 300 K, it is more difficult to confirm the relative effectiveness of rovibrational and electronic energy in this reaction. However, the data points in this region are consistent with the conclusion drawn regarding the toluene reactions that electronic and rovibrational energy are equally effective in promoting fragmentation. The steep translational energy dependencies and low asymptotic values of the charge-transfer product branching fractions, compared to toluene, indicate that translational energy is more effective in the ethylbenzene and *n*-propylbenzene systems for promoting fragmentation. Indeed, the NO^+ plus *n*-propylbenzene HT8P translational energy data have nearly the same energy dependence as the electronic energy data represented by the SIFT experiment. The similarity of the low translational energy HT8P and the flow tube data suggests that the unimolecular dissociation rate of the excited $C_9H_{12}^+$ charge-transfer product ions has increased compared to $C_8H_{10}^+$ and $C_7H_8^+$, i.e. kinetic and pressure (collisional stabilization of the charge-transfer product) shifts are the least apparent in the *n*-propylbenzene reactions. This conclusion is supported by the dissociation rate energy dependence for toluene[104] and propylbenzene.[105] For example, at a total energy of ca. 12.1 eV corresponding to reactions with O_2^+, the dissociation lifetime for $C_7H_8^+$ is approximately 20 μs whereas the dissociation lifetime for $C_9H_{12}^+$ is 0.1 μs. The latter time is approximately the lifetime between collisions with helium. Since helium is not a very efficient energy transfer agent, the fact that the product distribution are not disturbed for $C_9H_{12}^+$ is not surprising. Furthermore, the dissociation products of deuterated toluene reactions are found to be statistically scrambled whereas the fragment ions produced from deuterated ethylbenzene are not, especially at high ion internal energies, indicating more rapid dissociation for ethylbenzene ions.[16] Therefore, all of the data are consistent with the notion that the dissociation rate of the excited charge transfer product ions increases with increasing substituent alkyl chain length.

All of the alkylbenzene reactions discussed here proceed rapidly at approximately the thermal energy capture rate. The large reaction rate constants observed in these reactions are consistent with a complex formation mechanism. However, a predominance of the complex formation mechanism would express itself as an equivalence of electronic, rovibrational, and translational energy in promoting dissociative charge transfer. The flow drift tube (SIFDT) experiment shown in Figure 21 where approximately 0.5 eV of energy was put into translation did not produce as much fragmentation as the same amount of energy put into electronic (SIFT) or rovibrational (HTFA) energy. The HT8P data confirm this observation. This translational energy dependence suggests that complex formation is most significant at thermal and lower CM collision energies. Therefore, the reduced efficiency of translational energy in promoting dissociation observed in these reactions suggests that near-resonant charge transfer,[18] involving long-range interactions with minimal translational energy transfer, plays a role at low translational energies.

A systematic measurement of absolute charge-transfer cross sections at hyperthermal energies for these hydrocarbon sytems has not been made. However, as an example, cross sections of $\sim 25 \text{Å}^2$ are estimated for the $NO^+ +$ n-propylbenzene system at translational energies above 2 eV. This relatively small cross section compared with the size of the target molecule is an indication that hard-sphere type collisions with repulsive interactions should play a significant role at hyperthermal energies for this particular collision system. Additional evidence for short range charge transfer was observed in the HT8P $NO^+ +$ ethylbenzene and n-propylbenzene systems,[17] where the fragmentation branching fraction reflects threshold behavior similar to the line-of-centers functions observed in collision-induced dissociation (CID) experiments.[106] In CID experiments, the fragmentation cross section saturates at a value corresponding to an effective hard-sphere cross section. This saturation behavior resembles the flattening of the fragmentation branching fraction at high translational energies. However, the maximum degree of fragmentation observed at high translational energy appeared to be dependent on the initial energy available for reaction. For example, the amount of fragmentation observed for reactions of O_2^+ at high translational energies is greater than that observed for NO^+ which has a lower recombination energy. Indeed even the higher temperature O_2^+ HT8P data appears to yield more fragmentation than the lower internal energy 300 K data at high translational energies.[18] These differences could reflect differences in the hard-sphere cross section. However, another possibility is that long range near-resonant charge transfer is occurring in addition to short range hard sphere collisions at high translational energies, resulting in an increase in fragmentation with increasing internal energy available.

IV. SUMMARY OF INTERNAL ENERGY EFFECTS

As stated earlier, the reason for building the HTFA was to measure reactions at temperatures relevant to plasma environments such as the ionosphere. The data so far have demonstrated the importance of making high temperature measurements. However, it is not always possible to measure every reaction due to experiment and time constraints. Thus, it is useful to look for trends in the data so that better extrapolations of lower temperature data can be made for modeling applications. Trends are also important from a fundamental point of view.

The study of internal energy effects in the HTFA is a continuation of work that was started at lower temperatures using the variable temperature selected ion flow drift tube (VT-SIFDT). That data has been summarized previously and several trends were noted.[19] The HTFA comparisons taken as a function of translational energy allow us to verify those trends and look for new ones which become apparent due to the extended energy range. Data from other experiments that probe internal energy effects, sometimes with quantum state resolution, are now available and can be included in the comparison.

In the VT-SIFDT work, one of the clearest trends was for endothermic reactions where rotational and vibrational energy behaved identically.[19] This trend was also observed by Marquette, Rebrion and Rowe for the reactions of N^+ with H_2 by using para-hydrogen.[107] For exothermic reactions involving molecules with large rotational constants, differences were sometimes found between rotational energy and translational energy. Large rotational constants appeared to be a necessary but not sufficient condition for large effects to be found. In the VT-SIFDT work, it was not possible to distinguish whether rotational energy and translational energy were equivalent or if rotational energy had a negligible effect on reactivity for exothermic reactions of molecules with small rotational constants. The large energy range available in reactions studied in the HTFA, involving reactants with large vibrational constants, allows the conclusions to be tested and extended.

Table 2 lists all the reactions studied to date in the HTFA and a coded summary of the internal energy effects found. Only two endothermic reactions were studied (C^+ with H_2 and D_2) and the previous conclusion of equivalency of rotational and translational energy held. Therefore, all the studies of endothermic reactions made to date by this technique indicate that rotational and translational energy are equivalent. It was possible to compare the effects of rotational and translational energy for 14 exothermic reactions studied in the HTFA. For 11 of these, the equivalency of the two forms of energy held. For one to three reactions, rotational energy did not appear to change the reactivity. The range refers to the fact that the O^- reactions with NO and CO contain larger uncertainties which makes it difficult to conclude with certainty that no effect was found. In a very recent experiment, Anderson and coworkers have observed no change in the cross section of the reaction of NO^+ with ethanol when the rotational quantum state of NO^+ is changed.[108] Several studies have found that the rotations do not change the reactivity by more than 10% in the reaction of H_2^+ with H_2.[109–111] In other work, H_2 rotations hindered the clustering to CH_3^+ less than expected from theoretical calculations.[112] A good review of reactivity, studied as a function of rotational quantum state, is given elsewhere.[108]

Thus, it appears that in most ion-molecule reactions, rotational and translational energy are equivalent in controlling reactivity, at least in the low energy range where most of the data have been taken. This appears to be true for both the ion and neutral rotational energy, although the conclusion has been tested with only a few systems for ion rotations. In the higher energy range, the data are too sparse to make a conclusion. A recent work by Rempala and Ervin[113] indicates that in the high-energy regime, rotational energy is not effective in driving the endothermic reaction of S^- with H_2 at high energy. More work obviously needs to be done to examine the influence of rotations, particularly at higher kinetic energies.

There has been much more work on the effect of vibrational excitation on ion reactivity and much of the work up to 1992 has been summarized in two books.[114,115] Some of this work is outlined below. For diatomic and a few triatomic molecules, it has been possible to detect the vibration state by chemical means, the so-called monitor ion method.[69,81,96,116] The most detailed work on internal energy effects is often done using

Table 2. Summary of Internal Energy Effects Studied in the HTFA[a]

Reaction	*Internal energy effects*
$O^+ + N_2$	VE($v \geq 2$) $\gg$ KE; RE = 0
$O^+ + O_2$	VE>RE=KE; $k(v>0) = 2$–$3 \times k(v=0)$
$O^+ + NO$	RE = KE; $v \approx 0$
$N^+ + O_2$	VE > RE = KE,
$C^+ + H_2$	VE>RE = KE; VE makes the reactions exothermic
$C^+ + D_2$	VE>RE = KE; $k(v>0) = 6000 \times k(v=0)$
$Ar^+ + CO$	RE = KE; $k(v \geq 1) = 100 \times k(v=0) = k(v=1)$ or $k(v \geq 0) = 5 \times k(v=0)$
$Ar^+ + O_2$	RE = KE; $k(v \geq 1) = 100 \times k(v=0) = k(v=1)$ or $k(v \geq 0) = 5 \times k(v=0)$
$O^- + NO$	data scattered and inconclusive
$O^- + CO$	data scattered and inconclusive
$N_2^+ + O_2$	RE = KE; VE(N_2^+) = 0; For O_2, $k(v \geq 1) = 20 \times k(v=0) = k(v=1)$ or $k(v \geq 0) = 6 \times k(v=0)$
$O_2^+ + NO$	VE = RE = KE, all are flat
$Ar^+ + CO_2$	VE = RE = KE except at high energy IE < KE
$Ar^+ + SO_2$	IE > KE
$N_2^+ + CO_2$	IE < KE
$N_2^+ + SO_2$	VE = RE = KE for SO_2^+ channel; $N_2^+(v>0)$ promotes SO^+ channel
$O^- + CH_4$	RE = KE; VE for bends small: VE for stretches >0?
$O_2^+ + CH_4$	VE > RE = KE; VE promotes new channels
$X^+ + C_6H_6$	*k* independent of temperature; IE = EE for promoting fragmentation (KE not studied; $X^+ = NO^+$, O_2^+, N_4^+, O^+, Kr^+ ($^2P_{3/2}$), Kr^+ ($^2P_{1/2}$), N^+, N_2^+, Ar^+, F^+, Ne^+
$X^+ + C_{10}H_8$	*k* independent of temperature; IE = EE for promoting fragmentation (KE not studied; $X^+ = NO^+$, O_2^+, O^+, N^+, N_2^+, Ar^+, Ne^+
$X^+ + C_7H_8$	*k* independent of temperature; IE = EE>KE for promoting fragmentation; $X^+ = NO^+$, Xe_2^+, O_2^+, Xe^+, O^+, Kr^+ ($^2P_{3/2}$), N_2^+
$X^+ + C_8H_{10}$	*k* independent of temperature; IE = EE>KE for promoting fragmentation; $X^+ = NO^+$, Xe_2^+, O_2^+, Xe^+, O^+, Kr^+ ($^2P_{3/2}$), N_2^+
$X^+ + C_9H_{12}$	*k* independent of temperature; IE = EE>KE for promoting fragmentation; $X^+ = NO^+$, Xe_2^+, O_2^+, Xe^+, O^+, Kr^+ ($^2P_{3/2}$), N_2^+

[a]Abbreviations are used as follows: KE for translational energy, RE for Rotational Energy; VE for vibrational energy, IE for internal energy which includes both vibrational and rotational energy, and EE for electronic energy. Math symbols compare the effects of different types of energy on the rate constants, i.e. VE > RE = KE implies that RE and KE affect the rate constants equally and that vibrational energy is more effective than either; anything = 0 means no effect. For complete explanations see text for the individual reactions.

resonance enhanced multiphoton ionization (REMPI) to prepare ions in specific vibrational states. This technique has been used extensively by Zare and coworkers[117–123] and Anderson and coworkers[124–142] in guided-ion beams. Leone and Bierbaum[71,143–150] have used LIF detection of N_2^+ ions injected into a selected ion flow tube to study the effect of vibrational excitation.

A review of all the work described above is beyond the scope of this review and only generalities will be discussed. Of particular interest are the conclusions reached by Zare and coworkers,[117–123] who have studied reactions of state-selected ammonia ions, prepared with varying degrees of excitation in the symmetric, in-plane stretch and out-of-plane umbrella bend, as a function of CM collision energy with six different neutral species that varied in size from 2 to 13 atoms. Many of the reaction channels exhibited a strong dependence on both vibrational and translational energy. With regard to vibrational state selection, Zare and coworkers have concluded that in NH_3^+ reactions mode selectivity should be a general feature when the charge transfer channel is energetically open. Each reaction system that has an exothermic charge-transfer channel (deuterated ammonia (ND_3),[151] partially deuterated methylamine (CD_3NH_2),[118] and tetrahydrofuran (c-$(CH_2)_4O$)[118] showed varying degrees of mode selectivity, whereas those systems for which the charge-transfer channel is endothermic (D_2,[121] D_2O,[120] and CD_4[119]) were not influenced by the state preparation of the NH_3^+ reagent. The three NH_3^+ charge-transfer reactions investigated by Zare and coworkers involved a direct charge-transfer mechanism, i.e. the reactions did not proceed via long-lived complexes. Based on this mechanism, these authors postulated that the high degree of vibrational (Franck-Condon) overlap between the ammonia ion and the nascent neutral ammonia resulting from charge transfer drives mode selectivity for ion-molecule reactions involving the NH_3^+ ion. Anderson and coworkers[130] have studied the reaction of state-selected ammonia ions with methanol. This reaction system also has an active charge-transfer channel, and the reaction was determined to be mode selective. Similar Franck-Condon arguments are made in the work of Leone and Bierbaum on N_2^+ reactions.[145,147]

In all the work referenced above, little if any mention of the effect of neutral vibrations on reactivity was discussed. This stems from the fact that multiphoton ionization methods are able to produce vibrationally selected primary reactant ions in sufficient quantities to perform experiments. However, the neutral reactant is usually in large excess, and it is difficult to produce large quantities of vibrationally excited reacting species over a sufficiently large volume with quantum state resolution using laser excitation. However, there have been several studies using microwave discharges to study neutral vibrational effects.[14,15,152,153] Those studies worked, in part, because the ground vibrational state is unreactive or reacts slowly. Our approach has been to excite vibrations thermally and compare them to measurements made at the same translational energy at a colder internal temperature. Table 2 lists the vibrational effects derived from the HTFA data.

Only a few systems were studied in the HTFA that did not involve charge transfer for which vibrational effects could be derived. For the reactions of C^+ with H_2 (and D_2) and O^+ with N_2, vibrational energy was found to promote reactivity more than other channels. The latter conclusion had already been expounded by the NOAA group in the 1960s.[14,15] In the reaction of O^- with CH_4, bending excitation appeared to have no effect on reactivity, while stretches may have increased reactivity. In our VT-SIFDT work,[19] we found that for S_N2 reactions, the reactions fell into two classes. If the reaction involved methyl halides, vibrational excitation did not change the rate constant, but if the reactant was anything else, all forms of energy were equivalent. Larger systems more often behaved statistically. For smaller systems, conclusions seem very much dependent on the system. Thus, the data do not agree with the conclusion that state specific effects are only found for systems that have a charge-transfer channel. Some of the differences may stem from the size of the system and the higher energy range in the beam experiments.

Numerous systems involving charge transfer have been measured in the HTFA. Most reactions of small systems showed that vibrational excitation of the neutral increased reactivity more than did other forms of energy. In two reactions of NO, it was found that vibrational energy did not affect reactivity. For the two systems, N_2^+ with SO_2 and Ar^+ with CO_2, all forms of energy appeared equivalent, although in the latter reaction this conclusion applied only to the low energy regime. In the reactions of N_2^+ with CO_2 and Ar^+ with CO_2 at high energy, vibrational excitation hindered reactivity.

It appears that the effects of vibrations were correlated with the neutrals involved for several reagents. For two reactions involving NO, there was no measurable change in the rate constant as a function of vibrational level, although in one of the reactions our sensitivity was not great. CO_2 vibrations appeared to either hinder reactivity or be equal to other forms of energy. The latter case is an indication of a long-lived complex involving energy sharing. In a previous study of the reaction of O^+ with CO_2, we also found that vibrational energy was equal to other forms of energy.

Three charge transfer reactions involving O_2 all showed that vibrational excitation promoted reactivity in excess of other forms of energy. In these three reactions (with Ar^+, O^+, and N_2^+), the rate constants for the ground state are all close to 1×10^{-11} cm^3 s^{-1} at the sum of rotational and translational energy of 0.2–0.3 eV, which corresponds to the minimum value of the rate constant in the data. Derived rate constants for the vibrational enhancement also behaved similarly. If it is assumed that all excited states react at the same rate, the enhancement varies from about three for the O^+ reaction to about five for the other two reactions. For the reaction of O^+ with O_2, nearly resonant states have about a factor of five increase in the Frank-Condon factor for $v=1$ compared to $v=0$.[43] Thus, for the O_2 reactions, Franck-Condon arguments seem to be a major influence in the reactivity. Leone and Bierbaum have argued that Franck-Condon factors for molecules distorted by the interaction potential are the most relevant.[145,147] The Ar^+ reaction with CO behaves very similarly to the Ar^+ reaction with O_2. The N^+ reaction with O_2 has other channels and therefore is dynamically different, but does appear to show that vibrations increase the overall rate constant.

The reactions of N_2^+ and Ar^+ with SO_2 behave very differently. In the Ar^+ reaction, internal energy is more efficient at driving the reaction than translational energy. However, in this case we were not able to study the branching into SO^+ versus SO_2^+. In the N_2^+ reaction, we found that all forms of energy were equivalent for forming SO_2^+ and that N_2^+ vibrational excitation promoted SO^+ formation. Thus, the SO_2 reactions studied to date do not behave similarly as do the O_2, NO, and CO_2 reactions.

The most complicated of the simple systems is the reaction of O_2^+ with CH_4. An exothermic charge-transfer channel plays a prominent role at both elevated energy and temperature, although vibrational excitation is more effective than translational energy in promoting this channel. Vibrational energy was also found to increase the total rate constant and promote channels not previously observed at high translational energies. In essence, this is state-selected chemistry; i.e. only vibrational excited molecules produce certain products.

Reactions of atomic and diatomic ions with benzene, toluene, ethylbenzene, *n*-propylbenzene, and naphthalene represent the largest systems studied in the HTFA to date. All reactions were found to proceed near the capture rate at all temperatures studied. In general, these reactions proceed primarily by nondissociative and dissociative charge transfer except for the reactions involving NO^+, where the dissociation products observed in the HTFA experiments at high temperature are attributed to thermal decomposition of the charge-transfer product ions. Electronic and rovibrational energy are found to be nearly equivalent in promoting fragmentation whereas translational energy is found to be much less effective. The charge-transfer reactions were found to involve several mechanisms whose relative importance depends heavily on the initial conditions of the reaction partners. At thermal CM collision energies, complex formation and near-resonant charge transfer are important. Hard-sphere type collisions are significant at the highest CM collision energies studied (several eV), but there is also supporting evidence for near-resonant charge transfer at these collision energies.

The ability to predict the behavior of complex reaction systems is particularly important for modeling applications, which often require extrapolation of a limited amount of existing data to conditions of practical interest. In this review, we have attempted to identify trends in the data when possible. While the effect of rotational energy seems to be generally predictable, there are enough exceptions to warrant caution in making extrapolations. Furthermore, vibrational energy often displays state-specific effects both in overall reactivity and formation of new products. Therefore, it is still very difficult to predict reactivity at high temperature by extrapolating translational energy dependencies obtained at low temperature, and there continues to be a need to make measurements at true temperatures.

ACKNOWLEDGMENTS

We thank numerous colleagues who have worked on this apparatus: John Paulson, Robert Morris, Thomas Miller, Jeff Friedman, Peter Hierl, Itzhak Dotan, Melani Menendez-Barreto, John Seeley, John Williamson, Fred Dale, Paul Mundis, Susan Arnold, Tony Midey, Jane Van Doren, Berk Knighton, and Michael Berman. The authors thank Dick Zare, Scott Anderson, and Steve Leone for helpful discussions. We thank the Air Force Office of Scientific Research for the funds used to build and operate the apparatus.

REFERENCES

[1] Farrar, J. M.; Saunders Jr., W. H., (Eds.); *Techniques for the Study of Ion-Molecule Reactions*; John Wiley & Sons: New York, 1988.

[2] Ikezoe, Y.; Matsuoka, S.; Takebe, M.; Viggiano, A. A. *Gas Phase Ion-Molecule Reaction Rate Constants Through 1986*; Maruzen Company, Ltd.: Tokyo, 1987.

[3] Smith, M. A. Ion-molecule reaction dynamics at very low temperatures. In *Unimolecular and Bimolecular Ion-Molecule Reaction Dynamics*; Ng, C.-Y., Baer, T., Powis, I., Eds.; John Wiley and Sons: New York, 1994, 183.

[4] Smith, D.; Spanel, P. *Mass Spectrom. Rev.* **1995**, *14*, 255.

[5] Rebrion, C.; Marquette, J. B.; Rowe, B. R.; Clary, D. C. *Chem. Phys. Lett.* **1988**, *143*, 130.

[6] Adams, N. G.; Smith, D.; Clary, D. C. *Astrophys. J.* **1985**, *296*, L31.

[7] Troe, J. Statistical aspects of ion-molecule reactions. In *State-selected and state-to-state ion-molecule reaction dynamics: theory*; Baer, M., Ng, C.-Y., Eds.; John Wiley: New York, 1992, 485.

[8] Chen, A.; Johnsen, R.; Biondi, M. A. *J. Chem. Phys.* **1978**, *69*, 2688.

[9] Lindinger, W.; Fehsenfeld, F. C.; Schmeltekopf, A. L.; Ferguson, E. E. *J. Geophys. Res.* **1974**, *79*, 4753.

[10] Jursa, A. S., (Ed.); *Handbook of Geophysics and the Space Environment*; NTIS: Springfield VA, 1985.

[11] McFarland, M.; Albritton, D. L.; Fehsenfeld, F. C.; Ferguson, E. E.; Schmeltekopf, A. L. *J. Chem. Phys.* **1973**, *59*, 6610.

[12] McFarland, M.; Albritton, D. L.; Fehsenfeld, F. C.; Ferguson, E. E.; Schmeltekopf, A. L. *J. Chem. Phys.* **1973**, *59*, 6620.
[13] McFarland, M.; Albritton, D. L.; Fehsenfeld, F. C.; Ferguson, E. E.; Schmeltekopf, A. L. *J. Chem. Phys.* **1973**, *59*, 6629.
[14] Schmeltekopf, A. L.; Ferguson, E. E.; Fehsenfeld, F. C. *J. Chem. Phys.* **1968**, *48*, 2966.
[15] Schmeltekopf, A. L. *Planet Space Sci.* **1967**, *15*, 401.
[16] Ausloos, P. *J. Am. Chem. Soc.* **1982**, *104*, 5259.
[17] Williams, S.; Midey, A. J.; Arnold, S. T.; Morris, R. A.; Viggiano, A. A.; Chiu, Y.-H.; Levandier, D. J.; Dressler, R. A.; Berman, M. R. *J. Phys. Chem. A* **2000**, *104*, 10336.
[18] Dressler, R. A.; Levandier, D. J.; Williams, S.; Murad, E. *Comments At. Mol. Phys.* **1998**, *34*, 43.
[19] Viggiano, A. A.; Morris, R. A. *J. Phys. Chem.* **1996**, *100*, 19227.
[20] Graul, S. T.; Squires, R. R. *Mass Spectrom. Rev.* **1988**, *7*, 263.
[21] Ferguson, E. E.; Fehsenfeld, F. C.; Schmeltekopf, A. L. Flowing afterglow measurements of ion-neutral reactions. In *Adv. At. Mol Phys.*; Bates, D. R., Ed.; Academic: New York, 1969; Vol. 5, 1.
[22] Hierl, P. M.; Friedman, J. F.; Miller, T. M.; Dotan, I.; Mendendez-Barreto, M.; Seeley, J.; Williamson, J. S.; Dale, F.; Mundis, P. L.; Morris, R. A.; Paulson, J. F.; Viggiano, A. A. *Rev. Sci. Inst.* **1996**, *67*, 2142.
[23] Miller, T. M.; Friedman, J. F.; Menendez-Barreto, M.; Viggiano, A. A.; Morris, R. A.; Miller, A. E. S.; Paulson, J. F. *Phys. Scripta* **1994**, *T53*, 84.
[24] Gilbert, M. *Comb. and Flame* **1958**, *2*, 149.
[25] Takaishi, T.; Sensui, Y. *Trans. Faraday Soc.* **1963**, *59*, 2503.
[26] Poulter, K. F.; Rodgers, M.-J.; Nash, P. J.; Thompson, T. J.; Perkin, M. P. *Vacuum* **1983**, *33*, 311.
[27] Dressler, R. A.; Meyer, H.; Langford, A. O.; Bierbaum, V. M.; Leone, S. R. *J. Chem. Phys.* **1987**, *87*, 5578.
[28] Albritton, D. L.; Dotan, I.; Lindinger, W.; McFarland, M.; Tellinghuisen, J.; Fehsenfeld, F. C. *J. Chem. Phys.* **1977**, *66*, 410.
[29] Fahey, D. W.; Dotan, I.; Fehsenfeld, F. C.; Albritton, D. L.; Viehland, L. A. *J. Chem. Phys.* **1981**, *74*, 3320.
[30] Fahey, D. W.; Fehsenfeld, F. C.; Ferguson, E. E.; Viehland, L. A. *J. Chem. Phys.* **1981**, *75*, 669.
[31] Duncan, M. A.; Bierbaum, V. M.; Ellison, G. B.; Leone, S. R. *J. Chem. Phys.* **1983**, *79*, 5448.
[32] Anthony, E. B.; Schade, W.; Bastian, M. J.; Bierbaum, V. M.; Leone, S. R. *J. Chem. Phys.* **1997**, *106*, 5413.
[33] Amold, S. T.; Williams, S.; Dotan, I.; Midey, A. J.; Morris, R. A.; Viggiano, A. A. *J. Phys. Chem. A* **1999**, *103*, 8421.
[34] Midey, A. J.; Williams, S.; Amold, S. T.; Dotan, I.; Morris, R. A.; Viggiano, A. A. *Int. J. Mass Spectrom.* **2000**, *195*, 327.
[35] Ferguson, E. E.; Fehsenfeld, F. C.; Albritton, D. L. Ion chemistry of the Earth's atmosphere. In *Gas Phase Ion Chemistry*; Bowers, M. T., Ed.; Academic: San Diego, 1979; Vol. 1, 45.
[36] Ferguson, E. E. Ion-molecule reactions in the atmosphere. In *Kinetics of Ion-Molecule Reactions*; Ausloos, P., Ed.; Plenum Publishing Corp.: New York, 1979, 377.
[37] Viggiano, A. A.; Arnold, F. Ion chemistry and composition of the atmosphere. In *Atmospheric Electrodynamics*; Volland, H., Ed.; CRC Press: Boca Raton, 1995; Vol. 1, 1.
[38] Thomas, L. *Ann. Geophys.* **1983**, *1*, 61.
[39] Johnsen, R.; Biondi, M. A. *J. Chem. Phys.* **1973**, *59*, 3504.

[40] Rowe, B. R.; Fahey, D. W.; Fehsenfeld, F. C.; Albritton, D. L. *J. Chem. Phys.* **1980**, *73*, 194.
[41] Johnsen, R.; Brown, H. L.; Biondi, M. A. *J. Chem. Phys.* **1970**, *52*, 5080.
[42] Smith, D.; Adams, N. G.; Miller, T. M. *J. Chem. Phys.* **1978**, *69*, 308.
[43] Hierl, P. M.; Dotan, I.; Seeley, J. V.; Van Doren, J. M.; Morris, R. A.; Viggiano, A. A. *J. Chem. Phys.* **1997**, *106*, 3540.
[44] Ferguson, E. E. *Rev. Geophys. Space Phys.* **1974**, *12*, 703.
[45] Huber, K. P.; Herzberg, G. *Molecular Spectra and Molecular Structure. IV. Constants of Diatomic Molecules*; Van Nostrand Reinhold: New York, NY, 1979.
[46] Krupenie, P. H. *J. Phys. Chem. Ref. Data* **1972**, *1*, 423.
[47] Lias, S. G.; Bartmess, J. E.; Liebman, J. F.; Holmes, J. L.; Levin, R. D.; Mallard, W. G. *J. Phys. Chem. Ref. Data* **1988**, *17*, Supplement *1*, 1.
[48] Dotan, I.; Viggiano, A. A. *J. Chem. Phys.* **1999**, *110*, 4730.
[49] Graham, E.; Johnsen, R.; Biondi, M. A. *J. Geophys. Res.* **1975**, *80*, 2338.
[50] Le Garrec, J. L.; Lepage, V.; Rowe, B. R.; Ferguson, E. E. *Chem. Phys. Lett.* **1997**, *270*, 66.
[51] Ferguson, E. E. review. In *Interactions between Ions and Molecules*; Ausloos, P., Ed.; Plenum: New York, 1974, 320.
[52] Bundle, C. R.; Neumann, D.; Price, W. C.; Evans, D.; Potts, A. W.; Streets, D. G. *J. Chem. Phys.* **1970**, *53*, 705.
[53] Howorka, F.; Dotan, I.; Fehsenfeld, F. C.; Albritton, D. L. *J. Chem. Phys.* **1980**, *73*, 758.
[54] Dotan, I.; Hierl, P. M.; Morris, R. A.; Viggiano, A. A. *Int. J. Mass Spectrom. Ion Phys.* **1997**, *167/168*, 223.
[55] Viggiano, A. A.; Knighton, W. B.; Williams, S. *J. Chem. Phys.* **2001**, in preparation.
[56] Hierl, P. M.; Morris, R. A.; Viggiano, A. A. *J. Chem. Phys.* **1997**, *106*, 10145.
[57] Ervin, K. M.; Armentrout, P. B. *J. Chem. Phys.* **1986**, *84*, 6738.
[58] Twiddy, N. D.; Mohebati, A.; Tichy, M. *Int. J. Mass Spectrom. Ion Phys.* **1986**, *74*, 251.
[59] Fehsenfeld, F. C.; Dunkin, D. B.; Ferguson, E. E. *Astrophys. J.* **1974**, *188*, 43.
[60] Adams, N. G.; Smith, D. *Chem. Phys. Lett.* **1981**, *79*, 563.
[61] Johnsen, R.; Chen, A.; Biondi, M. *J. Chem. Phys.* **1980**, *73*, 3166.
[62] Liskow, D. H.; Bender, C. F.; Schaeffer, H. F. *J. Chem. Phys.* **1974**, *61*, 2507.
[63] Sakai, S.; Kato, S.; Morokuma, K. *J. Chem. Phys.* **1981**, *75*, 5398.
[64] Bauschlicher, C. W.; Shavitt, I. *Chem. Phys. Lett.* **1980**, *75*, 62.
[65] Peason, P. K.; Roueff, E. *J. Chem. Phys.* **1976**, *64*.
[66] Midey, A. J.; Viggiano, A. A. *J. Chem. Phys.* **1998**, *109*, 5257.
[67] Dotan, I.; Lindinger, W. *J. Chem. Phys.* **1982**, *76*, 4972.
[68] Alge, E.; Lindinger, W. *J. Geophys. Res.* **1981**, *86*, 871.
[69] Ferguson, E. E.; Richter, R.; Lindinger, W. *J. Chem. Phys.* **1988**, *89*, 1445.
[70] Koyano, I.; Tanaka, K.; Kato, T.; Suzuki, S. *Faraday Discuss. Chem. Soc.* **1987**, *84*, 265.
[71] Kato, S.; Frost, M. J.; Bierbaum, V. M.; Leone, S. R. *Can. J. Chem.* **1994**, *72*, 625.
[72] Schultz, R. H.; Armentrout, P. B. *J. Phys. Chem.* **1991**, *95*, 121.
[73] Lindinger, W.; Albritton, D. L.; Fehsenfeld, F. C.; Ferguson, E. E. *J. Geophys. Res.* **1975**, *80*, 3725.
[74] Dotan, I.; Midey, A. J.; Viggiano, A. A. *J. Am. Soc. Mass Spectrom.* **1999**, *10*, 815.
[75] Dotan, I.; Midey, A. J.; Viggiano, A. A. *J. Chem. Phys.* **2000**, In press.
[76] Dotan, I.; Lindinger, W. *J. Chem. Phys.* **1982**, *76*, 4972.
[77] Dotan, I.; Albritton, D. L.; Fehsenfeld, F. C. *J. Chem. Phys.* **1976**, *64*, 4334.
[78] Orth, R. G.; Futrell, J. H.; Nishimura, Y. *J. Chem. Phys.* **1981**, *75*, 3345.
[79] Viggiano, A. A.; Morris, R. A.; Miller, T. M.; Friedman, J. F.; Menedez-Barreto, M.; Paulson, J. F.; Michels, H. H.; Hobbs, R. H.; Montgomery, J. *J. Chem. Phys.* **1997**, *106*, 8455.
[80] Viggiano, A. A.; Dotan, I.; Morris, R. A. *J. Am. Chem. Soc.* **2000**, *122*, 353.

[81] Durup-Ferguson, M.; Bohringer, H.; Fahey, D. W.; Fehsenfeld, F. C.; Ferguson, E. E. *J. Chem. Phys.* **1984**, *81*, 2657.
[82] Rowe, B. R.; Dupeyrat, G.; Marquette, J. B.; Smith, D.; Adams, N. G.; Ferguson, E. E. *J. Chem. Phys.* **1984**, *80*, 241.
[83] Van Doren, J. M.; Barlow, S. E.; DePuy, C. H.; Bierbaum, V. M.; Dotan, I.; Ferguson, E. E. *J. Phys. Chem.* **1986**, *90*, 2772.
[84] Adams, N.; Smith, D.; Ferguson, E. E. *Int. J. Mass Spectrom. Ion Proc.* **1985**, *67*, 67.
[85] Dotan, I.; Davidson, J. A.; Fehsenfeld, F. C.; Albritton, D. L. *J. Geophys. Res.* **1978**, *83*, 1978.
[86] Lindinger, W.; Albritton, D. L.; Fehsenfeld, F. C. *J. Chem. Phys.* **1979**, *70*, 2038.
[87] Troe, J. *Int. J. Mass Spectrom. Ion Processes* **1987**, *80*, 17.
[88] Fisher, E. R.; Armentrout, P. B. *J. Chem. Phys.* **1991**, *94*, 1150.
[89] Alge, E.; Villinger, H.; Lindinger, W. *Plasma Chem. and Plasma Proc.* **1981**, *1*, 65.
[90] Lindsay, B. G.; Irvine, A. D.; Latimer, C. J. *Int. J. Mass Spectrom. Ion Processes* **1989**, *91*, 113.
[91] Tanaka, K.; Kato, T.; Koyano, I. *J. Chem. Phys.* **1986**, *84*, 750.
[92] Bohringer, H.; Arnold, F. *J. Chem. Phys.* **1986**, *84*, 2097.
[93] Kirchner, N. J.; Van Doren, J. M.; Bowers, M. T. *Int. J. Mass Spectrom. Ion Proc.* **1989**, *92*, 37.
[94] Troe, J. *Int. J. Mass Spectrom. Ion Proc.* **1987**, *80*, 17.
[95] Viggiano, A. A.; Morris, R. A.; Van Doren, J. M.; Paulson, J. F. *J. Chem. Phys.* **1992**, *96*, 275.
[96] Durup-Ferguson, M.; Bohringer, H.; Fahey, D. W.; Ferguson, E. E. *J. Chem. Phys.* **1983**, *79*, 265.
[97] Arnold, S. T.; Dotan, I.; Williams, S.; Viggiano, A. A.; Morris, R. A. *J. Phys. Chem. A* **2000**, *104*, 928.
[98] Su, T.; Chesnavich, W. J. *J. Chem. Phys.* **1982**, *76*, 5183.
[99] Su, T. *J. Chem. Phys.* **1988**, *89*, 5355.
[100] Eland, J. H. D.; Frey, R.; Schulte, H.; Brehm, B. *Intl. J. Mass Spec. and Ion Phys.* **1976**, *21*, 209.
[101] Jonsson, B. O.; Lindholm, E. *Arkiv Fysik* **1967**, *39*, 65.
[102] Gotkis, Y., Oleinikova, M., Naor, M., Lifshitz, C. *J. Phys. Chem.* **1993**, *97*, 12282.
[103] Ho, Y. P.; Dunbar, R. C.; Lifshitz, C. *J. Am. Chem. Soc.* **1995**, *117*, 6504.
[104] Huang, F.-S.; Dunbar, R. S. *Int. J. of Mass Spectrom. and Ion Process.* **1991**, *109*, 151.
[105] Hwang, W. G.; Moon, J. H.; Choe, J. C.; Kim, M. S. *J. Phys. Chem. A.* **1998**, *102*, 7512.
[106] Levine, R. D.; Bernstein, R. B. *Molecular Reaction Dynamics and Chemical Reactivity*; Oxford University Press: New York, 1987.
[107] Marquette, J. B.; Rebrion, C.; Rowe, B. R. *J. Chem. Phys.* **1988**, *89*, 2041.
[108] Green, R. J.; Qian, J.; Kim, H. T.; Anderson, S. L. *J. Chem. Phys.* **2000**, *113*.
[109] Softley, T. P.; Mackenzie, S. R.; Merkt, F.; Rolland, D. *Adv. Chem. Phys.* **1997**, *101*, 667.
[110] Sbar, N.; Dubrin, J. *J. Chem. Phys.* **1970**, *53*, 842 .
[111] Glenewinkel-Meyer, T.; Gerlich, D. *Isr. J. Chem.* **1997**, in press.
[112] Gerlich, D.; Kaefer, G. *Ap. J.* **1989**, *347*, 849.
[113] Rempala, K.; Ervin, K. M. *J. Chem. Phys.* **2000**, *112*, 4579.
[114] *State-Selected and State-to-State Ion-molecule Reaction Dynamics: Part 1. Experiment*; Ng, C.-Y.; Baer, M., Eds.; John Wiley: New York, 1992.
[115] *State-Selected and State-to-State Ion-molecule Reaction Dynamics: Part 2. Theory*; Baer, M.; Ng, C.-Y., Eds.; John Wiley: New York, 1992.
[116] Lindinger, W. *Int. J. of Mass Spectrom. Ion Proc.* **1987**, *80*, 115.
[117] Zare, R. N. *Science* **1998**, *279*, 1875.
[118] Poutsma, J. C.; Everest, M. A.; Flad, J. E.; Zare, R. N. *Appl. Phys. B.* **2000**.

[119] Everest, M. A.; Poutsma, J. C.; Flad, J. E.; Zare, R. N. *J. Chem. Phys.* **1999**, *111*, 2507.
[120] Everest, M. A.; Poutsma, J. C.; Zare, R. N. *J. Phys. Chem.* **1998**, *102*, 9593.
[121] Poutsma, J. C.; Everest, M. A.; Flad, J. E.; Jones Jr., G. C.; Zare, R. N. *Chem. Phys. Lett.* **1999**, *305*, 343.
[122] Guttler, R. D.; Jones Jr.; Posey, L. A.; Zare, R. N. *Science* **1994**, *266*.
[123] Conaway, W. E.; Ebata, T.; Zare, R. N. *J. Chem. Phys.* **1987**, *87*, 3447.
[124] Anderson, S. L. Multiphoton Ionization State Selection: Vibrational-Mode and Rotational State Control. In *State-Selected and State-to-State Ion-molecule Reaction Dynamics: Part 1. Experiment*; Ng, C.-Y. Baer, M., Eds.; John Wiley: New York, 1992.
[125] Tang, B.; Chui, Y.; Fu, H.; Anderson, S. L. *J. Chem. Phys.* **1991**, *95*, 3275.
[126] Chiu, Y.; Fu, H.; Huang, J.; Anderson, S. L. *J. Chem. Phys.* **1995**, *102*, 1199.
[127] Kim, H.-T.; Green, R. J.; Anderson, S. L. *J. Chem. Phys.* **2000**, *112*, 10831.
[128] Kim, H.-T.; Green, R. J.; Qian, J.; Anderson, S. L. *J. Chem. Phys.* **2000**, *112*, 5717.
[129] Qian, J.; Green, R. J.; Anderson, S. L. *J. Chem. Phys.* **1998**, *108*, 7173.
[130] Fu, H.; Qian, J.; Green, R. J.; Anderson, S. L. *J. Chem. Phys.* **1998**, *108*, 2395.
[131] Qian, J.; Fu, H.; Anderson, S. L. *J. Chem. Phys.* **1997**, *101*, 6504.
[132] Anderson, S. L. *Accts. of Chem. Res.* **1997**, *30*, 28.
[133] Chiu, Y.-h.; Fu, H.; Huang, J.-t.; Anderson, S. L. *J. Chem. Phys.* **1996**, *105*, 3089.
[134] Metayer-Zeitoun, C.; Alcarez, C.; Anderson, S. L.; Palm, H.; Dutuit, O. *J. Phys. Chem.* **1995**, *99*, 15523.
[135] Chiu, Y.-h.; Yang, B.; Fu, H.; Anderson, S. L. *J. Chem. Phys.* **1995**, *102*, 1188.
[136] Chiu, Y.-h.; Fu, H.; Huang, J.-t.; Anderson, S. L. *J. Chem. Phys.* **1994**, *101*.
[137] Chiu, Y.-h.; Yang, B.; Fu, H.; Anderson, S. L.; Schweizer, M.; Gerlich, D. *J. Chem. Phys.* **1992**, *96*, 5781.
[138] Yang, B.; Chiu, Y.-h.; Fu, H.; Anderson, S. L. *J. Chem. Phys.* **1991**, *95*, 3275.
[139] Yang, B.; Chiu, Y.-h.; Anderson, S. L. *J. Chem. Phys.* **1991**, *94*, 6459.
[140] Anderson, S. L. Vibrational Mode Effects in Polyatomic Ion Reactions. In *Gas Phase Ion Chemistry*; Jennings, K. R., Ed.; Kluwer Acad. Pub.: Dordrecht, 1991, 183.
[141] Orlando, T. M.; Yang, B.; Chiu, Y.-h.; Anderson, S. L. *J. Chem. Phys.* **1990**, *92*, 7356.
[142] Orlando, T. M.; Yang, B.; Anderson, S. L. *J. Chem. Phys.* **1989**, *90*, 1577.
[143] Kato, S.; Bierbaum, V. M.; Leone, S. R. *J. Phys. Chem. A* **1998**, *102*, 6659.
[144] Frost, M. J.; Kato, S.; Bierbaum, V. M.; Leone, S. R. *Chem. Phys.* **1998**, *231*, 145.
[145] Krishnamurthy, M.; Bierbaum, V. M.; Leone, S. R. *Chem. Phys. Lett.* **1997**, *281*, 49.
[146] Kato, S.; Gouw, J. A. d.; Lin, C.-D.; Bierbaum, V. M.; Leone, S. R. *Chem. Phys. Lett.* **1996**, *256*, 305.
[147] Kato, S.; Lin, J. A. d. G.-D.; Bierbaum, V. M.; Leone, S. R. *J. Chem. Phys.* **1996**, *105*, 5455.
[148] Gouw, J. A. d.; Ding, L. N.; Frost, M. J.; Kato, S.; Bierbaum, V. M.; Leone, S. R. *Chem. Phys. Lett.* **1995**, *240*, 362.
[149] Kato, S.; Frost, M. J.; Bierbaum, V. M.; Leone, S. R. *Rev. Sci. Instrum.* **1993**, *64*, 2808.
[150] Frost, M. J.; Kato, S.; Bierbaum, V. M.; Leone, S. R. *J. Chem. Phys.* **1994**, *100*, 6359.
[151] Guettler, R. D.; Jones, G. C.; Posey, L. A.; Zare, R. N. *Science* **1994**, *266*, 259.
[152] Viggiano, A. A.; Morris, R. A. *J. Chem. Phys.* **1993**, *99*, 3526.
[153] Jones, M.; Barlow, S. E.; Ellison, G. B.; Ferguson, E. E. *Chem. Phys. Lett.* **1986**, *130*, 218.
[154] Viggiano, A. A.; Paulson, J. F. *J. Chem. Phys.* **1983**, *79*, 2241.
[155] Viggiano, A. A.; Morris, R. A.; Deakyne, C. A.; Dale, F.; Paulson, J. F. *J. Phys. Chem.* **1990**, *94*, 8193.

FLOWING AFTERGLOW OPTICAL STUDIES OF ELECTRONIC STRUCTURES AND REACTIONS OF SMALL RARE GAS CLUSTER IONS

Masaharu Tsuji

OUTLINE

Advances in Gas-Phase Ion Chemistry
Volume 4, pages 137–177.

ABSTRACT

The application of the flowing afterglow methods to optical studies of formation and reactions of small rare gas cluster ions is discussed. Three representative topics are reviewed in this chapter. The first topic is the electronic structures and formation processes of rare gas heterocluster ions such as $ArKr^+$ and $NeKr_2^+$ at high heavier rare gas stagnation pressures. The second topic is the charge-transfer reaction of He_2^+ with NO at thermal energy. The charge-transfer mechanism of $NO^+(A^1\Pi)$ in the He_2^+/NO reaction is examined by comparing the observed vibrational and rotational distributions of $NO^+(A)$ with those calculated from various excitation models. The third topic is electron–ion and ion–ion recombination reactions of He_2^+. The internal state distributions of He_2^* in the $He_2^+/2e^-$ and $He_2^+/C_6F_5X^-$ (X = F,Cl) recombination reactions are determined and the recombination dynamics is discussed.

I. INTRODUCTION

The flowing afterglow (FA) technique was developed in 1963 by Ferguson and his coworkers of NOAA Laboratories in Boulder, Colorado for the study of ion–molecule reactions in the Earth's atmosphere.[1–4] During the past 37 years, this technique and the later-developed drift tube and selected ion flow tube techniques have been used to study a number of binary and ternary ion–molecule reactions of both positive and negative ions at thermal energies.[5,6] The application of the FA technique has been extended to Penning ionization by metastable $He(2^1S,2^3S)$, $Ne(^3P_{0,2})$, and $Ar(^3P_{0,2})$ atoms[7–10] and to electron–ion and ion–ion recombination reactions.[11–14] The main emphasis of these previous studies was placed upon the determination of rate constants and product distributions by mass spectrometric measurements of reactant and product ions.[1–16] Compared with the extensive mass spectrometric studies, little optical spectroscopic study had been carried out on the internal state distributions of products, before we initiated FA optical studies in 1977,[17] except for some pioneering studies of Penning ionization, charge-transfer (CT) reactions, and electron–ion recombination.[18–31] The determination of internal state distributions by FA optical studies has now been greatly increased, primarily through the efforts of Setser,[27–33] Yencha,[10,34–36] Miller and Bondybey,[37–39] Leone,[40–42] Adams and Smith,[43–51] Golde[52,53] and Tsuji,[54–162] and their coworkers. Setser's and Yencha's groups have studied Penning ionization and CT reactions of simple molecules and radicals by using UV and visible emission spectroscopy.[10,27–36] Miller and Bondybey[37–39] and Leone et al.[40–42] have used the laser-induced fluorescence (LIF) method for studying Penning ionization, CT reactions, and relaxation processes of $CO^+(A^2\Pi)$ and $N_2^+(A^2\Pi_u)$ by collisions with buffer rare gases. Leone et al.[41] have also used infrared chemiluminescence method for studying proton transfer reactions (e.g. F^-/HCl), associative detachment reactions (e.g. O^-/CO), and the N^+/O_2 heavy atom transfer reaction. Smith, Adams, Golde, and their coworkers[43–51] studied electron–ion recombination reactions of protonated ions and CO_2^+ by using emission spectroscopy and LIF method. There was no optical study on the ion–ion recombination reaction before our systematic studies except for a pioneering work of Smith et al.[51] for NO^+/NO_2^- by emission spectroscopy.

Since then, we have investigated various ionization and recombination processes in the FA using VUV, UV, and visible emission spectroscopy and the LIF method. There are two purposes in our FA optical studies. One is the detection of new ionic emissions, which are difficult to excite by such other ionization methods as electron-impact ionization and photoionization. In Table 1 are summarized ionic emissions identified in our laboratories and their main formation processes. The other is the determination of internal state distribution for understanding reaction dynamics at the microscopic level. In Table 2 are listed ionization and recombination processes studied by our group using a conventional FA technique coupled with emission spectroscopy or LIF detection. Among various reactions studied by our group, ion–ion recombination processes leading to rare gas halide excimers and NO* molecules have been reviewed previously.[163]

FA optical techniques have two advantages. One is the high intensities of fluorescence and LIF signals from products because of high densities and large volumes of active species in the flow tube. Therefore, high-resolution measurements, that can separate rovibrational structures of products, are possible in most cases, even though their formation rate coefficients are small ($<10^{-15}$ cm^3 s^{-1}). The other is that the reactions can be studied at thermal energy (~300 K), where the application of beam experiments is generally difficult. A disadvantage of the FA technique is that there are possibilities of secondary reactions and collisional deactivation of products due to high operating pressures of 0.1–5 Torr (1 Torr = 133.3 Pa). In general, initial rotational distributions are

Table 1. Ionic Emissions Identified at Kyushu University

Ion	*Transition*	*Reaction*	*References*
CS^+	$B^2\Sigma^+–A^2\Pi_i$	$He^+/CS_2,OCS$	54
	$B^2\Sigma^+–X^2\Sigma^+$	He*,Ne*/CS	55
PN^+	$B^2\Sigma^+–X^2\Sigma^+$	He*,Ne*/PN	56
SO^+	$A^2\Pi–X^2\Pi_r$	He^+/SO_2	57,58
S_2^+	$A^2\Pi_u–X^2\Pi_{g,r}$	He^*/S_2Cl_2	59,60
CX^+ (X = Cl,Br)	$A^1\Pi–X^1\Sigma^+$	He^+/CX_4	61–63
	$a^3\Pi_1–X^1\Sigma^+$	He^+/CX_4	61–64
SiX^+ (X = Cl,Br)	$a^3\Pi^+_{0,1}–X^1\Sigma^+$	He^+/SiX_4	63,65,66
MCl^+ (M = Ge,Sn)	$a^3\Sigma^+–X^1\Sigma^+$	He^+/MCl_4	63,67,68
BBr^+	$A^2\Pi–X^2\Sigma^+$	He^+/BBr_3	69
MH^+, MD^+ (M = Ge,Sn)	$a^3\Pi^+_{0,1}–X^1\Sigma^+$	$He^+/MH_4,MD_4$	70–74
$ArKr^+$	B 1/2–X 1/2	Ar^+/Kr	75
	C_1 3/2–A_1 3/2	Ar^+/Kr	76
	C_2 1/2–A_2 1/2	Ar^+/Kr	75
$KrXe^+$	B 1/2–X 1/2	Kr^+/Xe	77
	C_1 3/2–A_1 3/2	Kr^+/Xe	77
$HeRg_2^+$ (Rg = Ar,Kr)	B 1/2–X 1/2	$He^+/2Ar$	78,79
	B 1/2–A_2 1/2	$He^+/2Ar$	78,79
$NeRg_2^+$ (Rg = Ar,Kr,Xe)	B 1/2–X 1/2	$Ne^+/2Ar$	80,81
	C_1 3/2–A_1 3/2	$Ne^+/2Ar$	80,81
	B 1/2–A_2 1/2	$Ne^+/2Ar$	80,81
$ArKr_2^+$	B 1/2–X 1/2	$Ar^+/2Kr$	80
	C_1 3/2–A_1 3/2	$Ar^+/2Kr$	80
$ArXe_2^+$	B 1/2–X 1/2	$Ar^+/2Xe$	80
	C_1 3/2–A_1 3/2	$Ar^+/2Xe$	80
	B 1/2–A_2 1/2	$Ne^+/2Xe$	80
$KrXe_2^+$	B 1/2–X 1/2	$Kr^+/2Xe$	80
	C_1 3/2–A_1 3/2	$Kr^+/2Xe$	80

completely lost, when product ions in a ground state are probed by the LIF method. The vibrational and rotational relaxation of excited species often takes place for molecules with long radiative lifetimes (> 1 μs) and large quenching rate constants. This problem has been overcome by coupling a FA source to a low-pressure chamber through a small orifice by Leone et al.[40–42] and Sekiya et al.[161] Leone et al. probed nascent rovibrational distributions of ionic products in ion–molecule reactions and Penning ionization by using the LIF method, while our group determined them by using emission spectroscopy. The ionization processes that have been investigated by our group using such a low-pressure apparatus are listed in Table 3. We have also applied the low-pressure

Table 2. Ionization Processes Studied by a Conventional FA Apparatus at Kyushu University

Reactants	*Emitting product*	*Internal state distribution*[a]	*References*
1. Emission Spectra			
a. Excitation transfer			
$He^* + Ar^+, Kr^+, Xe^+$	$Ar^{+*}, Kr^{+*}, Xe^{+*}$	E	82–84
$Ne^* + Xe^+$	Xe^{+*}	E	85
b. Penning Ionization			
$He^* + N_2$	$N_2^+(B^2\Sigma_u^+)$	V,R	86
$He^* + O_2$	$O_2^+(A^2\Pi_u, c^4\Sigma_g^-)$	V,R	87
$He^* + NO$	$NO^+(A^1\Pi)$	R	88
$He^* + CO$	$CO^+(A^2\Pi, B^2\Sigma^+)$	V,R	86
$Ne^* + CO$	$CO^+(A^2\Pi)$	V	89
$He^*, Ne^* + HCl$, HBr	$HCl^+, HBr^+(A^2\Sigma^+)$	V,R	90
$He^*, Ne^* + CS$	$CS^+(B^2\Sigma^+)$	V	91
$He^*, Ne^* + PN$	$PN^+(B^2\Sigma^+)$	V	92
$He^* + CO_2$	$CO_2^+(A^2\Pi_u, B^2\Sigma_u^+)$	V	93
$He^*, Ne^* + OCS$	$OCS^+(A^2\Pi)$	R, $A^2\Pi_{1/2}/A^2\Pi_{3/2}$	94–97
$He^*, Ne^* + CS_2$	$CS_2^+(A^2\Pi_u)$	V	98
$He^*, Ne^* + N_2O$	$N_2O^+(A^2\Sigma^+)$	V	95,98–100
$He^* + H_2O$, D_2O	$H_2O^+, D_2O^+(A^2A_1)$	V	101
$He^*, Ne^* + ICN$	$ICN^+(A^2\Sigma^+)$	V	98
$He^* + S_2Cl_2$	$S_2^+(A^2\Pi_u)$	V	60
$He^*, Ne^* + SiCl_4$	$SiCl_4^+(C^2T_2)$	V	102
$He^*, Ne^*, Ar^* + X\text{-}(C{\equiv}C)_2\text{-}X$, $X = H, CH_3, C_2H_5$	$X\text{-}(C{\equiv}C)_2\text{-}X^+(A)$	V	103
c. Charge-Transfer Reactions			
$He^+ + Ar(^3P_2)$	Ar^{+*}	E	82
$He_2^+ + Kr$, Xe	Kr^+, $Xe^+(^2P_{1/2,3/2})$	$^2P_{1/2}/^2P_{3/2}$	104
$He^+ + N_2$	$N_2^+(B^2\Sigma_u^+, C^2\Sigma_u^+, D'^2\Pi_g)$	E,V,R	105–108
$He_2^+, Ne_2^+ + N_2$	$N_2^+(B^2\Sigma_u^+)$	V,R	86,109
$He^+ + O_2$	$O_2^+(c^4\Sigma_g^-)$	V,R	87
$He_2^+ + O_2$	$O_2^+(A^2\Pi_u, c^4\Sigma_g^-)$	V,R	87
$He_2^+ + NO$	$NO^+(A^1\Pi)$	V,R	88
$He_2^+ + CO$	$CO^+(A^2\Pi, B^2\Sigma^+)$	V,R	86
$Ar^+ + CS$	$CS^+(B^2\Sigma^+)$	V,R	110
$Ar^+ + PN$	$PN^+(B^2\Sigma^+)$	V,R	110
$Ar^+ + HBr$, DBr	$HBr^+, DBr^+(A^2\Sigma^+)$	V,R	111
$He^+ + N_2O$	$N_2O^+(A^2\Sigma^+), N_2^+(B^2\Sigma_u^+)$	V,R	99,100
$He^+ + CO_2$	$CO^+(A^2\Pi)$	V	112
$He_2^+ + CO_2$	$CO_2^+(A^2\Pi_u)$	V	113
$He^+ + CS_2$, OCS	$CS^+(B^2\Sigma^+)$	V	114
$Ar_2^+ + CS_2$	$CS_2^+(A^2\Pi_u)$	V	115
$CO^+ + CS_2$	$CS_2^+(A^2\Pi_u)$	V	116
He^+, $Ne^+ + OCS$	$CO^+(A^2\Pi)$	V	117

(continued)

Table 2. Continued

Reactants	Emitting product	Internal state distribution[a]	References
$Ar^+ + OCS$	$OCS^+(A^2\Pi)$	V	96,97
$N_2^+ + OCS$	$OCS^+(A^2\Pi)$	V	118
$He^+ + H_2O$, D_2O	OH^+,$OD^+(A^3\Pi)$	V,R	119
$He^+ + SO_2$, $SOCl_2$, $SOBr_2$, SO_2Cl_2, SO_2FCl	$SO^+(A^2\Pi)$	V	120
$He^+ + SeOCl_2$	$SeO^+(A^2\Pi)$		121
$He^+ + C_2H_2$	$CH^+(A^1\Pi)$	V,R	122
$He^+ + CCl_4$,$CFCl_3$	$CCl^+(A^1\Pi,a^3\Pi)$	E	61,123
$Ar^+ + SiCl_4$	$SiCl_4^+(C^2T_2)$	V	102
$He^+ + SiBr_4$	$SiBr^+(a^3\Pi^+_{0,1})$	V	66
$He^+ + GeCl_4$,	$GeCl^+(a^3\Sigma^+)$	V	68
$He^+ + SnCl_4$,	$SnCl^+(a^3\Sigma^+)$	V	67
$He^+ + SiH_4$	$SiH^+(A^1\Pi)$	V,R	124,125
$He^+ + GeH_4$,GeD_4	GeH^+,$GeD^+(a^3\Pi^+_{0,1})$	V	74,126
$Ar_2^+ + X\text{-}(C\equiv C)_2\text{-}X$	$X\text{-}(C\equiv C)_2\text{-}X^+(A)$	V	103
d. Chemiluminescent Reactions			
$C^+ + O_2$,CO_2,NO_2,N_2O	$CO^+(A^2\Pi)$,$CN(A^2\Pi_i,B^2\Sigma^+)$	V	127–129
$C^+ + CS_2$,OCS	$CS^+(A^2\Pi)$	V	130,131
e. Electron–Ion Recombination			
$He^+ + 2e^-$	He*	E	132
$He_2^+ + 2e^-$	He_2*	E	121
$CO_2^+ + e^-$	$CO(d^3\Delta_i,e^3\Sigma^- a'^3\Sigma^+)$	E,V,R	133,134
$CH_n^+ + e^-$	$CH(A^2\Delta,B^2\Sigma^-,C^2\Sigma^+)$	E,V	135
f. Ion–Ion Recombination			
$He^+ + C_6F_5X^-$ (X = F,Cl,Br,I,CF_3)	He*	E	136
$He_2^+ + C_6F_5X^-$ (X = F,Cl)	He_2*	E,R	121,137
$CO_2^+ + C_6F_6^-$	$CO(d^3\Delta_i,a'^3\Sigma^+)$	E,V	138
Ar^+,Kr^+,$Xe^+ + SF_6^-$	ArF*,KrF*,XeF*	E,V	139–142
Ar^+,Kr^+,$Xe^+ + C_6F_6^-$	ArF*,KrF*,XeF*	E	142,143
Ar^+,Kr^+,$Xe^+ + C_6F_5Cl^-$	ArCl*,KrCl*,XeCl*	E	121,143
Ar^+,Kr^+,$Xe^+ + Cl^- + He$	ArCl*,KrCl*,XeCl*	E,V	141,144–147
$Xe^+ + Br^- + He$	XeBr*	E	148
$NO^+(X: v'') + SF_6^-$	NO*	E,V,R	149,150
$NO^+ + C_6F_5X^-$(X = F,Cl,Br,I,CF_3)	NO*	E,V,R	151–154
2. LIF Spectra			
a. Penning Ionization			
$He^* + N_2$	$N_2^+(X^2\Sigma_g^+)$	V	155
$Ne^* + N_2$	$N_2^+(X^2\Sigma_g^+)$	V	156
He*,$Ne^* + CO_2$	$CO_2^+(X^2\Pi_g)$	$X^2\Pi_{1/2}/X^2\Pi_{3/2}$	121
He*,Ne*,$Ar^* + OCS$	$OCS^+(X^2\Pi)$	$X^2\Pi_{1/2}/X^2\Pi_{3/2}$	157
He*,$Ne^* + N_2O$	$N_2O^+(X^2\Pi)$		121

(continued)

Table 2. Continued

Reactants	*Emitting product*	*Internal state distribution*[a]	*References*
b. Charge-Transfer Reactions			
He^+, $He_2^+ + N_2$	$N_2^+(X^2\Sigma_g^+)$	V	155
$He^+ + N_2O$	$N_2^+(X^2\Sigma_g^+)$	V	100
c. Ion–Molecule Reactions			
$C^+ + C_2H_4, C_2H_6, C_3H_6, C_3H_8$	$CH(X^2\Pi)$	V	158
$C^+ + CH_3OH$, CH_3OD	$CH(X^2\Pi)$	V	159

[a]Electronic state (E), vibrational state (V) and/or rotational state (R) distributions were determined.

Table 3. Ionization Processes Studied by Using the FA Apparatus Coupled with a Low Pressure Chamber

a. Penning ionization			
$He^* + O_2$	$O_2^+(A^2\Pi_u)$	V, R	160
$He^* + CO$	$CO^+(A^2\Pi)$	V	161
$He^* + HCl$, HBr	HCl^+, $HBr^+(A^2\Sigma^+)$	V, R	162
b. Charge-transfer reactions			
$He^+ + N_2$	$N_2^+(B^2\Sigma_u^+, C^2\Sigma_u^+, D'^2\Pi_g)$	E	107
$He_2^+ + N_2$	$N_2^+(B^2\Sigma_u^+)$	V, R	121
$He^+ + H_2O$, D_2O	OH^+, $OD^+(A^3\Pi)$	V, R	119
$He^+ + C_2H_2$	$CH^+(A^1\Pi)$	V, R	122
$He^+ + SiH_4$	$SiH^+(A^1\Pi)$	V, R	124,125
$He^+ + CS_2$	$CS^+(B^2\Sigma^+)$	V	121
$He^+ + OCS$	$CO^+(A^2\Pi)$	V, R	121
$He_2^+ + CO$	$CO^+(A^2\Pi, B^2\Sigma^+)$	V, R	121
$Ar^+ + H_2O$	$H_2O^+(A^2A_1)$	V	121

apparatus to studies of Penning ionization of HCl, HBr, and O_2 by metastable He(2^3S) atoms,[160,162] because these ionization processes have captured considerable attention since Richardson et al.[27,28] found non-Franck Condon (FC) like vibrational distributions of $HCl^+(A^2\Sigma^+)$, $HBr^+(A^2\Sigma^+)$, and $O_2^+(A^2\Pi_u)$ in their FA optical studies at high buffer gas pressures. A comparison between FA and low-pressure data suggested that rovibrational distributions of these ions were partly relaxed in FA due to collisions with buffer gases. However, our low-pressure data demonstrated that these Penning-ionization reactions proceed through non-FC like mechanisms due to strong interactions in the entrance and/or exit channels.

There has been continuing interest in electronic structures and formation mechanisms of small rare gas cluster ions in recent years.[164–171] We have recently applied the FA technique to studies of electronic structures and elementary reactions of small rare gas cluster ions, as shown in Tables 1 and 2. In this chapter, our recent results for the following reactions will be summarized: e.g.,

$$Ar^{+} + Kr \rightarrow ArKr^{+*} + h\nu, \quad \text{(Radiative association)} \tag{1}$$

$$Ar^{+} + 2Kr \rightarrow ArKr^{+*} + Kr, \quad \text{(Three-body association)} \tag{2}$$

$$He_2^{+} + NO \rightarrow NO^{+}(A^1\Pi) + 2He, \quad \text{(Charge-transfer reaction)} \tag{3}$$

$$He_2^{+} + 2e^{-} \rightarrow He_2^{*} + e^{-}, \quad \text{(Collisional radiative recombination)} \tag{4}$$

$$He_2^{+} + C_6F_6^{-} \rightarrow He_2^{*} + C_6F_6. \quad \text{(Ion–ion recombination)} \tag{5}$$

II. FLOWING AFTERGLOW METHODS FOR STUDYING FORMATION AND REACTIONS OF SMALL RARE GAS CLUSTER IONS

The FA apparatus used for studying association processes 1 and 2 and CT reaction 3 is essentially identical to that described in a previous review.[172] In brief, the FA cell consisted of a stainless steel main flow tube (60 mm internal diameter, 50 cm long) and a quartz flow tube (11 mm internal diameter, 40 cm long). They were continuously evacuated using a high-capacity (10 $m^3\,min^{-1}$) mechanical booster pump in order to reduce a loss of active species by collisions with the inside wall of the quartz flow tube. The carrier rare gas (flow rates 5000–20000 sccm for He and 1000–5000 sccm for Ar) flowed along the flow tube at a linear flow velocity of about 10^4 cm s^{-1} for He and 5×10^3 cm s^{-1} for Ar. Metastable atoms and atomic ions of rare gases and electrons were generated by a microwave discharge in high-purity rare gases operated at rare gas pressures of 0.1–5 Torr. When we used a dc discharge, no ionic active species were generated in an Ar flow and the $[He^{+}]/[He(2^3S)]$ ratio was much lower than that obtained using a microwave discharge. Thus, the combination of the microwave discharge with the high-capacity pumping system was necessary to generate and maintain high densities of active ionic species in a flow system. A sample gas was introduced through a stainless steel nozzle (0.4–0.6 mm internal diameter) 10 or 20 cm downstream from the center of the discharge. The stagnation pressure of the sample gas just before the nozzle was measured using a high-pressure gauge. Radiative association 1 and CT reaction 3 were studied without applying a high stagnation pressure, while high stagnation pressures of 1–5 atm were necessary for the study of three-body association 2. The pressure of sample gas near the orifice was expected to be close to the stagnation pressure. Emission spectra at a few mm downstream from an inlet of the sample gas were measured. VUV emissions in the 120–200 nm region were observed through a MgF window using a VUV monochromator attached directly to the main flow tube. On the other hand, UV and visible emissions in the 200–1000 nm region were collected through a quartz window and focused on the inlet slit of a conventional 1 or 1.25 m monochromator using two quartz lenses.

In a rare gas FA, metastable atoms, atomic ions, and dimer ions are present as rare gas active species [e.g. $He(2^3S)$, $He^+(2^2S_{1/2})$, and He_2^+ in a He flow and $Ar(^3P_{0,2})$, $Ar^+(^2P_{1/2,3/2})$, and Ar_2^+ in an Ar flow]. The contribution of active ionic species to the observed emissions was examined by applying an electrostatic potential (-50 to ~ 50 V) to a pair of ion-collector grids placed between the discharge region and the reaction zone. The overlap of emissions due to ionic reactions with those due to neutral reactions has often made detailed analysis of each reaction difficult. This problem was overcome using an ion modulation technique, by which emissions from ionic reactions can be detected exclusively.[94,172]

Figure 1 shows the FA cell used for studying the $He_2^+/2e^-$ and $He_2^+/C_6F_6^-$ reactions 4 and 5, as an example of the FA apparatus used for studying electron–ion and ion–ion recombination reactions. The positive He_2^+ ions were formed by the three-body reaction 6:

$$He^+ + 2He \xrightarrow{k_6} He_2^+ + He, \quad [k_6 = 6.8 \times 10^{-32}\ cm^6\ s^{-1}(Ref.15)]. \tag{6}$$

Therefore, high buffer gas pressures above 1 Torr and a long distance between the discharge region and the reaction zone were necessary to obtain high $[He_2^+]/[He^+]$ ratios. When studying the $He_2^+/2e^-$ recombination reaction 4, the emission spectrum was directly measured through one of the quartz observation windows without adding an attaching gas. On the other hand, C_6F_6 was added to the FA, 20 cm downstream from the center of the discharge, when $He_2^+/C_6F_6^-$ recombination reaction 5 was studied. There are two regions in the FA above and below the inlet of C_6F_6, which are denoted by regions A and B in Figure 1, respectively. The active species in region A were $He(2^3S)$, He^+, and electrons formed directly in the

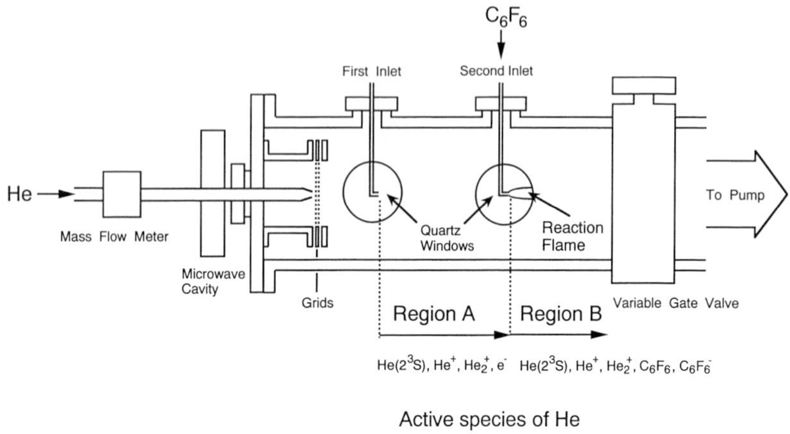

Figure 1. The FA cell for studying $He_2^+/2e^-$ and $He_2^+/C_6F_6^-$ recombination reactions.

discharge, and He_2^+ formed by the secondary reaction 6. In region B, the $C_6F_6^-$ anions were formed by a fast non-dissociative thermal electron attachment to C_6F_6:

$$e^- + C_6F_6 \xrightarrow{k_7} C_6F_6^- \quad [k_6 = 2.1 \times 10^{-7}\ \text{cm}^3\ \text{s}^{-1}\ (\text{Ref.173})]. \tag{7}$$

Thus, the active species in region B were He(2^3S), He^+, He_2^+, C_6F_6, and $C_6F_6^-$. The electron density 10 cm downstream from the discharge was measured with a single Langmuir probe to be 3.2×10^9–2.6×10^{10} cm^{-3} in a He pressure range 0.2–1.4 Torr. Since the electrons were completely scavenged through process 7, the density of anion was expected to be nearly the same as that of the electron density. Emission spectra from region B were observed to probe the $He_2^+/C_6F_6^-$ recombination reaction.

III. ELECTRONIC STRUCTURES AND FORMATION PROCESSES OF RARE GAS HETEROCLUSTER IONS

Although emission spectra of rare gas heterodimer ions had been observed earlier in various mixtures of rare gases,[174–177] the origin of the bands had been erroneously attributed to neutral heterodimer until a later systematic study of Tanaka et al.[165] Tanaka et al. found that an electronic discharge in a binary mixture of rare gases provided several discrete emission band groups confined to a narrow wavelength region. These were characteristic of the pair of rare gases and closely related in energy to the ionization energy difference between the two atoms. Such spectra were observed for nine of the ten binary mixtures of He, Ne, Ar, Kr, and Xe. These bands were ascribed to a bound–bound intramolecular CT transition $Rg^{+}Rg' \rightarrow RgRg'^{+}$, in which Rg is the lighter and Rg′ the heavier element in the mixture. The validity of their assignment was confirmed by later high-resolution rotational analyses of $HeAr^{+}$, $HeNe^{+}$, and $ArKr^{+}$.[178–180]

The potential-energy diagram of $RgRg'^{+}$ (Rg = Ne, Ar, Kr) is shown in Figure 2(a). There are three upper states (B 1/2, C_1 3/2, C_2 1/2) derived from the $Rg^{+}(^2P_{1/2}) + Rg'(^1S_0)$ dissociation limit and three lower states (X 1/2, A_1 3/2, A_2 1/2) correlating with the $Rg(^1S_0) + Rg'^{+}(^2P_{1/2,3/2})$ dissociation limits. Among them, five parallel transitions ($\Delta\Omega = 0$), denoted by A ~ E in Figure 2(a), generally have appreciable intensities. A single upper state (B 1/2) is derived from the $He^{+}(^2S_{1/2}) + Rg'(^1S_0)$ dissociation limit for $HeRg'^{+}$, as shown in Figure 2(b). Therefore, only two parallel transitions [B 1/2 → X 1/2 (A band) and B 1/2 → A_2 1/2 (D band)] have been observed for $HeRg'^{+}$.[80] Although all the five or two transitions have been observed for $NeRg'^{+}$ (Rg′ = Ar, Kr, Xe) and $HeRg'^{+}$ (Rg′ = Ar, Kr, Xe), respectively,[165] only the A band had been observed for $ArKr^{+}$ and $KrXe^{+}$.[177,180] We have recently succeeded in detecting the B, C, and D bands of $ArKr^{+}$ and the B and C bands of $KrXe^{+}$ by the Ar and Kr in FA

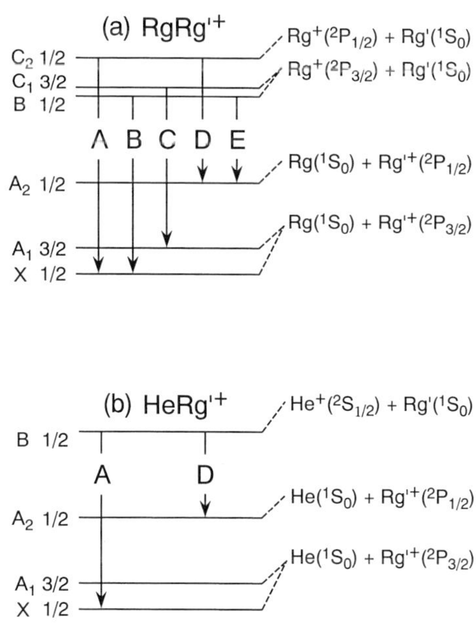

Figure 2. Energy-level diagrams of (a) $RgRg'^{+}$(Rg = Ne, Ar, Kr) and (b) $HeRg'^{+}$.

reactions of Kr and Xe, respectively.[75–77] Here, results for the $ArKr^+$ bands are represented as an example of heterodimer bands. In many reaction systems, new continuous bands, which could be assigned to heterotrimer bands, appeared near the heterodimer bands at high stagnation pressures of Rg′.[78–81] Their possible electronic transitions, dominant excitation processes, and geometries have already been discussed.

A. Emission Spectra of Heterodimer Ions

When Kr was added to an Ar afterglow, three new visible emission systems were observed in the 675–710, 712–720, and 935–990 nm regions, as shown in Figures 3(a)–(c). They were ascribed to the B $1/2 \rightarrow$ X $1/2$, C_1 $3/2 \rightarrow A_1$ $3/2$, and C_2 $1/2 \rightarrow A_2$ $1/2$ transitions of $ArKr^+$, corresponding to the B, C, and D bands in Figure 2, respectively. As expected these bands disappeared when ionic active species were removed from the reaction zone. This indicated that the $Ar^+(^2P_{1/2,3/2})$ ions participated in the formation of $ArKr^{+*}$. Dominant excitation processes of $ArKr^{+*}$ were expected to be two-body radiative associations 8a and 8b at low Kr pressures and three-body clustering reactions 9a and 9b at high Kr pressures in view of the electronic-state correlation between $Ar^+(^2P_J)$ and $ArKr^+(\Omega)$ (see Figure 2):

$$Ar^+(^2P_{1/2}) + Kr \rightarrow ArKr^+(C_2\ 1/2) + h\nu, \tag{8a}$$

$$Ar^+(^2P_{3/2}) + Kr \rightarrow ArKr^+(B\ 1/2,\ C_1\ 3/2) + h\nu, \tag{8b}$$

$$Ar^+(^2P_{1/2}) + 2Kr \rightarrow ArKr^+(C_2\ 1/2) + Kr, \tag{9a}$$

$$Ar^+(^2P_{3/2}) + 2Kr \rightarrow ArKr^+(B\ 1/2,\ C_1 3/2) + Kr. \tag{9b}$$

The molecular constants of the C_1 $3/2$, B $1/2$, A_2 $1/2$, and A_1 $3/2$ states obtained from a vibrational analysis of the B, C, and D bands are summarized in Table 4 along with reported data of the C_2 $1/2$ and X $1/2$ states.[180] In the vibrational analysis of the B and D bands, vibrational constants of the C_2 $1/2$ and X $1/2$ states were fixed with the reliable values obtained by Holland et al.[180] from the rotational analysis of the A band.

Since the B $1/2 \rightarrow$ X $1/2$ and C_2 $1/2 \rightarrow$ X $1/2$ systems share the same lower state, the v'' numbering was beyond question. Less certain, however, was the numbering in the upper B $1/2$ state. It is highly likely that the $(0, v'')$ bands with the most favorable Franck-Condon (FC) factors appear with the highest intensity, because a significant difference is expected in the

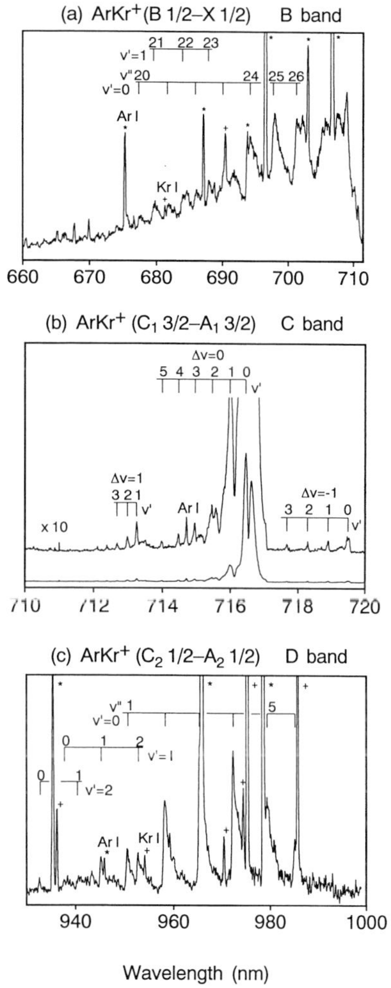

Figure 3. The B, C, and D bands of $ArKr^+$ produced from the Ar afterglow reactions of Kr. Lines marked with * and + are ArI and KrI atomic lines respectively.

equilibrium internuclear distance between the B 1/2 and X 1/2 states according to the ab initio calculation of Bender and Winter.[166] Therefore, the v' numbering of the B $1/2 \rightarrow$ X 1/2 transition was derived on the reasonable assumption that the lowest observed upper level is $v' = 0$. There was no clear minimum in the calculated potential curve of the B 1/2 state.[166] However, the observation of the B $1/2 \rightarrow$ X 1/2 transition implies that the

Table 4. Molecular Constants and Dissociation Energies of $ArKr^+$

States	T_0 (cm^{-1})	ω_e (cm^{-1})	$\omega_e x_e$ (cm^{-1})	D_0 (cm^{-1})	r_e (Å)	*References*
$C_2 1/2$	19440.6	64.225(11)[a]	2.84	740	3.733	180
$C_1 3/2$	(18406.9)[b]	67.29(77)	2.19(17)	342	~3.4	75,166
B 1/2	18103.4(23)	44.7(7)[c]	–	645	–	76,166
A_2 1/2	8832.1	93.7(12)	2.44(17)	1093	~3.6	76,166
A_1 3/2	(4453.5)[d]	57.40(64)	2.17(14)	101	~3.8	75,166
X 1/2	0	244.14(92)	4.066(39)	4554	2.45	180

[a]Values in parentheses represent one standard deviation (σ).
[b]Estimated from T_0 values of A_1 3/2 and the transition energy of C_1 3/2-A_1 3/2.
[c]$\Delta G(1/2) = \omega_e - 2\omega_e x_e$.
[d]Estimated from equation 12.

B 1/2 state has a shallow minimum. The B 1/2 → X 1/2 transition was observed from only the lowest two $v' = 0$ and 1 levels. Therefore, only the $T_0(= \nu_0)$ value of the B 1/2 → X 1/2 transition and the $\Delta G(1/2)$ value of the B 1/2 state, given in Table 4, could be determined.

The C band consists of the $\Delta v = -1$, 0, and 1 sequences from $v' = 0$–5. The $\Delta v = 0$, and 1 sequences are degraded to red, while the $\Delta v = -1$ sequence is degraded to blue. These opposite degradation suggests that the relations, $B_{v'} < B_{v''=v'}$ and $B_{v'} > B_{v''=v'+1}$, hold in this system. The molecular constants of the A_1 3/2 and C_1 3/2 states were obtained from the R heads of the $\Delta v = 0$ and 1 sequences.

The (0, v'') bands with low v'' values appear for the D band, because the equilibrium internuclear distance of the A_2 1/2 state ($r_{e''} \approx 3.6$ Å)[166] is slightly shorter than that of the C_2 1/2 state ($r_{e'} = 3.733$ Å).[180] The ν_e value of the C_2 1/2 → A_2 1/2 transition and the vibrational constants of the lower A_2 1/2 state were determined from a least-squares fit to eleven R heads. The T_e value of the A_2 1/2 state, given in Table 4, was obtained using the ν_e value of the C_2 1/2 → A_2 1/2 transition and the known T_e value of the C_2 1/2 state.[180]

Prior to our FA studies of the B, C, and D bands of $ArKr^+$ cluster ion, only the A band had been known. However, the B 1/2 → A_2 1/2 transition (the E band), which was predicted to occur at 1130 nm by Tanaka et al.,[165] could not be detected due to a low sensitivity of the optical detection system in the near infrared region. It is clear from Figure 2(a) that the following relation holds among the wavenumbers of the A, B, D, and E bands:

$$\nu(A) - \nu(D) + \nu(E) - \nu(B) = 0. \tag{10}$$

By using the ν_0 values of the B and D bands and the known ν_0 value of the A band,[180] the ν_0 value of the E band was calculated to be 9271.4 cm^{-1} from equation 10. From this value, the wavelength of the (0, 0) band of the B 1/2 → A_2 1/2 transition was precisely predicted to be 1078 nm.

Figure 4 shows the Morse potentials of the C_2 1/2, B 1/2, A_2 1/2, and X 1/2 states of $ArKr^+$ calculated using molecular constants given in Table 4. According to ab initio calculations of $ArKr^+$ by Bender and Winter[166] and Ma et al.,[169] the potential curve of $ArKr^+$ in the ground X 1/2 state is strongly bound, while the A_2 1/2, C_1 3/2, and C_2 1/2 states appear to have shallow minima at large internuclear distances, and the A_1 3/2 and B 1/2 states have no clear minima. The theoretical results of Bender and Winter[166] are thus inconsistent with our finding that the A_1 3/2 and B 1/2 states are weakly bound. Therefore, more detailed ab initio calculations are required to obtain accurate $ArKr^+$ potentials.

Although we have estimated the ν_e value of the C_1 3/2 → A_1 3/2 transition of $ArKr^+$, the T_e values of the C_1 3/2 and A_1 3/2 states could not been determined. We attempted to estimate them using the following relationships between the atomic and molecular splittings of $RgRg'^+$.[181]

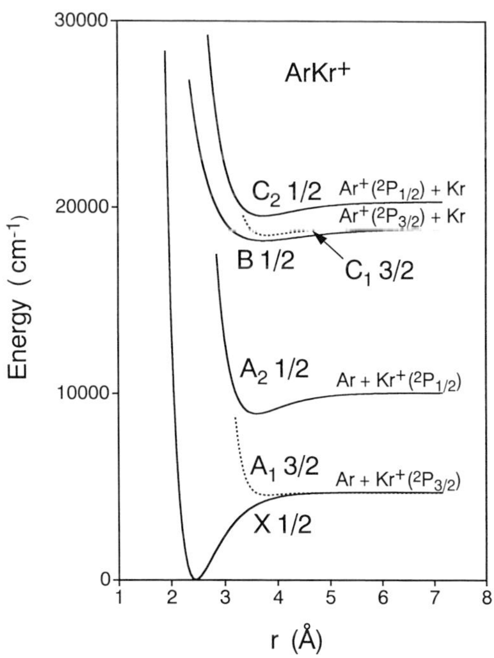

Figure 4. Morse potentials for the C_2 1/2, B 1/2, A_2 1/2, and X 1/2 states of $ArKr^+$, deduced from the data in Table 4.

The energies of the magnetic interaction of orbital and spin angular momenta are −A and +A/2 in the atomic ion, while they are −A/2, 0, and +A/2 in the molecular ion. Here, A is a spin-orbit coupling constant in the ground state of atomic ion. An additional off-diagonal matrix element arises in the molecular ion from the microscopic spin-orbit operator

$$\langle {}^2\Pi_{1/2}|H_{so}|{}^2\Sigma_{1/2}\rangle = -A/\sqrt{2}, \tag{11}$$

pushing apart the two 1/2 states. Figure 5 illustrates this for the configuration of $Ar + Kr^+$. According to simple perturbation theory, T'_{A1}, the energy of the A_1 3/2 state is given from the following relations involving the lower A_2 1/2, X 1/2, and A_1 3/2 states:

$$T'_{A1} = \left\{T'_{A2} + T'_X + 2A \pm \sqrt{(T'_{A2} + T'_X)^2 - 2A^2}\right\}/2. \tag{12}$$

Here, T'_{A2} and T'_X, are energies of perturbed (observed) A_2 1/2 and X 1/2 states, respectively. In order to obtain the real root of the T'_{A1} value from equation 12, the following relation must be satisfied:

$$T'_{A2} - T'_X \geq -\sqrt{2}A. \tag{13}$$

Similar relations hold for the upper C_2 1/2, B 1/2, and C_1 3/2 states. When the potential energies of the C_1 3/2 and A_1 3/2 states of the $NeRg'^+$ (Rg′ = Ar, Kr, Xe) and $ArXe^+$ systems were estimated using equation 12, a good agreement between the calculated and observed energies was found for $NeRg'^+$ and $ArXe^+$. These results justified the use of equation 11 as the off-diagonal matrix elements for $NeRg'^+$ and $ArXe^+$. Equation 12 was

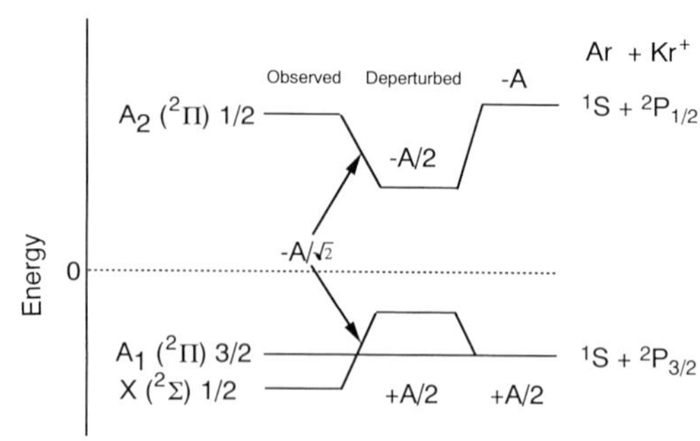

Figure 5. Energy-level diagram of the $Ar + Kr^+$ configuration.

also applied to $ArKr^+$ for the determination of the T_e values of the C_1 3/2 and A_1 3/2 states. Since the A values of Ar^+ and Kr^+ in the ground state are −954.7 and −3580.7 cm^{-1}, respectively, the energy difference must be larger than 1350.1 cm^{-1} for the C_2 1/2 and B 1/2 states and 5063.9 cm^{-1} for the A_2 1/2 and X 1/2 states according to equation 13. The observed energy difference of the A_2 1/2 and X 1/2 states (8832.1 cm^{-1}) satisfies equation 13, and the T_0 value of the A_1 3/2 state was estimated to be 4453.5 cm^{-1}. On the other hand, the observed energy difference between the C_2 1/2 and B 1/2 states (1337.2 cm^{-1}) does not satisfy equation 13, so that no real root of the T_0 value was obtained for the C_1 3/2 state. For $ArXe^+$ and $NeRg'^+$, whose energy splittings can well be explained by the spin-orbit coupling constants of the atomic ions, there is a good agreement between the molecular and atomic splittings. On the other hand, there is a large discrepancy between the energy separation between the A_2 1/2 and X 1/2 states of $ArKr^+$ (8832.1 cm^{-1}) and the spin-orbit splitting of $Kr^+(^2P_J$: 5371.0 cm^{-1}). This implies that the energy splittings among the C_2 1/2, B 1/2, and C_1 3/2 states of $ArKr^+$ cannot be explained only by the spin-orbit coupling constant of $Kr^+(^2P_J)$. The T_0 value of A_1 3/2 was estimated from equation 12, while the T_0 value of the C_1 3/2 state was evaluated from the T_0 value of the A_1 3/2 state and the ν_0 value of the C_1 3/2 → A_1 3/2 transition. Further detailed experimental and theoretical studies are required in order to determine accurate term values of the C_1 3/2 and A_1 3/2 states.

B. Emission Spectra of Heterotrimer Ions

Emission spectra of heavier rare gases in He, Ne, Ar, and Kr afterglows were measured at higher stagnation pressures than those used for the detection of the heterodimer bands. For example, Figures 6(a)–6(c) show emission spectra obtained from the reactions of Kr in Ne afterglows at Kr stagnation pressures of 0.5, 1.0, and 2.1 atm. Four known $NeKr^+$ bands (A, B, D, and E) appear at low Kr stagnation pressures. A new continuous band appears in the 159–215 nm region at high Kr stagnation pressures which consists of two components. The first component in the 159–184 nm region [F in Figure 6(c)] appears near the A and B bands of $NeKr^+$, while the second one with a typical Gaussian type envelope is found in the 184–215 nm region [S in Figure 6(c)]. Continuum F appears at higher Kr stagnation pressure than $NeKr^+$(B band) and its intensity increases more rapidly than that of the band. Also, the intensity of continuum S increases more rapidly than that of continuum F.

Similar continuous bands were observed near or at the longer wavelength side of the heterodimer bands in all the reaction systems studied except for

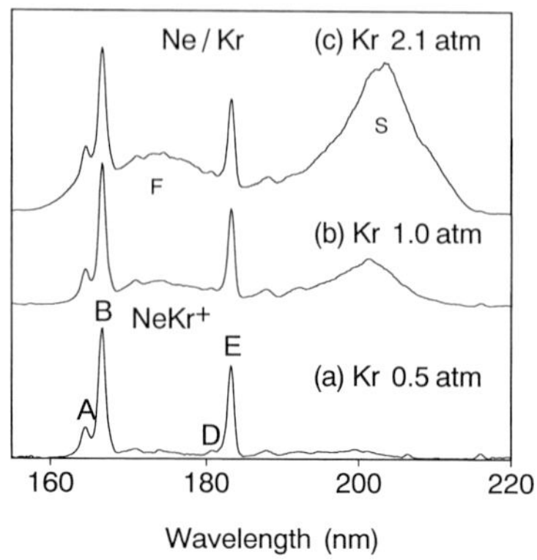

Figure 6. Emission spectra obtained from the Ne afterglow reactions of Kr at various Kr stagnation pressures.

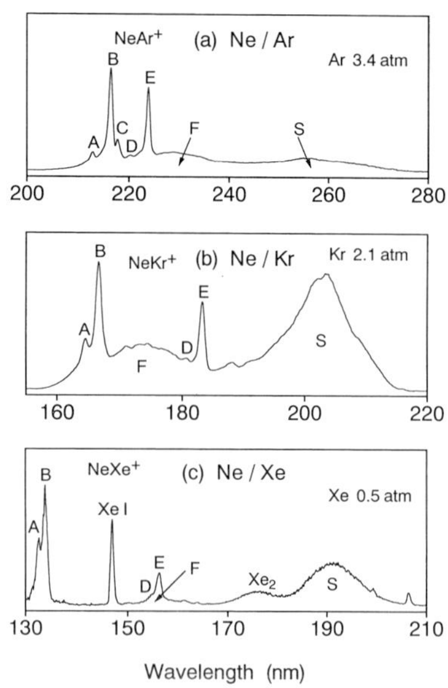

Figure 7. Emission spectra obtained from the Ne^+/Rg' ($Rg' = Ar$, Kr, and Xe) systems.

the He^+/Ne system.[78–81] For example, continua observed in the Ne^+/Rg' ($Rg' = Ar$, Kr, Xe) systems are shown in Figure 7. The following general aspects were found for the continua.

(a) In most cases, the continuous bands consist of two components and the onset wavelengths of the shorter-wavelength components are close to the

B 1/2 → X 1/2 (A band) transition of $HeRg'^{+}$ and the B 1/2 → X 1/2 (B band) transition of $RgRg'^{+}$ (Rg = Ne, Ar, Kr). The observed wavelengths depend on both Rg and Rg′.

(b) Although the continua appear at high stagnation pressures of Rg′, the onset stagnation pressure becomes low with the increasing polarization of the Rg′ atom. The second continuum increases more rapidly than the first one with increasing stagnation pressure of Rg′. The effects of the addition of a third rare gas as a foreign gas suggested that the collisional relaxation participated in the formation of the second continua.

These features were very similar to those of the well-known bound-free transitions of Rg_2^* and Rg_2X^* excimers (X = halogen), which appear at high rare-gas pressures in the longer-wavelength region of Rg* atomic line and the main RgX* excimers.[182–189] Therefore, the continuous bands were ascribed to the bound-free transitions of $RgRg'^{+}_2$:

$$Rg^+Rg'_2 \rightarrow RgRg'^{+}_2 + h\nu. \tag{14}$$

All the $RgRg'^{+}_2$ bands disappeared when the active ionic species were removed from the discharge flows. This implies that $He^+(^2S_{1/2})$ in the He flow and $Rg^+(^2P_{1/2})$ and/or $Rg^+(^2P_{3/2})$ in the Ne, Ar, and Kr flows are responsible for the formation of $RgRg'^{+*}_2$. Since the $[Ne^+(^2P_{1/2})]/[Ne^+(^2P_{3/2})]$ ratio was small (~0.1), the lower $Ne^+(^2P_{3/2})$ spin-orbit component was expected to be more important than the upper $Ne^+(^2P_{1/2})$ one for the formation of the $NeRg'^{+*}_2$. The relative contribution of the two spin-orbit components of Ar^+ or Kr^+ to the formation of $RgRg'^{+*}_2$ (Rg = Ar, Kr) was examined by adding N_2 or O_2 into the Ar or Kr FA, respectively. Since reaction rate constants of the $Ar^+(^2P_{1/2})/N_2$ and $Kr^+(^2P_{3/2})/O_2$ reactions are much larger than those of the $Ar^+(^2P_{3/2})/N_2$ and $Kr^+(^2P_{1/2})/O_2$ ones, N_2 and O_2 can be used as filter gases of $Ar^+(^2P_{1/2})$ and $Kr^+(^2P_{3/2})$, respectively.[190,191] The effects of N_2 addition for the Ar^+/Kr and Ar^+/Xe systems and O_2 addition for the Kr^+/Xe system implied that the lower $Rg^+(^2P_{3/2})$ component was responsible for the formation of the $RgRg'^{+*}_2$ observed in the Ar^+/Kr, Ar^+/Xe, and Kr^+/Xe systems. Thus, the first and second components were attributed to the B 1/2–X 1/2 and B 1/2–A_2 1/2 transitions in the cases of $HeRg'^{+}_2$, and the B 1/2–X 1/2 and/or C_1 3/2–A_1 3/2 transitions and the B 1/2–A_2 1/2 transition in the cases of $RgRg'^{+}_2$ (Rg = Ne, Ar, Kr), respectively. The emission intensity of the second continuum increased more rapidly than that of the first with increasing stagnation pressure of Rg′ or a foreign gas. This was explained by the fact that the first continuum occurs from high vibrationally excited levels of $HeRg'^{+}_2$(B 1/2) and $RgRg'^{+}_2$(B 1/2), while the second continuum arises predominantly from low vibrationally excited levels formed by collisional relaxation of the high vibrational levels of these states. Summing up the above results, the major excitation and relaxation processes of $Rg^+Rg'_2{}^*$ cluster ions in high and low vibrational levels is as follows:

$$Rg^+Rg'(B\ 1/2, C_1\ 3/2) + Rg' \rightarrow Rg^+Rg'_2(B\ 1/2, C_1\ 3/2\!:\text{high } v') + h\nu, \tag{15a}$$

$$Rg^+Rg'(B\ 1/2, C_1\ 3/2) + 2Rg' \rightarrow Rg^+Rg'_2(B\ 1/2, C_1\ 3/2\!:\text{high } v') + Rg', \tag{15b}$$

$$Rg^+Rg'_2(B\ 1/2, C_1\ 3/2\!:\text{high } v') + Rg' \rightarrow Rg^+Rg'_2(B\ 1/2, C_1\ 3/2\!:\text{low } v') + Rg'. \tag{15c}$$

As an example of the heterotrimer ions, the energy diagram of $NeKr_2^+$ is shown in Figure 8. This diagram is obtained assuming that the B and C_1 states and the X and A_1 states of $NeKr_2^+$ are degenerated. According to recent ab initio calculations of the ground states of $RgRg_2'^+$ by Hiraoka et al.,[171] their stable geometry is the symmetrical triangle [Figure 9(a)] and the equilibrium internuclear distance of Rg′–Rg′ in $RgRg_2'^+$ is nearly the same as that of known $Rg_2'^+$ dimer ions in the ground state. Therefore, they concluded that the Rg atom acts as a guest atom of $Rg_2'^+$. The dissociation energies of Ar_2^+ (1.26 eV), Kr_2^+ (1.13 eV), and Xe_2^+ (1.02 eV) are much larger

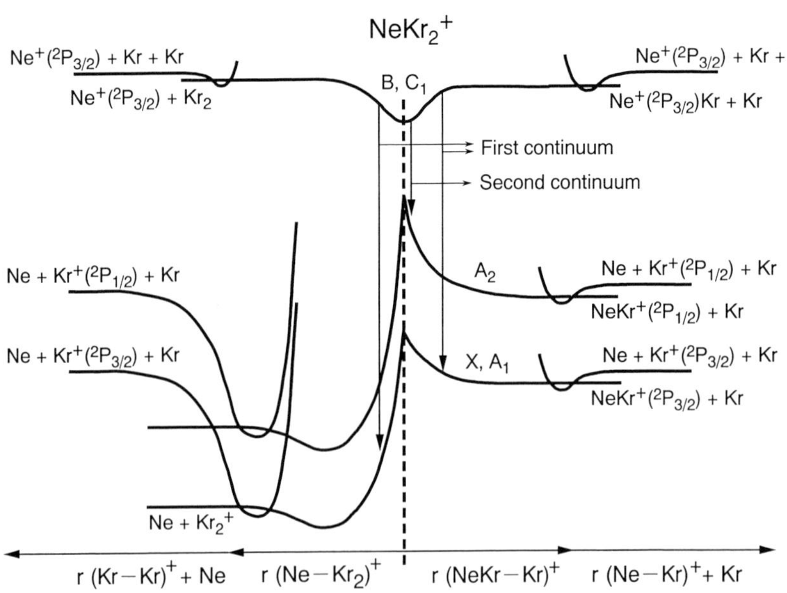

Figure 8. The potential energy diagram of the Ne^+/Kr/Kr system.

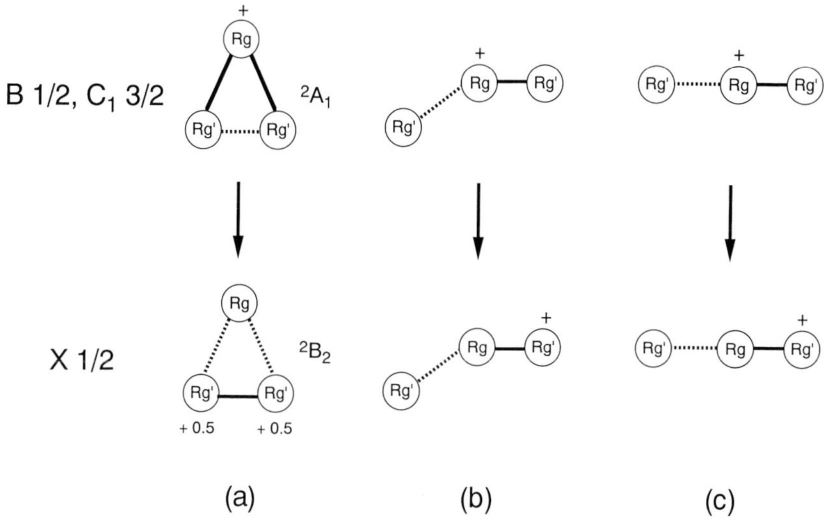

Figure 9. Possible structural configurations of $RgRg'^+_2$ in the upper $Rg^+Rg'_2$ states.

than those of Ar_2 (12 meV), Kr_2 (17 meV), and Xe_2 (24 meV) in the ground states. Therefore, significant blue shifts from the associated $RgRg'^+$ bands will occur for the $Rg^+Rg'_2 \rightarrow RgRg'^+_2$ transitions from highly vibrationally excited levels near the $Rg^+ + Rg'_2$ dissociation limit, if the geometries of $Rg^+Rg'_2$ are similar to those of $RgRg'^+_2$. Although slight blue shifts from the related B band of heterodimer bands were found for the onset wavelengths of the first continua of $NeAr^+_2$ and $NeKr^+_2$, such significant blue shifts were not observed. These facts led us to conclude that the geometries of $Rg^+Rg'_2$ are significantly different from those of $RgRg'^+_2$ in the ground state. Since the onset wavelengths of the first continua of $RgRg'^+_2$ were located close to those of the related $RgRg'^+$ bands, the major chromophore of $RgRg'^+_2$ will be $RgRg'^+$ and the second Rg′ atom will act as a guest atom of Rg^+Rg' in the upper $Rg^+Rg'_2$ states.

Assuming that Rg^+-Rg′ is a chromophore and the second Rg′ behaves as a guest atom, possible geometries of $Rg^+Rg'_2$ in the emitting excited states are asymmetrical triangle structure (b) and linear structure (c), as shown in Figure 9. If $Rg^+Rg'_2$ has the linear geometry (c), their maximum transition energies of the first continua will be nearly the same as those of the correlated $RgRg'^+$ bands. On the other hand, if $Rg^+Rg'_2$ has bent geometry (b), blue shifts of the first continua due to a bound character of Rg'^+_2 will occur for the $RgRg'^+_2$ bands. Small blue shifts from the B bands of the heterodimer were actually observed for the $HeRg'^+_2$ (Rg′ = Ar, Kr),

$NeRg'^+_2$ ($Rg' = Ar$, Kr), and $ArXe_2^+$ bands. This implies that a minor bound-free transition leading to $Rg + Rg'^+_2$ products (left side route in Figure 8) as well as a major bound-free transition leading to $RgRg'^+ + Rg'$ products (right side route in Figure 8) occurs in these systems. Consequently, the asymmetrical triangle structure (b) was expected to be more important in most cases as a possible geometry of $Rg^+Rg'_2$. When a charge-transfer type of electronic transition from Rg^+ to Rg' in Rg'-Rg^+–Rg' occurs, Rg'^+-Rg–Rg′ is formed. The lower Rg'^+-Rg–Rg′ states are much more unstable than the upper Rg'-Rg^+–Rg' states because of a high Rg–Rg′ repulsion at short range. Therefore, bound-free transitions will take place at short $(Rg'\text{-}Rg)^+$–Rg' distances.

C. Formation Processes of Heterodimer Ions

Since the rotational constant of $HeNe^+$ is largest among rare gas heterodimer ions,[192] rotational fine structures in the electronic transition of $HeNe^+$ can be easily separated. Therefore, reliable data for the rotational distribution of $HeNe^+(B\ 1/2)$ can be obtained. When emission spectra obtained from the He afterglow reactions of Ne are measured at various He and Ne pressures, the $HeNe^+(B\ 1/2–X\ 1/2)$ emission system is observed, as shown in Figure 10. Possible excitation processes of $HeNe^+(B\ 1/2)$ are as follows:

$$\begin{aligned} &He^+ + Ne \xrightarrow{k_{16}} HeNe^+(B\ 1/2) + h\nu, \\ &k_{16} = 2 \times 10^{-17}\ cm^3\ s^{-1}(Ref.164) \end{aligned} \tag{16}$$

$$\begin{aligned} &He^+ + Ne + Ne \xrightarrow{k_{17a}} HeNe^+(B\ 1/2) + Ne, \\ &k_{17a} = 2.4 \times 10^{-32}\ cm^6\ s^{-1}(Ref.164) \end{aligned} \tag{17a}$$

$$He^+ + Ne + He \rightarrow HeNe^+(B\ 1/2) + He. \tag{17b}$$

Although rate constant of process 17b has not been measured, it will be the same order as that of process 17a with a somewhat smaller stabilization efficiency. The contribution of process 16 is expected to be much smaller than that of processes 17a and 17b, because the k_{16} value is too small to probe the product $HeNe^+(B\ 1/2)$ ion under our experimental conditions. The vibrational and rotational distributions were determined by a spectral simulation. No significant changes in rovibrational distributions were found under the conditions, where process 17a or 17b was dominant. The $N_0 : N_1$ ratio of $HeNe^+(B\ 1/2)$ was 100: ~10 and the rotational temperatures of the $v' = 0$ and 1 levels were 70–110 and 40–70 K, respectively.

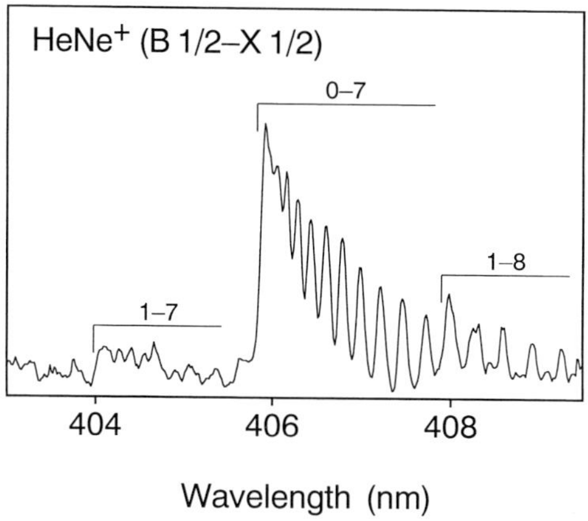

Figure 10. Emission spectrum of $HeNe^+$(B 1/2–X 1/2) produced from Ne reactions in a He afterglow.

The average vibrational and rotational energies of $HeNe^+$(B 1/2) were evaluated to be only 9.9–20 and 47–72 cm^{-1}, respectively. This implied that little excess energy is released as vibrational and rotational energies of $HeNe^+$(B 1/2) in processes 17a and 17b. The average fractions of the total available energy deposited into vibrational and rotational modes of $HeNe^+$(B 1/2), $\langle f_v \rangle$ and $\langle f_r \rangle$, were determined to be only about 0.014–0.033 and 0.097–0.12, respectively. These results imply that most of the excess energy is released as the relative translational energy of the products of these clustering reactions.

$HeAr^+$(B 1/2–X 1/2) emissions resulting from the following reactions were observed from Ar reactions in a He afterglow:

$$He^+ + Ar \rightarrow HeAr^+(B\ 1/2) + h\nu, \tag{18}$$

$$He^+ + Ar + He \rightarrow HeAr^+(B\ 1/2) + He, \tag{19a}$$

$$He^+ + Ar + Ar \rightarrow HeAr^+(B\ 1/2) + Ar. \tag{19b}$$

The relative contribution of processes 18 and 19a was examined by comparing the dependence of the emission intensity of $HeAr^+$(B 1/2–X 1/2) on the He pressure with that of $N_2^+(C^2\Sigma_u^+ - X^2\Sigma_g^+)$ resulting from the two-body CT reaction:

$$He^+ + N_2 \rightarrow N_2^+(C^2\Sigma_u^+) + He. \tag{20}$$

Measurements were carried out at low Ar and N_2 pressures (below 0.05 Torr), where process 19b was expected to be insignificant. The dependence of $HeAr^+$(B 1/2–X 1/2) on the He pressure was similar to that of $N_2^+(C^2\Sigma_u^+ - X^2\Sigma_g^+)$ below ~0.75 Torr, while a significant enhancement of the emission intensity was found for $HeAr^+$(B 1/2–X 1/2) above that pressure. This result indicated that process 18 is dominant at low He pressures below ~0.75 Torr, while process 19a is significant above this value. The rate constant of 18 was determined to be 4.0×10^{-13} cm^3 s^{-1}, indicating that the formation rate of $HeAr^+$(B 1/2) with a large binding energy (0.16 eV) is much faster than that of $HeNe^+$(B 1/2) with a small binding energy (0.036 eV).

The vibrational distributions of $HeAr^+$(B 1/2) were determined by a spectral simulation, because it was difficult to resolve the vibrational structure of the $HeAr^+$(B 1/2–X 1/2) transition. Since no appreciable change in the spectral features was found by changing the rotational temperatures of $HeAr^+$(B 1/2:$v'=0$–8) in the 100–300 K range, they were fixed at 300 K. The vibrational distributions of $HeAr^+$(B 1/2) in the He^+/Ar two-body radiative association and He^+/Ar/He and He^+/Ar/Ar three-body asssociation are shown in Figure 11. There are no significant differences in the vibrational distributions of $HeAr^+$(B 1/2) in the three reactions, although the vibrational populations of $v'=0$ and 1 in the He^+/Ar/Ar reaction are larger than those in the other two reactions. The $\langle E_v \rangle$ values for the He^+/Ar, He^+/Ar/He, and He^+/Ar/Ar reactions were

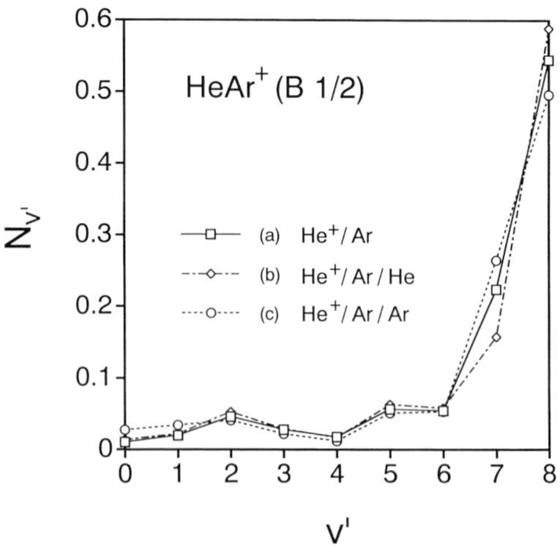

Figure 11. Vibrational distributions of $HeAr^+$ produced from the He^+/Ar, He^+/2Ar, and He^+/Ar/He reactions.

estimated to be 975, 965, and 951 cm^{-1}, respectively. The average fractions of available energies deposited into vibrational energy of $HeAr^+$(B 1/2), $\langle f_v \rangle$, were nearly constant for the three reactions (0.61–0.62), indicating that much higher fractions of the total excess energies are released as vibrational energy of $HeAr^+$(B 1/2) than those in the case of $HeNe^+$(B 1/2).

IV. CHARGE-TRANSFER REACTIONS OF He_2^+ IONS

Since the available recombination energy (RE) of He_2^+ has a large width of 18.3–20.3 eV due to a repulsive nature of $He_2(X^1\Sigma_g^+)$ ground state, more resonant CT reactions are possible than for atomic ions such as He^+ with a constant RE of 24.59 eV.[193] Figure 12 shows an energy-level diagram of RE He_2^+ and emitting ionic states of NO, N_2, and CO. The $NO^+(A^1\Pi)$, $N_2^+(B^2\Sigma_u^+)$, and $CO^+(B^2\Sigma^+)$ states are located within the RE of He_2^+, while the $CO^+(A^2\Pi_i)$ state is located outside of it. The CT reactions of He_2^+ with N_2 and CO have been studied by observing optical emissions from product ions in a low energy ion-beam experiment[194] and in a helium FA:[86]

$$He_2^+ + N_2 \rightarrow N_2^+(B^2\Sigma_u^+) + 2He, \tag{21}$$

$$He_2^+ + CO \rightarrow CO^+(A^2\Pi_{i,}\ B^2\Sigma^+) + 2He. \tag{22}$$

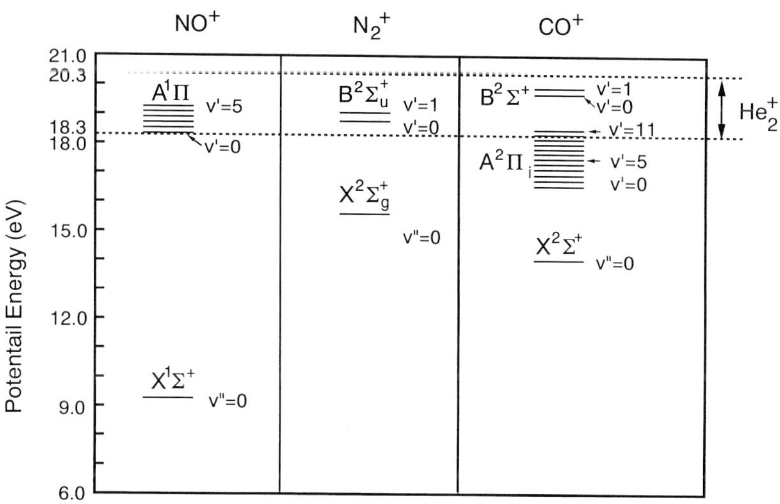

Figure 12. Energy-level diagram of $He_2^+(X^2\Sigma_u^+)$, $NO^+(A^1\Pi)$, $N_2^+(B^2\Sigma_u^+)$, and $CO^+(B^2\Sigma^+)$. The dotted lines delineate the range of the He_2^+ recombination energy.

Bearman et al.[194] found that vibrational distributions of the product ions in the above processes at 11 eV (lab) were consistent with a vertical FC-like model. Although Endoh et al.[86] also found FC-like vibrational distributions for $N_2^+(B)$ and $CO^+(B)$, a higher vibrational excitation than that predicted from a FC vertical model was found for $CO^+(A)$. They found a slight degree of rotational excitation in $N_2^+(B: v'=0)$ and $CO^+(B: v'=0)$, and the effective rotational temperatures were estimated to be 900 ± 60 and 890 ± 100 K, respectively.

We have recently studied the He_2^+/NO reaction at thermal energy by observing the $NO^+(A^1\Pi\text{-}X^1\Sigma^+)$ emissions in the VUV region.[88] The ionization mechanism was determined from the observed vibrational and rotational distributions of $NO^+(A)$.

A. Excitation Processes of $NO^+(A^1\Pi)$ in a He Afterglow

Figure 13(a) shows an emission spectrum of NO in a helium afterglow at a He pressure of 1.5 Torr, where active species were $He(2^3S)$, He^+, and He_2^+. Strong $NO^+(A^1\Pi\text{–}X^1\Sigma^+)$ emission from $v'=0\text{–}5$ were observed. When He^+ and He_2^+ ions were trapped using ion-collector grids, the $NO^+(A\text{–}X)$ emission reduced in intensity by about 20–30%, as shown in Figure 13(b). Figure 13(c) shows an emission spectrum resulting from ionic reactions, which is obtained by subtracting Figure 13(b) from Figure 13(a). The dependence of $[He^+]$ and $[He_2^+]$ on the He pressure was monitored using the $N_2^+(C\text{–}X)$ and $CO_2^+(B\text{–}X)$ emissions resulting from the following reference reactions:[107,113]

$$He^+ + N_2 \rightarrow N_2^+ + (C^2\Sigma_u^+) + He, \tag{23}$$

$$He_2^+ + CO_2 \rightarrow CO_2^+(B^2\Sigma_u^+) + 2He. \tag{24}$$

The effects of ion collection and the dependence of emission intensity of $NO^+(A\text{–}X)$ on $[He^+]$ and $[He_2^+]$ suggested that $NO^+(A\text{–}X)$ emissions in Figures 13(b) and 13(c) arise from the following Penning ionization and CT reactions, respectively:

$$He(2^3S) + NO \rightarrow NO^+(A^1\Pi) + He + e^-, \tag{25}$$

$$He_2^+ + NO \rightarrow NO^+(A^1\Pi) + 2He. \tag{26}$$

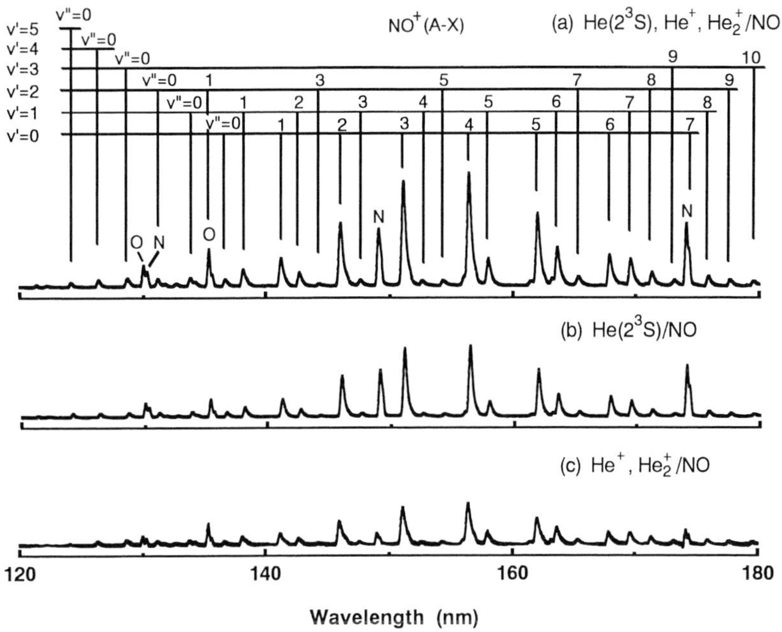

Figure 13. Emission spectra resulting from the (a) He(2^3S), He^+, He_2^+/NO, (b) He(2^3S)/NO, and (c) He^+, He_2^+/NO reactions, respectively. Spectrum (c) was obtained by subtracting spectrum (b) from spectrum (a).

B. Vibrational and Rotational Distributions of $NO^+(A^1\Pi)$

The relative rovibrational distributions of $NO^+(A^1\Pi)$ in the He_2^+/NO and He(2^3S)/NO reactions were determined by a computer simulation of NO^+(A–X). The rovibrational distributions of NO^+(A) were independent of buffer He gas pressure in the 0.5–1.5 Torr range and NO pressure in the 1–20 m Torr range. It was therefore concluded that collisional relaxation was insignificant and the observed rovibrational distributions reflected the nascent populations. The observed vibrational distribution of NO^+(A) is given in Table 5 together with the other related data. From the observed rovibrational distributions of NO^+(A), the average vibrational and rotational energies deposited into NO^+(A) were determined to be $\langle E_v \rangle = 0.22 \pm 0.02$ eV and $\langle E_r \rangle = 0.10 \pm 0.1$ eV in the He_2^+/NO reaction and $\langle E_v \rangle = 0.17$ eV and $\langle E_r \rangle = 0.063 \pm 0.005$ eV in the He(2^3S)/NO Penning ionization.

The vibrational distributions of NO^+(A) in electron-impact excitation and photoionization agree with FC factors for the NO(X) → NO^+(A)

Table 5. Vibrational Distributions of $NO^+(A)$ in the He_2^+/NO Charge-Transfer, Electron-Impact Ionization, Photoionization, $He(2^3S)/NO$ Penning Ionization, Franck-Condon Factors for the $NO(X) \rightarrow NO^+(A)$ Ionization

$N_{v'}$	*References*	$NO^+(A)$					
		N_0	N_1	N_2	N_3	N_4	N_5
He_2^+/NO	88	0.495	0.205	0.107	0.082	0.089	0.022
Electron impact	195	0.66	0.27	0.067	0.023		
Photoionization (HeI: 58.4 nm)	196	0.712	0.198	0.068	0.023		
$He(2^3S)/NO$	29	0.575	0.200	0.081	0.061	0.065	0.017
Franck-Condon factors	88	0.719	0.229	0.044	0.0068	0.00094	0.00012

ionization, as shown in Table 5. The $NO^+(A)$ ion in the He_2^+/NO reaction is more vibrationally excited than predicted from FC factors for the $NO(X) \rightarrow NO^+(A)$ ionization. It is also slightly more excited than in the $He(2^3S)/NO$ Penning ionization, which also proceeds through non-FC like ionization. This shows that the He_2^+/NO reaction proceeds through non-FC like ionization, even though the energy of $NO^+(A)$ is located within the RE of He_2^+ (Figure 12). On the basis of the present finding, the ionization mechanism of NO leading to the resonant $A^1\Pi$ state is different from that of N_2 and CO leading to the resonant $B^2\Sigma^+$ state. This shows that closed shell N_2 and CO molecules are not perturbed by an access of He_2^+, while an open shell NO molecule is perturbed before an electron jump.

The rotational temperature of $NO^+(A)$ in the He_2^+/NO reaction was 1170 ± 100 K, which was essentially independent of v'. The similar rotational excitation for $v' = 0$–5 levels was explained by the fact that the CT reaction occurs near resonantly because the RE of He_2^+ has a large latitude of 2 eV. Only 0.10 ± 0.01 eV of the RE, which amounts to only 0.5% of the RE, is converted into the rotational energy of $NO^+(A)$. The rotational temperatures of $NO^+(A)$ were slightly higher than those of $N_2^+(B{:}\, v' = 0)$ and $CO^+(B{:}\, v' = 0)$.[86] A higher rotational excitation is consistent with the fact that the He_2^+/NO reaction proceeds through an intimate collision, where more conversion of the RE of He_2^+ into the rotational energy of the product $NO^+(A)$ ion becomes possible via decomposition of a non-linear complex.

C. Charge-Transfer Mechanisms

The following four models are considered to explain the observed non-FC vibrational distribution: (Model 1) *distorted FC model*: In this model, the target NO molecule is distorted by an approach of the reactant He_2^+ ion due to charge-permanent-dipole and charge-induced-dipole interactions between the ion-dipole pair. (Model 2) *the effect of vibrational wavefunction of* $He_2^+(X^2\Sigma_u^+: v''=0)$: Since the potential energies of $NO^+(A: v'=0\text{–}5)$ lie in the effective RE region of He_2^+, the probability density of vibrational wavefunction of $He_2^+(X^2\Sigma_u^+: v''=0)$, $P_{RE}(r-r_e)=\Psi_0^2(r-r_e)$, may affect the observed $N_{v'}$ distribution. This effect was taken into account by combining FC or distorted FC factors for vertical ionization $(FCF)_{v'}$ with the probability distribution of RE, $P_{RE}(r-r_e)$. (Model 3) *the effect of energy resonance*: The energy-resonant requirement was included by multiplying an exponential function as proposed by Anderson et al.[197] (Model 4) *the effect of the perturbation of* He_2^+: In the He_2^+/NO reaction, a significant distortion of He_2^+ would occur because the binding energy of He_2^+(X: 2.37 eV) is smaller than that of NO(X: 6.50 eV). This effect was taken into account by calculating distorted $P_{RE}(r-r_e)$ values from an equilibrium internuclear distance of $He_2^+(X: r_e=1.08\ \text{Å})$ to $r_e=2.0$ Å.

None of the above four models reproduced the observed $N_{v'}$ distribution well. Although the $N_{v'}$ population decreases rapidly for $v'=0\text{–}2$, it becomes rather flat above that. The deviation from FC factors is relatively small for the low $v'=0\text{–}2$ levels, while it is very large for the high $v'=3\text{–}5$ levels. Thus, it is reasonable to assume that the $N_{v'}$ distribution is bimodal. There exist both singlet and triplet entrance $He_2^+ + NO$ surfaces, which can interact with the exit $NO^+(A: v') + 2He$ potentials. Therefore, one possible explanation for the bimodal distribution is the different entrance surfaces, which preferentially provide the low and high vibrational levels. The other possible explanation is different reaction dynamics for the low and high vibrational levels.

Two types of reaction mechanisms have been proposed at low energy for the $A^+ + BC \rightarrow BC^+ + A$ CT reactions.[198] One is Demkov-type (direct) mechanism, where the transition occurs by the avoided crossing along the r(B–C) coordinate. The probability for this transition will be governed by the energy defects and the FC factors between the reactant and product states at infinite separation. The other is Landau-Zener type (intimate) mechanism, where the transition occurs by an avoided crossing along the R(A–BC) coordinate. In this case, the properties of the reactants at infinite separation are expected to be lost, so that the probability for this transition would not reflect such properties as the energy defects and FC factors for infinite separation. This type of reaction proceeds through intimate coupling via a collision complex. In the He_2^+/NO reaction, the low $NO^+(A: v'=0\text{–}2)$ levels may be formed approximately via a Demkov-type mechanism, because the deviation from FC factors is relatively small. On the other hand, the high $NO^+(A: v'=3\text{–}5)$ levels are probably produced via a Landau-Zener type mechanism, because the deviation from the FC factors for infinite separation is very large. In order to determine the origin of the bimodal nature, detailed theoretical calculations of potential surfaces will be necessary.

V. ELECTRON–ION AND ION–ION RECOMBINATION REACTIONS OF He_2^+

The He_2^+ dimer ion is an ideal molecule for studying recombination processes by emission spectroscopy because there are many emitting excited states for He_2*.[192] Two pioneering FA optical studies have been carried out on the electron–He_2^+ recombination process leading to He_2* by Schmeltekopf and Broida[23] and Collins and Robertson.[19,20] Schmeltekopf and Broida[23] observed several He_2* emission systems resulting from the collisional radiative recombination:

$$He_2^+(X) + 2e^- \rightarrow He_2^* + e^-. \tag{27}$$

They found that the rotational populations of high and low rotational levels of the $e^3\Pi$ state were expressed by double Boltzmann temperatures of 903 and 492 K at 1.4 Torr and 878 and 278 K at 250 Torr, respectively. Collins and Robertson[19,20] examined the axial variation and pressure dependence of the intensity of the He_2* emission and He_2^+ concentration, and identified the dominant populating mechanism of He_2* to be recombination process 27.

We have recently made further detailed FA optical studies of process 27 and the following ion–ion recombination processes:

$$He_2^+ + C_6F_6^- \rightarrow He_2^* + C_6F_6, \tag{28a}$$

$$He_2^+ + C_6F_5Cl^- \rightarrow He_2^* + C_6F_5Cl. \tag{28b}$$

A. $He_2^+/2e^-$ Recombination

When emission spectra resulting from collisional radiative recombination 27 were observed, ten singlet and twenty triplet systems of He_2* bands were identified in the 350–1000 nm region including those previously observed by Collins and Robertson[20] in the UV and visible region. Although all of the He_2* states below the recombination energy of He_2^+ (22.22 eV) can be formed in the $He_2^+/2e^-$ reaction, He_2* states in the

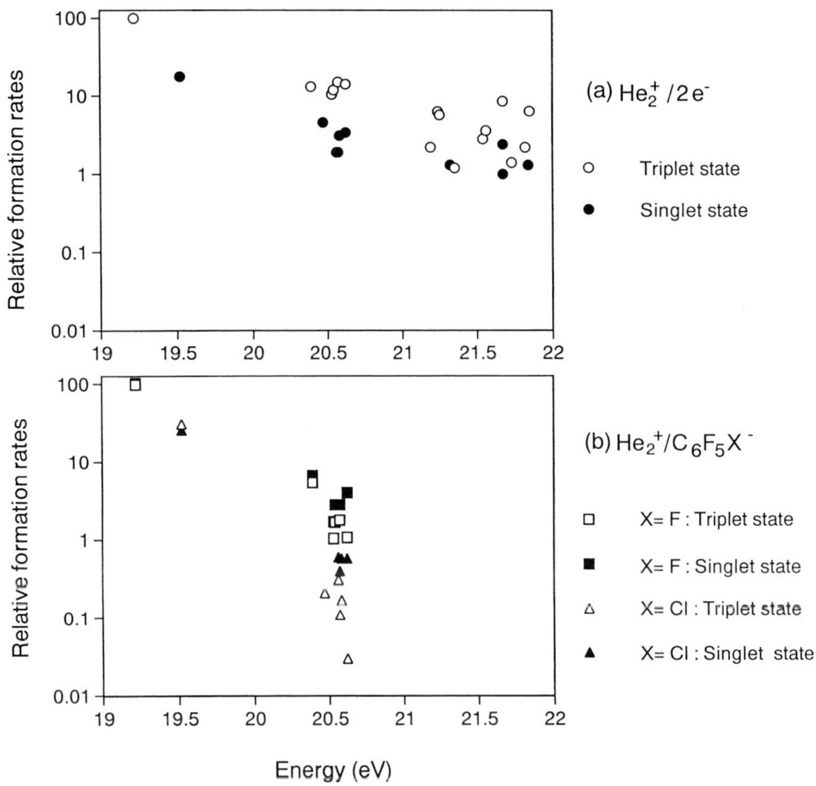

Figure 14. Relative formation rates of He_2* in the (a) $He_2^+/2e^-$ and (b) $He_2^+/C_6F_5X^-$ recombination reactions.

19.2–21.9 eV region were observed. The separate relative formation rates of singlet and triplet states are shown in Figure 14(a). The relative formation rates are largest for the lowest $c^3\Sigma^+$ and $C^1\Sigma^+$ states, and decrease with increasing excitation energy of He_2* for both the singlet and triplet states. The total formation ratio of the singlet/triplet states was about 1/4, which was close to a statistical ratio of 1/3. No vibrational excitation was observed for all the He_2* bands. The lack of vibrational excitation was explained by the large FC factors for the $He_2^+ \rightarrow He_2$* recombination due to similar equilibrium internuclear distances of He_2^+ (1.081 Å) and the observed He_2* states (1.069–1.097 Å).[192]

The rotational distribution of $He_2(d^3\Sigma_u^+)$, and its average rotational energy and the fraction of the total excess energy, are listed in Table 6.

Table 6. Rotational Energy Disposal in the $He_2^+/2e^-$ and $He_2^+/C_6F_5X^-$ ($X=F,Cl$) Recombination Reactions

Reaction	*State*	E_{state} (eV)	E_{excess} (eV)	T_{rot} (K)		*Obs.* E_r(eV)	*Obs.* $\langle f_r \rangle$	*Calc.*[a] E_r(eV)	*Calc.*[a] $\langle f_r \rangle$
$He_2^+/2e^-$	$He_2(d^3\Sigma_u^+)$	20.4	1.8	450	800	0.07	0.04	0.22	0.12
$He_2^+/C_6F_6^-$	$He_2(c^3\Sigma_g^+)$	19.2	2.5	1000		0.08	0.03	0.07	0.03
$He_2^+/C_6F_6^-$	$He_2(d^3\Sigma_u^+)$	20.4	1.3	500	1200	0.10	0.08	0.04	0.03
$He_2^+/C_6F_6^-$	$He_2(C^1\Sigma_g^+)$	19.5	2.2	1000		0.09	0.04	0.06	0.03
$He_2^+/C_6F_5Cl^-$	$He_2(c^3\Sigma_g^+)$	19.2	2.3	950		0.07	0.03	0.07	0.03
$He_2^+/C_6F_5Cl^-$	$He_2(d^3\Sigma_u^+)$	20.4	1.1	500	1500	0.13	0.12	0.03	0.03

[a]Calculated from statistic prior distributions.

The rotational distribution of $He_2(d^3\Sigma_u^+)$ was independent of the He pressure in the 1.7–20 Torr range, indicating that collisional relaxation was insignificant under our experimental conditions. The rotational distribution of $He_2(d^3\Sigma_u^+)$ was reproduced by double Boltzmann rotational temperatures of 450 and 800 K. On the basis of this finding, more than two excitation mechanisms will be involved in process 29. In Table 6 are also shown calculated rotational energy and its fraction of the total excess energy. These values were estimated from a statistical prior distribution,[199] assuming a long lived $[He_2{}^* - e^-]$ complex:

$$He_2^+ + 2e^- \rightarrow [He_2^* - e^-] \rightarrow He_2^* + e^-. \tag{29}$$

The observed E_r and $\langle f_r \rangle$ values are about one-third of the calculated ones, indicating that most of the energy is released as relative translational energy of products. It is known that $N_2^+(B^2\Sigma_u^+)$ is rotationally excited under low-energy electron-impact ionization due to multiplepole interactions between N_2 and an incident electron.[200] Similar multiplepole interactions between $He_2{}^*$ and electrons in the exit channels may result in the slight rotational excitation of $He_2{}^*$.

B. $He_2^+/C_6F_5X^-$ ($X=F$, Cl) Recombination

When C_6F_5X ($X=F$ or Cl) was added to the FA, several $He_2{}^*$ bands with high excitation energies of 21.19–21.84 eV disappeared and some $He_2{}^*$ systems with low excitation energies of 19.22–20.62 eV remained, as shown in Figure 15 for the case of the $He_2^+/C_6F_6^-$ reaction. Although the formation of the $He_2{}^*$ states with excitation below 21.70 and 21.48 eV are energetically

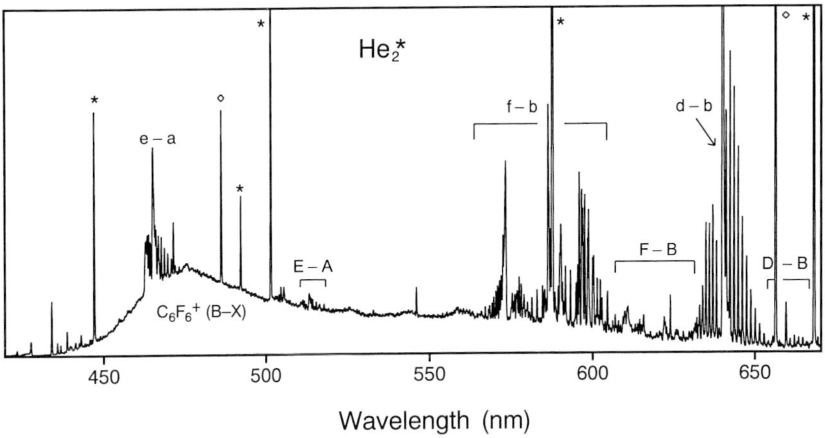

Figure 15. He_2* emissions resulting from the $He_2^+/C_6F_6^-$ ion–ion neutralization reaction of in a He afterglow. Lines marked with * and ◇ are He* and H*(impurity) lines, respectively. The broad $C_6F_6^+$(B–X) emission in the 430–530 nm region arises from the He(2^3S)/C_6F_6 and He_2^+/C_6F_6 reactions.

allowed in the $He_2^+/C_6F_6^-$ and $He_2^+/C_6F_5Cl^-$ reactions, respectively, only those in the 19.2–20.6 eV region were detected. The relative formation rates in reactions 28a and 28b are shown in Figure 14b. The relative formation rates are largest for the lowest $c^3\Sigma^+$ and $C^1\Sigma^+$ states, and decreases more rapidly than those in process 27 with increasing excitation energy of He_2* for both the singlet and triplet states. The sum of the formation rates of the $c^3\Sigma^+$ and $C^1\Sigma^+$ states accounts for 92 and 86% of the total formation rates for $C_6F_6^-$ and $C_6F_5Cl^-$, respectively. The total formation ratios of the singlet/triplet states were about 1/3.5 for $C_6F_6^-$ and 1/4 for $C_6F_5Cl^-$, which is close to a statistical ratio of 1/3.

The rotational temperatures of He_2* and their average rotational energies are given in Table 6. The rotational distributions of $He_2(d^3\Sigma_u^+)$ in the $He_2^+/C_6F_5X^-$ reactions were expressed by double Boltzmann temperatures, as in the case of the $He_2^+/2e^-$ reaction, while that of the $He_2(c^3\Sigma_g^+)$ state was reproduced by a single Boltzmann temperature. The observed rotational distributions of He_2(c,d) in processes 28a and 28b were close to those predicted from statistical calculations assuming a long-lived [He_2^+–$C_6F_5X^-$] complex.[199] Therefore, the He_2(c,d) may be formed by a long-lived ion-pair complex. The $\langle f_v \rangle$ and $\langle f_r \rangle$ values in both the $He_2^+/2e^-$ and $He_2^+/C_6F_5X^-$ reactions were 0 and 0.03–0.04, respectively. The low $\langle f_v \rangle$ and $\langle f_r \rangle$ values suggest that, at these recombination energies, most of the excess energy is released as relative translation energies of products and vibrational energy of C_6F_5X due to a strong mutually attractive force in the entrance channels and different equilibrium geometries between $C_6F_5X^-$ and C_6F_5X.

The observed He_2* states are formed by promotion of an electron from the 1sσ* orbital to the 3sσ, 3pσ, 3pπ, 3dσ, 3dπ, or 3dδ bonding orbital. No evidence for the formation of the upper $n = 4$ states was found in the 20.79–21.38 eV range, though their formation is energetically accessible. The lack of these upper He_2* states is probably due to the fact that the interparticle distance leading to such high energy states (15.7–44.4 Å) is too large for an efficient electron jump.

There was a great similarity between the $He_2^+/C_6F_6^-$ and $He^+/C_6F_6^-$ neutralization reactions leading to He_2* and He*, respectively.[136,138] In both reactions, favorable product channels are the formation of the $n = 3$ triplet states, and their relative emission rates generally decrease with increasing excitation energy of the neutral products. The neutralization reactions of He_2^+ and He^+ with $C_6F_6^-$ proceed through an electron jump from the HOMO orbital of $C_6F_6^-$ to a vacant orbital of He_2^+ or He^+, respectively. Our results indicate that the interactions of the HOMO orbital of $C_6F_6^-$ with the 3sσ, 3pσ, 3pπ, 3dσ, 3dπ, or 3dδ bonding molecular orbitals of He_2^+ at the crossing points are as large as those with the 3s, 3p, and 3d atomic orbitals of He^+.

We have initiated a theoretical analysis of the $He_2^+/C_6F_6^-$ reaction. The energies of electronic configurations of low-lying A–F and a–f states of He_2* have been calculated using MRCI(4,28) and CASSCF(4,28) methods. Good agreement between the observed and calculated excitation energies of He_2* was found when polarization Rydberg-type orbitals were added to the cc-pV5Z basis functions. The $C_6F_6^-$ potentials were optimized using B3LYP/6–31G* method. The $C_6F_6^-$ anion has two optimum geometries (C_{2v} and D_2). The former structure is more stable than the latter one by 17.1 kcal. Figure 16 shows ab initio entrance He_2^+–$C_6F_6^-$ ($^3A_1$,$^1A_1$) potentials determined in a parallel theoretical study; it was difficult to probe the other low-lying states in our experiments. It can be seen that He_2(d, D) states are formed through direct curve crossings between the entrance potentials and those for He_2(d, D)–C_6F_6 at a large internuclear distance of about 19 Å, while the A state is produced through avoided crossings between the entrance potentials and that for He_2(A)–C_6F_6 at shorter distance of about 12 Å. There are no direct avoided crossings between the entrance potentials and He_2(b, B)–C_6F_6 states. Therefore, these states will be produced through second curve crossings between intermediate He_2(A)–C_6F_6 potentials and those for He_2(b, B)–C_6F_6, if they are involved in the exit channels. A high performance computer having an extremely large memory was required to calculate potential energies of other exit He_2*–C_6F_6 channels.

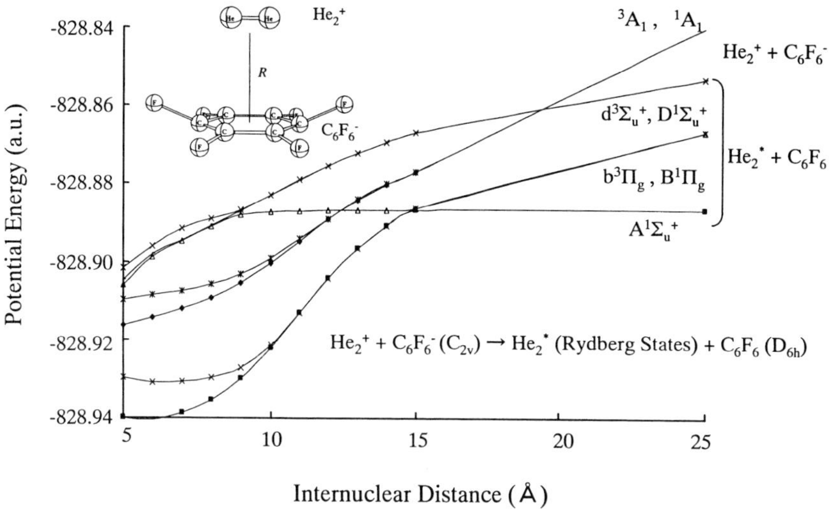

Figure 16. Ab initio potential energy curves for $He_2^+ + C_6F_6^-$ and a few low lying $He_2^* + C_6F_6$ states obtained in a parallel theoretical approach.

VI. SUMMARY AND CONCLUSIONS

The FA optical technique has been applied to studies of electronic structures and reactions of small rare gas cluster ions. From vibrational analyses of new emission systems of $ArKr^+$ and $KrXe^+$, molecular constants of related states were determined. A high-resolution rotational analysis of heterodimer bands using Fourier transform spectrometer and LIF method as well as detailed theoretical work will be necessary for better understanding of electronic structures of heterodimer ions. Since heterodimer bands are exclusively produced from association processes of correlated spin-orbit states of Rg^+, these bands will be useful in probing spin-orbit states of $Rg^+(^2P_J)$ in the FA. New continuous bands converging to heterodimer bands were observed at high stagnation pressures of heavier rare gases. They were ascribed to bound-free transitions of heterotrimer bands. In order to confirm the validity of this assignment, detailed ab initio calculations, as reported for the rare gas halide Rg_2X^* excimers,[184,186] will be required. Since the heterotrimer bands have typical bound-free character, these continuous bands are new promising candidates for excimer lasers in the VUV region generated by an electrical discharge of rare gas mixtures. In our FA study, definitive information on the excitation mechanism of the heterotrimer ions has not been obtained. Therefore, further kinetic studies, including measurements of radiative lifetimes of excited ions, will be required for the clarification of the excitation and decay mechanisms of rare gas heterocluster ions.

The nascent vibrational and rotational distributions of $NO^+(A^1\Pi)$ resulting from He_2^+/NO CT reaction at thermal energy were determined by analyzing the $NO^+(A^1\Pi-X^1\Sigma^+)$ emission in a He flowing afterglow. This was the first example that the product vibrational distribution gave a non-FC distribution, even though the product ionic levels lie within the resonance region. We are planning to extend this study to thermal energy CT reactions of heterocluster ions such as $HeNe^+$ and $HeAr^+$ with open shell molecules.

Electronic state distributions of He_2^* in the $He_2^+/2e^-$ and $He_2^+/C_6F_5X^-$ recombination reactions were determined. The electronic state populations of He_2^* decrease with their increasing excitation energies He_2^*. Collins and Robertson[19,20] predicted that Saha equilibrium is established between ions and electrons in FA. Although information on reionization rates and radiative and non-radiative decay rates of He_2^* are necessary for the understanding of such an equilibrium, no such experimental data have yet been obtained for He_2^*. Therefore, kinetic data for He_2^* obtained using the LIF technique are necessary for detailed analysis of the recombination dynamics for He_2^+. We found that the electronic state populations of He_2^* in the $He_2^+/C_6F_5X^-$ reactions rapidly decrease with increasing excitation energy of He_2^*. No vibrational excitation was found, though a slight rotational excitation was observed. I hope that the theoretical approach of our group, which is now in progress, will provide more dynamical information on the $He_2^+/C_6F_5X^-$ recombination.

ACKNOWLEDGMENTS

The author would like to thank his co-workers, especially Prof. Y. Nishimura, Prof. Y. Sakai, Dr. K. Mogi, Mr. M. Tanaka, Mr. H. Ishimi, Miss E. Oda, Mr. M. Hisano, and Mr. T. Tanoue, for their contribution to the studies reviewed here. This study was supported by a Grant-in-Aid for Scientific Research from the Japanese Ministry of Education, Science, Sports, and Culture (Nos. 06453026 and 09440201), the Iwatani Naoji Foundation, the Showa Shell Petroleum Foundation for Environmental Research, the Mitsubishi Foundation, and Kyushu University Interdisciplinary Progams in Education and Projects in Research Development (2001).

REFERENCES

[1] Ferguson, E. E.; Fehsenfeld, F. C.; Schmeltekopf, A. L. *Adv. At. Mol. Phys.* **1969**, *5*, 1.

[2] Fehsenfeld, F. C. *Int. J. Mass Spectrom. Ion Phys.* **1975**, *16*, 151.

[3] Ferguson, E. E.; Fehsenfeld, F. C.; Schmeltekopf, A. L. In *Gas Phase Ion Chemistry*; Bowers, M. T., Ed.; Academic Press: Florida,1979; Vol. 1, Chapter 24, p 309.

[4] Ferguson, E. E. In *Swarms of Ions and Electrons in Gases*; Lindinger, W.; Märk, T. D.; Howorka, F., Eds.; Springer: New York, 1984, p 126.
[5] McFarland, M.; Albritton, D. L.; Fehsenfeld, F. C.; Ferguson, E. E.; Schmeltekopf, A. L. *J. Chem. Phys.* **1973**, *59*, 6610.
[6] Smith, D.; Adams, N. G. In *Gas Phase Ion Chemistry*; Bowers, M. T. Ed.; Academic Press: Florida, 1979; Vol. 1, Chapter 1, p 1.
[7] Schmeltekopf, A. L.; Fehsenfeld, F. C. *J. Chem. Phys.* **1970**, *53*, 3173.
[8] Bolden, R. C.; Hemsworth, R. S.; Shaw, M. J.; Twiddy, N. D. *J. Phys. B* **1970**, *3*, 61.
[9] Jones, M. T.; Dreiling, T. D.; Setser, D. W.; N. McDonald, N. *J. Phys. Chem.* **1985**, *89*, 4501.
[10] Yencha, A. J. In *Electron Spectroscopy: Theory, Techniques and Applications*; Brundle, C. R.; Baker, A. D., Eds.; Academic Press: New York, 1984; Vol. 5, p 197.
[11] Smith, D.; Adams, N. G. In *Physics of Ion–Ion and Electron–Ion Collisions*; Brouillard, F.; McGowan, J. W., Eds.; Plenum: New York, 1981, p 501.
[12] Mitchell, J. B. A. In *Atomic Processes in Electron–Ion and Ion–Ion Collisions*; Brouillard, F., Ed.; Plenum Press: New York, 1988, p 185.
[13] Rowe, B. R. In *Dissociative Recombination: Theory, Experiment, and Application*; Rowe, B. R.; Brian, J.; Mitchell, A.; Canosa, A., Eds.; World Scientific: Singapore, 1993, p 113.
[14] Mitchell, J. B. A.; Rebrion-Rowe, C. *Int. Rev. Phys. Chem.* **1997**, *16*, 201.
[15] Ikezoe, Y.; Matsuoka, S.; Takebe, M.; Viggiano, A. *Gas Phase Ion Molecule Reaction Rate Constants through 1986*; Maruzen: Tokyo, 1987, and references therein.
[16] Adams, N. G.; Smith, D. In *Rate Coefficients in Astrochemistry*; Millar, T. J.; Williams, D. A., Eds.; Kluwer: Dordrecht, 1988, p 173.
[17] Tsuji, M.; Tsuji, K.; Fukutome, H.; Nishimura, Y. *Chem. Lett.* **1977**, 673.
[18] Collins, C. B.; Robertson, W. W. *J. Chem. Phys.* **1964**, *40*, 701.
[19] Collins, C. B.; Robertson, W. W. *J. Chem. Phys.* **1964**, *40*, 2202.
[20] Collins, C. B.; Robertson, W. W. *J. Chem. Phys.* **1964**, *40*, 2208.
[21] Robertson, W. W. *J. Chem. Phys.* **1966**, *44*, 2456.
[22] Schmeltekopf, A. L.; Ferguson, E. E.; Fehsenfeld, F. C. *J. Chem. Phys.* **1968**, *48*, 2966.
[23] Schmeltekopf, A. L.; Broida, H. P. *J. Chem. Phys.* **1963**, *39*, 1261.
[24] Wauchop, T. S.; Broida, H. P. *J. Quant. Spectrosc. Radiat. Transfer* **1972**, *12*, 371.
[25] Wauchop, T. S.; Broida, H. P. *J. Chem. Phys.* **1972**, *56*, 330.
[26] Taieb, G.; Broida, H. P. *Chem. Phys.* **1977**, *21*, 313.
[27] Richardson, W. C.; Setser, D. W.; Albritton, D. L.; Schmeltekopf, A. L. *Chem. Phys. Lett.* **1971**, *12*, 349.
[28] Richardson, W. C.; Setser, D. W. *J. Chem. Phys.* **1973**, *58*, 1809.
[29] Coxon, J. A.; Clyne, M. A. A.; Setser, D. W. *Chem. Phys.* **1975**, *7*, 255.
[30] Piper, L. G.; Gundel, L.; Velazco, J. E.; Setser, D. W. *J. Chem. Phys.* **1975**, *62*, 3883.
[31] Coxon, J. A.; Marcoux, P. J.; Setser, D. W. *Chem. Phys.* **1976**, *17*, 403.
[32] Chang, R. S. F.; Setser, D. W.; Taylor, G. W. *Chem. Phys.* **1978**, *25*, 201.
[33] Marcoux, P. J.; van Swaay, M.; Setser, D. W. *J. Phys.Chem.* **1979**, *83*, 3168.
[34] Wu, K. T.; Yencha, A. J. *Chem. Phys.* **1978**, *32*, 247.
[35] Wu, K. T.; Yencha, A. J. *Chem. Phys.* **1980**, *49*, 127.
[36] Wu, K. T.; Mogner, H.; Yencha, A. J. *Chem. Phys.* **1982**, *68*, 285.
[37] Miller, T. A.; Bondybey, V. E. *J. Chim. Phys-Chim.* Biol. **1980**, *77*, 695.
[38] Miller, T. A.; Bondybey, V. E. *Appl. Spectrosc. Rev.* **1982**, *18*, 105.
[39] Miller, T. A. *Ann. Rev. Phys. Chem.* **1982**, *33*, 257, and references therein.
[40] Guyer, D. R.; Hüwel, L.; Leone, S. R. *J. Chem. Phys.* **1983**, *79*, 1259.
[41] Bierbaum, V.; Ellison, G. B.; Leone, S. R. In *Gas Phase Ion Chemistry*; Bowers, M. T., Ed.; Academic Press: Florida,1979; Vol. 3, Chapter 17, p 1, and references therein.
[42] Sonnenfroh, D. M.; Leone, S. R. *J. Chem. Phys.* **1989**, *90*, 1677.

[43] Adams, N. G.; Herd, C. R.; Geoghegan, M.; Smith, D.; Canosa, A.; Gomet, J. C.; Rowe, B. R.; Queffelec, J. L.; Moriais, M. *J. Chem. Phys.* **1991**, *94*, 4852.
[44] Adams, N. G. *Adv. Gas Phase Ion Chem.* **1992**, *1*, 271.
[45] Adams, N. G. In *Dissociative Recombination: Theory, Experiment, and Application*; Rowe, B. R.; Brian, J.; Mitchell, A.; Canosa, A., Eds.; World Scientific: Singapore, 1993, p 99.
[46] Foley, B. L.; Adams, N. G.; Lee, H. S. *J. Phys. Chem.* **1993**, *97*, 5218.
[47] Adams, N. G. *Int. J. Mass Spectrom. Ion Processes* **1994**, *132*, 1.
[48] Adams, N. G.; Babcock, L. M. *J. Phys. Chem.* **1994**, *98*, 4564.
[49] Adams, N. G.; Babcock, L. M. *Astrophys. J.* **1994**, *434*, 184.
[50] Butler, J. M.; Babcock, L. M.; Adams, N. G. *Mol. Phys.* **1997**, *91*, 81.
[51] Smith, D.; Adams, N. G.; Church, M. J. *J. Phys. B* **1978**, *11*, 4041.
[52] Gougousi, T.; Johnsen, R.; Golde, M. F. *J. Chem. Phys.* **1997**, *107*, 2430.
[53] Skrzypkowski, M. P.; Gougousi, T.; Johnsen, R.; Golde, M. F. *J. Chem. Phys.* **1998**, *108*, 8400.
[54] Tsuji, M.; Obase, H.; Matsuo, M.; Endoh, M.; Nishimura, Y. *Chem. Phys.* **1980**, *50*, 195.
[55] Tsuji, M.; Obase, H.; Nishimura, Y. *J. Chem. Phys.* **1980**, *73*, 2575.
[56] Obase, H.; Tsuji, M.; Nishimura, Y. *Chem. Phys. Lett.* **1981**, *81*, 119.
[57] Tsuji, M.; Yamagiwa, C.; Endoh, M.; Nishimura, Y. *Chem. Phys. Lett.* **1980**, *73*, 407.
[58] Murakami, I.; Tsuji, M.; Nishimura, Y. *Chem. Phys. Lett.* **1982**, *92*, 131.
[59] Tsuji, M.; Murakami, I.; Nishimura, Y. *Chem. Phys. Lett.* **1980**, *75*, 536.
[60] Tsuji, M.; Murakami, I.; Nishimura, Y. *J. Chem. Phys.* **1981**, *75*, 5373.
[61] Tsuji, M.; Mizuguchi, T.; Shinohara, K.; Nishimura, Y. *Can. J. Phys.* **1983**, *61*, 838.
[62] Tsuji, M.; Kuhn, R.; Maier, J. P.; Nishitani, S.; Shinohara, K.; Obase, H.; Nishimura, Y. *Chem. Phys. Lett.* **1985**, *119*, 473.
[63] Nishimura, Y.; Mizuguchi, T.; Tsuji, M.; Obara, S.; Morokuma, K. *J. Chem. Phys.* **1983**, *78*, 7260.
[64] Tsuji, M.; Shinohara, K.; Mizuguchi, T.; Nishimura, Y. *Can. J. Phys.* **1983**, *61*, 251.
[65] Tsuji, M.; Mizuguchi, T.; Nishimura, Y. *Can. J. Phys.* **1981**, *59*, 985.
[66] Tsuji, M.; Shinohara, K.; Nishitani, S.; Mizuguchi, T.; Nishimura, Y. *Can. J. Phys.* **1984**, *62*, 353.
[67] Tsuji, M.; Nishimura, Y.; Mizuguchi, T. *Chem. Phys. Lett.* **1981**, *83*, 483.
[68] Tsuji, M.; Mizuguchi, T.; Nishimura, Y. *Chem. Phys. Lett.* **1981**, *84*, 318.
[69] Yamaguchi, S.; Tsuji, M.; Nishimura, Y. *Chem. Phys. Lett.* **1987**, *138*, 29.
[70] Tsuji, M.; Shimada, S.; Nishimura, Y. *Chem. Phys. Lett.* **1982**, *89*, 75.
[71] Tsuji, M.; Yamaguchi, S.; Shimada, S.; Nishimura, Y. *J. Mol. Spectrosc.* **1984**, *103*, 498.
[72] Yamaguchi, S.; Obase, H.; Tsuji, M.; Nishimura, Y. *Chem. Phys. Lett.* **1985**, *119*, 477.
[73] Yamaguchi, S.; Obase, H.; Tsuji, M.; Nishimura, Y. *Can. J. Phys.* **1986**, *64*, 700.
[74] Yamaguchi, S.; Tsuji, M.; Nishimura, Y. *Can. J. Phys.* **1986**, *64*, 1374.
[75] Tsuji, M.; Tanaka, M.; Nishimura, Y. *Chem. Phys. Lett.* **1997**, *266*, 246.
[76] Tsuji, M.; Tanaka, M.; Nishimura, Y. *Chem. Phys. Lett.* **1996**, *256*, 623.
[77] Tsuji, M.; Tanaka, M.; Nishimura, Y. *Chem. Phys. Lett.* **1996**, *262*, 349.
[78] Tsuji, M.; Tanaka, M.; Nakamura, M.; Nishimura, Y. *Chem. Lett.* **1995**, 895.
[79] Tsuji, M.; Tanaka, M.; Oda, E.; Ishimi, H.; Nishimura, Y. *Chem. Phys.* **1998**, *236*, 319.
[80] Tsuji, M.; Tanaka, M.; Nishimura, Y. *J. Chem. Phys.* **1997**, *107*, 4852.
[81] Tsuji, M.; Tanaka, M.; Nakamura, M.; Nishimura, Y. *Chem. Lett.* **1996**, 367.
[82] Tsuji, M.; Kaneko, N.; Furusawa, M.; Muraoka, T.; Nishimura, Y. *J. Chem. Phys.* **1993**, *98*, 8565.
[83] Tsuji, M.; Kaneko, N.; Nishimura, Y. *J. Chem. Phys.* **1993**, *99*, 4539.
[84] Tsuji, M.; Kaneko, N.; Ishimi, H.; Nishimura, Y. *J. Chem. Phys.* **1994**, *101*, 8726.
[85] Tsuji, M.; Kaneko, N.; Nishimura, Y. *J. Chem. Phys.* **1994**, *101*, 7451.
[86] Endoh, M.; Tsuji, M.; Nishimura, Y. *J. Chem. Phys.* **1983**, 79, 5368.

[87] Endoh, Ph.D. Thesis, Grad. Sch. Eng. Sci., Kyushu Univ. 1984.
[88] Tsuji, M.; Ishimi, H.; Hisano, M.; Oota, H.; Nishimura, Y. *J. Chem. Phys.* **1999**, *110*, 9064.
[89] Tsuji, M.; Obase, H.; Nishimura, Y. *Rep. Inst. Adv. Mat. Study, Kyushu Univ.* **1988**, *2*, 101.
[90] Tsuji, M.; Maier, J. P.; Obase, H.; Nishimura, Y. *Chem. Phys.* **1986**, *110*, 17.
[91] Obase, H.; Tsuji, M.; Nishimura, Y. *Chem. Phys.* **1981**, *57*, 89.
[92] Obase, H.; Tsuji, M.; Nishimura, Y. *Chem. Phys.* **1983**, *74*, 89.
[93] Endoh, M.; Tsuji, M.; Nishimura, Y. *J. Chem. Phys.* **1982**, *77*, 4027.
[94] Tsuji, M.; Matsuo, M.; Nishimura, Y. *Int. J. Mass Spectrom. Ion Phys.* **1980**, *34*, 273.
[95] Tsuji, M.; Endoh, M.; Obase, H.; Nishimura, Y. *Rep. Inst. Ind. Sci. Kyushu Univ.* **1983**, *74*, 109.
[96] Sekiya, H.; Tsuji, M.; Nishimura, Y. *Chem. Phys. Lett.* **1983**, *100*, 494.
[97] Tsuji, M.; Maier, J. P.; Obase, H.; Sekiya, H.; Nishimura, Y. *Chem. Phys. Lett.* **1987**, *137*, 421.
[98] Tsuji, M.; Maier, J. P. *Chem. Phys.* **1988**, *126*, 435.
[99] Tsuji, M.; Tsuji, K.; Nishimura, Y. *Int. J. Mass Spectrom. Ion Phys.* **1979**, *30*, 175.
[100] Sekiya, H.; Endoh, M.; Tsuji, M.; Nishimura, Y. *Nippon Kagaku Kaishi* **1984**, 1498.
[101] Tsuji, M.; Maier, J. P.; Obase, H.; Nishimura, Y. *Chem. Phys. Lett.* **1988**, *147*, 619.
[102] Tsuji, M.; Nakamura, M.; Mizuguchi, T.; Kaneko, N.; Nishimura, Y. *Rep. Inst. Adv. Mat. Study, Kyushu Univ.* **1995**, *9*, 7.
[103] Tsuji, M.; Maier, J. P. *Chem. Phys.* **1985**, *97*, 397.
[104] Tsuji, M.; Furusawa, M.; Nishimura, Y. *Chem. Phys. Lett.* **1990**, *169*, 49.
[105] Sekiya, H.; Tsuji, M.; Nishimura, Y. *Rep. Inst. Adv. Mat. Study, Kyushu Univ.* **1987**, *1*, 51.
[106] Sekiya, H.; Tsuji, M.; Nishimura, Y. *Rep. Inst. Ind. Scc. Kyushu Univ.* **1987**, *82*, 125.
[107] Sekiya, H.; Tsuji, M.; Nishimura, Y. *J. Chem. Phys.* **1987**, *87*, 325.
[108] Tsuji, M.; Hamagami, T.; Tanoue, T.; Nishimura, Y. *Rep. Inst. Adv. Mat. Study, Kyushu Univ.* **1999**, *13*, 37.
[109] Obase, H.; Tsuji, M.; Nishimura, Y. *Chem. Phys. Lett.* **1984**, *105*, 214.
[110] Obase, H.; Tsuji, M.; Nishimura, Y. *Chem. Phys.* **1984**, *87*, 93.
[111] Obase, H.; Tsuji, M.; Nishimura, Y. *Chem. Phys.* **1985**, *99*, 111.
[112] Tsuji, M.; Endoh, M.; Susuki, T.; Mizukami, K.; Nishimura, Y. *J. Chem. Phys.* **1984**, *81*, 3559.
[113] Endoh, M.; Tsuji, M.; Nishimura, Y. *Chem. Phys.* **1983**, *82*, 67.
[114] Tsuji, M.; Shimada, S.; Mizukami, K.; Obase, H.; Nishimura, Y. *Rep. Inst. Adv. Mat. Study, Kyushu Univ.* **1996**, *10*, 13.
[115] Endoh, M.; Tsuji, M.; Nishimura, Y. *Chem. Phys. Lett.* **1984**, *109*, 35.
[116] Tsuji, M.; Mizukami, K.; Sekiya, H.; Obase, H.; Shimada, S.; Nishimura, Y. *Chem. Phys. Lett.* **1984**, *107*, 389.
[117] Sekiya, H.; Tsuji, M.; Nishimura, Y. *Bull. Chem. Soc. Jpn.* **1984**, *57*, 3329.
[118] Obase, H.; Tsuji, M.; Nishimura, Y. *Chem. Phys. Lett.* **1987**, *141*, 133.
[119] Yamaguchi, S.; Tsuji, M.; Nishimura, Y. *J. Chem. Phys.* **1988**, *88*, 3111.
[120] Tsuji, M.; Murakami, I.; Okano, S.; Nishimura, Y. *Rep. Grad. School Eng. Sci. Kyushu Univ.* **1999**, *21*, 1.
[121] Tsuji, M. **2001**, unpublished results.
[122] Yamaguchi, S.; Tsuji, M.; Nishimura, Y. *J. Chem. Phys.* **1987**, *87*, 1637.
[123] Tsuji, M.; Furusawa, M.; Mizuguchi, T.; Muraoka, T.; Nishimura, Y. *J. Chem. Phys.* **1992**, *97*, 245.
[124] Tsuji, M.; Kobarai, K.; Yamaguchi, S.; Nishimura, Y. *Chem. Phys. Lett.* **1989**, *158*, 470.
[125] Yamaguchi, S.; Tsuji, M.; Obase, H.; Sekiya, H.; Nishimura, Y. *J. Chem. Phys.* **1987**, *86*, 4952.
[126] Tsuji, M.; Kobarai, K.; Nishimura, Y. *J. Chem. Phys.* **1990**, *93*, 3133.

[127] Tsuji, M.; Susuki, T.; Endoh, M.; Nishimura, Y. *Chem. Phys. Lett.* **1982**, *86*, 411.
[128] Tsuji, M.; Susuki, T.; Mizukami, K.; Nishimura, Y. *J. Chem. Phys.* **1985**, *83*, 1677.
[129] Tsuji, M.; Nagano, I.; Susuki, T.; Mizukami, K.; Obase, H.; Nishimura, Y. *J. Phys. Chem.* **1986**, *90*, 3998.
[130] Mizukami, K.; Obase, H.; Tsuji, M.; Nishimura, Y. *Chem. Phys. Lett.* **1985**, *116*, 510.
[131] Tsuji, M.; Mizukami, K.; Obase, H.; Nishimura, Y. *J. Phys. Chem.* **1988**, *92*, 1163.
[132] Tsuji, M.; Nakamura, M.; Oda, E.; Hisano, M.; Nishimura, Y. *Jpn. J. Appl. Phys.* **1998**, *37*, 5775.
[133] Tsuji, M.; Nakamura, M.; Nishimura, Y.; Obase, H. *J. Chem. Phys.* **1995**, *103*, 1413.
[134] Tsuji, M.; Nakamura, M.; Nishimura, Y.; Obase, H. *J. Chem. Phys.* **1998**, *108*, 8031.
[135] Tsuji, M.; Kobarai, K.; Kouno, H.; Obase, H.; Nishimura, Y. *J. Chem. Phys.* **1991**, *94*, 1127.
[136] Tsuji, M.; Nakamura, M.; Nishimura, Y.; Oda, E.; Oota, H.; Hisano, M. *J. Chem. Phys.* **1999**, *110*, 2903.
[137] Tsuji, M.; Oda, E.; Tanaka, M.; Nakamura, M.; Nishimura, Y. *Chem. Lett.* **1997**, 465.
[138] Tsuji, M.; Oda, E.; Nakamura, M.; Nishimura, Y. *Chem. Lett.* **1998**, 413.
[139] Tsuji, M.; Furusawa, M.; Nishimura, Y. *Chem. Phys. Lett.* **1990**, *166*, 363.
[140] Tsuji, M.; Furusawa, M.; Nishimura, Y. *J. Chem. Phys.* **1990**, *92*, 6502.
[141] Tsuji, M.; Muraoka, T.; Ide, M.; Ujita, H.; Nishimura, Y. *J. Chem. Phys.* **1994**, *101*, 2880.
[142] Tsuji, M.; Ishimi, H.; Nishimura, Y. *Chem. Lett.* **1996**, 515.
[143] Tsuji, M. *Houshasen Kagaku* **1996**, *62*, 16.
[144] Tsuji, M.; Furusawa, M.; Kouno, H.; Nishimura, Y. *J. Chem. Phys.* **1991**, *94*, 4291.
[145] Tsuji, M.; Muraoka, T.; Kouno, H.; Nishimura, Y. *J. Chem. Phys.* **1992**, *97*, 1079.
[146] Tsuji, M.; Ide, M.; Muraoka, T.; Nishimura, Y. *J. Chem. Phys.* **1993**, *99*, 1710.
[147] Tsuji, M.; Ide, M.; Muraoka, T.; Nishimura, Y. *J. Chem. Phys.* **1994**, *101*, 328.
[148] Tsuji, M.; Ide, M.; Oda, E.; Nishimura, Y. *J. Chem. Phys.* **1998**, *109*, 3374.
[149] Tsuji, M.; Ishimi, H.; Nakamura, M.; Nishimura, Y.; Obase, H. *J. Chem. Phys.* **1995**, *102*, 2479.
[150] Tsuji, M.; Ishimi, H.; Nishimura, Y. *Chem. Lett.* **1995**, 873.
[151] Tsuji, M.; Ishimi, H.; Nishimura, Y.; Obase, H. *J. Chem. Phys.* **1995**, *102*, 6013.
[152] Tsuji, M.; Ishimi, H.; Nishimura, Y.; Obase, H. *Int. J. Mass Spectrom. Ion Processes* **1995**, *145*, 165.
[153] Tsuji, M.; Ishimi, H.; Nishimura, Y.; Obase, H. *J. Chem. Phys.* **1996**, *105*, 2701.
[154] Tsuji, M.; Ishimi, H.; Oota, H.; Hisano, M.; Nishimura, Y. *Bull. Chem. Soc. Jpn.* **1999**, *72*, 227.
[155] Sekiya, H.; Hirayama, T.; Endoh, M.; Tsuji, M.; Nishimura, Y. *Chem. Phys.* **1986**, *101*, 291.
[156] Tsuji, M.; Ochsner, M.; Maier, J. P.; Obase, H.; Sekiya, H.; Nishimura, Y. *Chem. Phys. Lett.* **1985**, *120*, 17.
[157] Ochsner, M.; Tsuji, M.; and Maier, J. P. *Chem. Phys. Lett.* **1985**, *115*, 373.
[158] Tsuji, M.; Nagano, I.; Nishiyama, N.; Obase, H.; Sekiya, H.; Nishimura, Y. *J. Phys. Chem.* **1986**, *90*, 3106.
[159] Tsuji, M.; Nagano, I.; Nishiyama, N.; Sekiya, H.; Nishimura, Y. *Laser Chem.* **1987**, *7*, 333.
[160] Tsuji, M.; Obase, H.; Endoh, M.; Yamaguchi, S.; Yamaguchi, K.; Kobarai, K.; Nishimura, Y. *J. Chem. Phys.* **1988**, *89*, 6753.
[161] Sekiya, H.; Tsuji, M.; Nishimura, Y. *Chem. Lett.* **1986**, 1997.
[162] Obase, H.; Tsuji, M.; Nishimura, Y. *J. Chem. Phys.* **1987**, *87*, 2695.
[163] Tsuji, M. *Trends in Physical Chemistry*, Science Information: Trivandrum, Vol. 5, 25, 1995.
[164] Johnsen, R. *Phys. Rev. A* **1983**, *28*, 1460.
[165] Tanaka, Y.; Yoshino, K.; Freeman, E. *J. Chem. Phys.* **1975**, *62*, 4484.

[166] Bender, C. F.; Winter, N. W. *Appl. Phys. Letters* **1978**, *33*, 29.
[167] Balasubramnian, K.; Liao, M. Z.; Lin, S. H. *Chem. Phys. Lett.* **1987**, *138*, 49.
[168] Lopéz, G. E. *J. Comput. Chem.* **1995**, *6*, 758.
[169] Ma, N. L.; Li, W-K.; Ng, C. Y. *J. Chem. Phys.* **1993**, *99*, 3617.
[170] Bieske, E. J.; Maier, J. P. *Chem. Rev.* **1993**, *93*, 2603.
[171] Hiraoka, K.; Shimizu, A.; Minamitsu, A.; Nasu, M.; Wasada, H.; Yamabe, S. *J. Chem. Phys.* **1998**, *108*, 6689.
[172] Tsuji, M. In *Techniques of Chemistry: Techniques for the Study of Ion Molecule Reactions; Chapter IX. Spectroscopic Probes*; Farrar, J. M.; Saunders, Jr.; W. Eds.; John Wiley & Sons, 1988, p 489.
[173] Shimamori, H.; Tatsumi, Y.; Sunagawa, T. *J. Chem. Phys.* **1993**, *99*, 7787.
[174] Druyvestern, M. J. *Nature* **1931**, *128*, 1076.
[175] Jongerius, H .M.; van Koeveringe, J. J. L.; Oskam, H. J. *Physica (Utr.)* **1959**, *25*, 406.
[176] Friedl, L. W. *Z. Naturforsch*, **1959**, *14a*, 848.
[177] Kugler, E. *Ann. Phys. (Leipz.)* **1964**, *14*, 137.
[178] Dabrowski, I.; Herzberg, G. *J. Mol. Spectrosc.* **1978**, *73*, 183.
[179] Dabrowski, I.; Herzberg, G.; Yoshino, K. *J. Mol. Spectrosc.* **1981**, *89*, 491.
[180] Holland, F.; Huber, K. P.; Hoy, A. R.; Lipson, R. H. *J. Mol. Spectrosc.* **1991**, *145*, 164.
[181] Huber, K. P.; Lipson, R. H. *J. Mol. Spectrosc.* **1986**, *119*, 433.
[182] Brodmann, R.; Zimmerer, G. *J. Phys. B* **1977**, *10*, 3395.
[183] Mangano, J. A.; Jacob, J. H.; Rokni, M.; Hawryluk, A. *Appl. Phys. Lett.* **1977**, *31*, 26.
[184] Huestis, D. L.; Schotter, N. E. *J. Chem. Phys.* **1978**, *69*, 3100.
[185] Lorents, D. C.; Huestis, D. L.; McCusker, M. V.; Nakano, H. H.; Hill, R. W. *J. Chem. Phys.* **1978**, *68*, 4657.
[186] Wadt, W. R.; Hay, P. J.; *J. Chem. Phys.* **1978**, *68*, 3850.
[187] Bonified, T.D.; Rambow, F. H. K.; Walters, G. K.; McCusker, M. V.; Lorents, D. C.; Gutcheck, R. A. *Chem. Phys. Lett.* **1980**, *69*, 290.
[188] Verkhovtseva, E. T.; Bondarenko, E. A.; Doronin, Yu. S. *Chem. Phys. Lett.* **1987**, *140*, 181.
[189] Leichner, P. K.; Ericson, R. *Phys. Rev. A* **1974**, *9*, 251.
[190] Rakshit, A. B.; Warneck, P. *J. Chem. Phys.* **1980**, *73*, 2673.
[191] Giles, K.; Adams, N. G.; Smith, D. *J. Phys. B* **1989**, *22*, 873.
[192] Huber, K. P.; Herzberg, G. *Constants of Diatomic Molecules*, Van Nostrand Reinhold: New York, 1979.
[193] Leventhal, J. J. In *Gas Phase Ion Chemistry*; Bowers, M. T., Ed.; Academic Press: Florida,1979; Vol. 3, Chapter 24, p 309.
[194] Bearman, G. H.; Earl, J. D.; Pieper, R. J.; Harris, H. H.; Leventhal, J. J. *Phys. Rev. A* **1976**, *13*, 1734.
[195] Maier II, W. B.; Holland, R. F. *J. Chem. Phys.* **1971**, *54*, 2693.
[196] Edqvist, O.; Lindholm, E.; Selin, L. E.; Sjogren, H.; Asbrink, L. *Ark. Fysik.* **1970**, *40*, 439.
[197] Anderson, S. L.; Turner, T.; Mahan, B. H.; Lee, Y. T. *J. Chem. Phys.* **1982**, *77*, 1842.
[198] Kato, T. *J. Chem. Phys.* **1984**, *80*, 6105.
[199] Levine, R. D.; Kinsey, J. L. In *Atom Molecule Collision Theory*; Bernstein, R. B., Ed.; Plenum Press: New York, 1979, p 693.
[200] Tokue, I.; Wang, J.; Ito, Y. *Bull. Chem. Soc. Jpn.* **1993**, *66*, 69.

MERGED-BEAMS STUDIES OF ELECTRON–MOLECULAR ION INTERACTIONS IN ION STORAGE RINGS

Mats Larsson

OUTLINE

Advances in Gas-Phase Ion Chemistry
Volume 4, pages 179–211.

ABSTRACT

The development of ion storage rings for atomic and molecular physics has proved to be very fruitful for the study of electron–molecular ion interactions. The long storage times accessible in the rings make it possible to prepare vibrationally cold molecular ions for further use in merged electron–ion beams experiments. In recent years, it has been possible to obtain very cold electron beams in storage rings, something which has increased the resolution in the collision experiments. In the present work, we review the most recent developments and use HD^+, H_3^+, O_2^+, and H_3O^+ to exemplify the potential of storage rings experiments.

I. INTRODUCTION

The interaction of a free electron with a molecular ion is likely to lead to dissociation of the molecular ion, but the time scale of the dissociation process can, in principle, range from tens of femtoseconds to billions of years or even longer depending on the energy of the electron. The interaction of a low-energy electron (kinetic energy, ε, $\ll 1$ eV) with a molecular ion results in dissociative recombination (DR), a process which has a very large cross section and can be described as

$$AB^+ + e^-(\varepsilon) \rightarrow AB^{**} \rightarrow A + B + KER \tag{1}$$

where A and B are atoms or molecules, AB^{**} is the neutral resonant state into which the electron is captured, and KER is the kinetic release that goes into the reaction products. The process is very effective because the molecular ion can stabilize the electron capture by dissociating into neutral products on a time scale sufficiently short to compete favorably with autoionization. It is the dominant neutralizing process in plasmas where the gas temperature is a few thousand degrees or less, and the electron temperature is lower than few eV (i.e., $T_e < 20{,}000-30{,}000$ K).

When the electron energy is increased, the molecular ion can also be dissociated directly into neutral and charged products:

$$AB^+ + e^-(\varepsilon) \rightarrow A^+ + B + e^- + KER \tag{2}$$

$$AB^+ + e^-(\varepsilon) \rightarrow A^+ + B^- + KER \tag{3}$$

Process 2 is called dissociative excitation and process 3 resonant ion-pair formation. If the electron energy is larger than the ionization energy of the molecular ion, the following processes can occur

$$AB^+ + e^-(\varepsilon) \rightarrow A^+ + B^+ + 2e^- + KER \tag{4}$$

$$AB^+ + e^-(\varepsilon) \rightarrow AB^{2+} + 2e^- \tag{5}$$

where process 4 is called dissociative ionization. The doubly charged molecular ion formed in process 5 can be extremely long-lived, but owing to the fact that its vibrational levels are located above the dissociation limit, it must eventually decay by dissociation. In principle, this can take a very long time, maybe billions of years; in practice, a doubly charged molecular ion will have a short lifetime because of its high reactivity.

Finally, if the molecular ion is negatively charged, electron impact detachment (without or with dissociation) occurs if the electron energy is larger than the electron affinity:

$$AB^- + e^-(\varepsilon) \rightarrow AB + 2e^- \tag{6}$$

$$AB^- + e^-(\varepsilon) \rightarrow A + B + 2e^- + KER \tag{7}$$

Another possible process is dissociative excitation (cf. reaction 2):

$$AB^- + e^-(\varepsilon) \rightarrow A^- + B + e^- + KER \tag{8}$$

Of the processes 1–8 listed above, dissociative recombination has attracted by far the most attention. The development of ion storage rings for atomic and molecular physics during the 1990s has greatly contributed to the development of studies of dissociative recombination,[1,2] but the process has been intensely studied also during the decades following its original discovery.[3,4] The various experimental and theoretical methods developed and applied to the study of dissociative recombination are described in several review articles[5–7] and in proceedings from conferences dedicated to dissociative recombination.[8–11] A comprehensive review of dissociative electron–molecular ion recombination studies in storage rings have recently been published;[12] however, the production of the book in which the review appears was delayed, and the review includes only results available up to August 1998. Recent results are also discussed in Ref. 13.

The present article will be restricted to discussions of dissociative recombination and resonant ion-pair formation (process 3). Dissociative excitation, dissociative ionization, and electron impact detachment of negative ions have been reviewed recently.[14,15] The field of merged-beams experiments in atomic and molecular physics has recently been reviewed,[16] and the reader is referred to this excellent article for a detailed discussion of the merged-beams technique.

Because of the recent review of electron–ion recombination,[12] the present article will focus on some important selected examples rather than attempting to be comprehensive.

II. ION STORAGE RINGS

Storage rings for ions and electrons were first developed for experiments in particle physics. However, during the late 1980s and 1990s, several heavy-ion storage rings became available for atomic and molecular physics.[1,17] Figure 1 shows the storage ring CRYRING at the Manne Siegbahn Laboratory, Stockholm University, as an example of all storage ring facilities used for atomic and molecular physics. It consists of ion sources, acceleration devices, and a ring system, 52 m in circumference, in which a beam of charged particles is confined by magnetic fields. CRYRING is also a synchrotron; thus the ions can be further accelerated in the ring while the magnets are ramped synchronously to keep the beam orbit constant. The pressure in the entire ring is kept below 10^{-11} torr and this is one of the keys to experiments with molecular ions. Even at a respectable vacuum of 10^{-9} torr, the beam would disappear in a fraction of a second by collisions with rest gas molecules. Even at the ultra-high vacuum in CRYRING, the beam lifetime is determined primarily by rest gas collisions and can range from 10 seconds to hours (for fully stripped atomic ions). The ions make up to one million revolutions

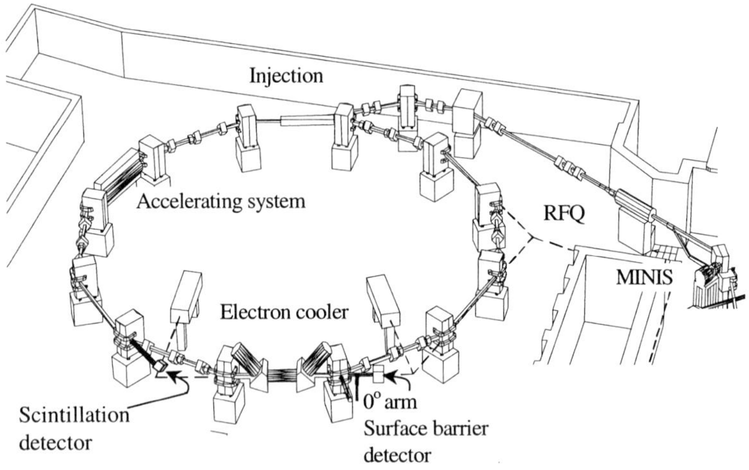

Figure 1. Schematic view of CRYRING. Molecular ions are created in the ion source MINIS, accelerated and mass selected. In some cases they are further accelerated by the Radio Frequency Quadrupole (RFQ), and injected into the ring. The accelerating system is used to further increase the ion energy. Reaction products from the electron cooler section exit the ring and hit detectors located on the 0° arm. The scintillation detector, which detects neutral particles arising from collisions of the stored beam with rest gas molecules, is used as a beam monitor.

per second, which means that they travel a very long distance during an injection cycle (on the order of the circumference of the Earth or more). Thus, focusing elements play an essential role in keeping the particles in stable orbits. Nevertheless, six out of the 12 straight sections in CRYRING are left without focusing or correction magnets. Some of these sections are used for experiments. The other six straight sections contain quadrupole magnets (four poles) as focusing elements and sextupole magnets (six poles) for second-order corrections to the beam orbit. Storage rings with similar characteristics as CRYRING are available at the University of Aarhus, Denmark (ASTRID), the Max-Planck-Institute for Nuclear Physics, Heidelberg, Germany (TSR), and in Tokyo (TARN II).

The electron cooler, which is positioned in a dispersive-free straight section, serves two different purposes. First, it reduces the phase space volume occupied by the ion beam by beam cooling. Coulomb collisions between electrons and ions give rise to a friction force which slows the velocity of those ions travelling with non-zero velocity in the reference frame of the electron beam. Optimal cooling is obtained when the ion and electron beams have the same velocity. Second, the electron beam acts as a target in the studies of collisions between ions and electrons.

The principal advantages of using an ion storage ring for the study of electron–molecular ion interaction can be summarized in the following points.

- Vibrationally hot molecular ions that are injected into the ring can cool internally after a few seconds of storage owing to emission of infrared radiation. This is the case for heteronuclear diatomic molecules and infrared active modes in polyatomic molecules.
- Lighter molecular ions (H_2^+, HD^+, H_3^+, HeH^+, etc.) can be effectively cooled translationally by means of the electron cooler. During each passage through the electron cooler, heat is transferred from the ion beam to the continuously renewed electron beam, which reduces the momentum spread in and the diameter of the stored ion beam. A phase-spaced cooled ion beam is a signature of good alignment of the electron beam with respect to the ion beam, and it has advantages when particle imaging of the recombination reaction products is performed.
- The electron beam is much larger ($\varnothing = 40$ mm in CRYRING) than the ion beam and to a good approximation homogeneous, which means that absolute cross sections can be obtained without measurement of the beam overlap integral. This is usually one of the major challenges in merged-beams experiments.
- The confinement by means of dipole (bending) magnets allows high beam energy, from a few to tens of MeV, which leads to low cross sections for electron capture from rest gas molecules (Rg). Thus, the

process $AB^+ + Rg \rightarrow A + B$ is negligible for light ions and manageable for heavier systems ($m_{AB} > 10$ amu).

- Energy resolving surface barrier detectors can be used, and because of the high ion beam energy, those rest gas collisions that have high cross section at high beam energy, such as charge stripping, lead to reaction products that are easily separated from those of dissociative recombination. The detection efficiency of surface barrier detectors is 100% at the beam energies used in storage rings.
- The merged-beams geometry in the electron cooler section gives the possibility to study electron–molecular ion interactions at meV collision energies. The spread of relative energies of electrons and ions can be made very small and depends essentially only on the quality of the electron beam.

III. HYDROGEN IONS

Dissociative recombination of the simplest diatomic and triatomic molecules, H_2^+ and H_3^+, has received considerable attention. Because of the rapid conversion of H_2^+ to H_3^+ in a hydrogen plasma, it was not until the development of a merged electron–ion beams technique that the study of dissociative recombination of H_2^+ at low collision energies became possible.[18] Its simple electronic structure made it suitable for theoretical studies, and in the early 1990s, accurate calculations by means of the multichannel quantum-defect theory (MQDT) were available.[19] The results were in good agreement with single-pass merged-beams results,[20–22] and showed that the $(2p\sigma_u)^2\ ^1\Sigma_g^+$ resonant state of H_2 totally dominates recombination of H_2^+ ($v=0$) at low electron energies. The rotational motion of the H_2^+ ion was not considered in the MQDT calculations, but was incorporated in a later study.[23] This clearly indicated that rotation had a larger effect on the cross section than previously anticipated. Such was the situation for dissociative recombination of H_2^+ when the ion storage ring experiments started.

Dissociative recombination of H_3^+ has been studied for many years by means of both afterglow and beam techniques, and the DR rate coefficient for H_3^+ has generated considerable controversy. The history of H_3^+ recombination has been reviewed many times,[24–27] most recently[28] in connection with the Royal Society meeting in London on H_3^+, and will therefore not be repeated here. Because of the recent discoveries of H_3^+ in dark[29] and diffuse[30,31] interstellar molecular clouds, and because of the indications that chemical models of molecular clouds could have bistable solutions,[32] the DR rate coefficient for H_3^+ in its zeroth vibrational level has been brought into focus again. Table 1 shows the more recent

Table 1. Recent Measured H_3^+ Dissociative Recombination Rate Coefficients at $T_e = 300$ K

$\alpha(H_3^+, \nu=0)$ (10^{-7} cm^{3}s^{-1})	*Method*	*Reference*
1.6	Infrared absorption spectroscopy	33
1.15	Ion storage ring; CRYRING	34
0.78	FALP	35
0.1–0.2	FALP	36
~0.1	FALP	27
1.2	Single-pass merged beams	37

measured rate coefficients for recombination of H_3^+ ($\nu=0$). As follows from Table 1, there is not yet consensus regarding the rate coefficient, although the spread is not larger than a factor of ten. A detailed discussion of the results can be found in Ref. 28.

A. HD^+ and H_2^+

Most of the work at ion storage rings has been performed with HD^+ rather than its isotopomer H_2^+ for the simple reason that HD^+ is easily cooled vibrationally by means of emissions of infrared radiation, whereas H_2^+ in principle should not cool at all on the time scales relevant in storage ring experiments (i.e., on time scales of one minute or less; the term "in principle" is used because in practice the interaction of H_2^+ with cold electrons leads to a reduction of the population of higher vibrational levels). A detailed account of experiments on HD^+ in storage rings are given in Ref. 12, so only a brief summary will be given here before we describe some of the most recent work on HD^+.

In the first ion storage ring experiment,[38] it was shown that recombination of HD^+ has peaks in the cross section at "high" electron energies (~9 eV), which are due to Rydberg states converging to HD^+ ($2p\sigma_u$). The mechanism is now well understood and has been treated theoretically by means of MQDT.[39,40] Rotational coupling has also been included in MQDT and a detailed comparison with theory has been made.[40,41] Resonances at low electron energies,[39–41] arising from the interference between direct electron capture to the $(2p\sigma_u)^2\ {}^1\Sigma_g^+$ resonant state (the direct mechanism) and electron capture to Rydberg states predissociated by the resonant state (the indirect mechanism), were observed at both CRYRING and TARN II. An essential feature, making it possible to observe these resonances, was that both rings were equipped with electron coolers utilizing the adiabatic expansion technique,[42] which could produce electron beams with a transverse energy spread of 10 meV. Whereas the agreement between the experimental results is very good (in the initial TARN II experiment only relative cross sections were presented; however, later absolute cross sections were measured also[43]), the theoretical results are larger by a factor of 50%[41] or more.[40] More recent theoretical attempts have not improved the agreement.[43] A paper giving a comparison of the experimental cross sections for recombination of HD^+ ($\nu=0$), measured at ASTRID, CRYRING, and TSR, is in preparation;[44] the preliminary results show a very good agreement.

Until recently, no experiment had been performed in which the recombination cross section of excited vibrational levels in HD^+ or H_2^+ were measured. This is naturally a considerable challenge because the vibrational population of excited levels must be controlled or monitored. When the TSR was modified to extract part of the stored beam, it became possible to use the Coulomb explosion imaging technique[45] to measure the nuclear arrangement of the HD^+ as a function of time after injection. The Coulomb explosion image gave direct information about the vibrational population of the HD^+ ions stored in the TSR and, by simultaneously measuring also the recombination events in the electron cooler, it was possible to obtain information about the rate of recombination of vibrationally excited levels with respect to that of $\nu=0$.[46,47] A comparison with theory reveals good agreement for all levels except $\nu=5$, for which theory predicts a value a factor of 250 larger than experiment.[46,47] The reason for this discrepancy is not well understood.

The kinetic energy going into the particles A and B in equation 1 is dependent on the state of excitation of A and B. The higher their state of excitation, the less energy is available as kinetic energy. The argument suggests that the state of excitation can be measured by measuring how much kinetic energy is shared between A and B. In practice, this can be done by recording the image of the neutral dissociation products at some distance from the electron cooler. At ion storage rings, this is done by an imaging device consisting of several multichannel plates (MCPs) linked to a phosphor screen and a charged-coupled-device (CCD) camera.[48,49] The transverse separation, D (i.e., the projected distance on the surface of the imaging detector), between the dissociation products can be expressed in terms of the kinetic energy release KER; the ion beam energy, E_b; the masses of the neutral fragments, m_A and m_B, respectively; the distance between the location of the recombination event and the imaging detector, L; and the angle θ between the molecular axis and the electron beam during dissociation:

$$D = \sqrt{\mathrm{KER}/E_b}\,\frac{m_A + m_B}{\sqrt{m_A m_B}}\,L\sin\theta \tag{9}$$

This technique makes it possible to address the question of the state of excitation of hydrogen and deuterium following dissociative recombination, i.e., to determine n and n' in the reaction

$$HD^+ + e^- \rightarrow H(n) + D(n') \tag{10}$$

Figure 2 shows the energetics of HD^+ $X^2\Sigma_g^+$ and the relevant dissociation limits of HD. When the electron energy is nominally 0 eV, only the H(1s) + D(1s) and H(1s) + D(n = 2) (or vice versa) limits are energetically available. However, DR at low electron energy proceeds through the $(2p\sigma_u)^2\ {}^1\Sigma_g^+$ state, and there is no way dissociation of this state can produce ground state products. Thus, as long as the electron energy is smaller than 1 eV, H(1s) + D(n = 2) is the only decay channel. This has been verified experimentally by means of imaging experiments.[50–52] When the electron energy is increased so that H(1s) + D($n \geq 3$) channels become available, it has been shown experimentally that part of the dissociation flux is redirected to the new channels.[53] This occurs in a "democratic" way so that no single channel becomes dominant.

An interesting question is whether part of the flux is redirected to the ion-pair limit $H^+ + D^-$ (or $H^- + D^+$), which is located just above H(1s) + D(n = 4) (see Figure 2). The imaging technique cannot reveal any

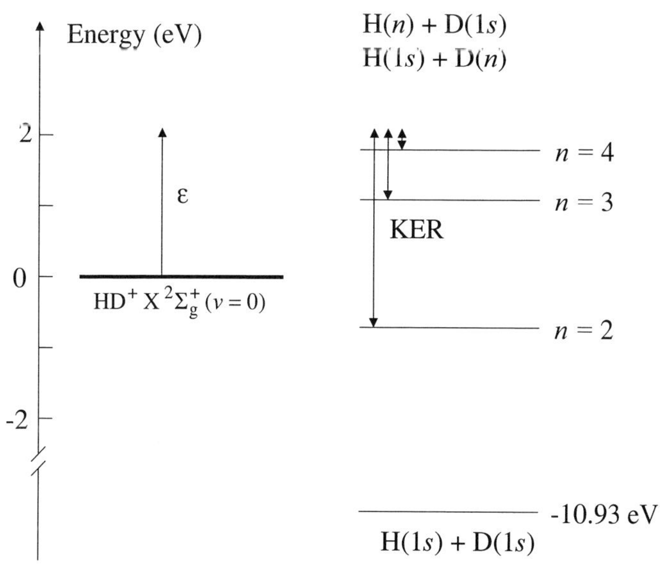

Figure 2. Energy level diagram for HD^+. The electron energy, ε, is represented by the length of the arrow.

information about this. Instead one must monitor the formation of ion pairs. In the CRYRING, the problem was solved by recording the production of D^- as a function of electron energy.[54,55] Figure 3 shows the first dipole magnet in CRYRING following the electron cooler. The dipole magnet serves two purposes; first, it bends the circulating ion beam so that it stays on a closed orbit around the ring; second, it acts as a charge state analyzer, so that H^- and D^- ions are separated from the main beam. Unfortunately, due to geometric constraints as shown in Figure 3, the H^- ions could not be detected. Figure 4 shows the experimental cross section for resonant ion pair formation (RIP), $HD^+ + e^- \rightarrow H^+ + D^-$, as a function of electron energy. The cross section rises sharply at 1.92 eV, in perfect agreement with the expected position at 1.913 eV. Since there is no competing process that can give rise to the formation of D^-, the signal-to-noise is infinite in principle. In practice, D^- is formed also when the electron energy is lower than 1.92 eV. The reason for this is that, at both edges of the

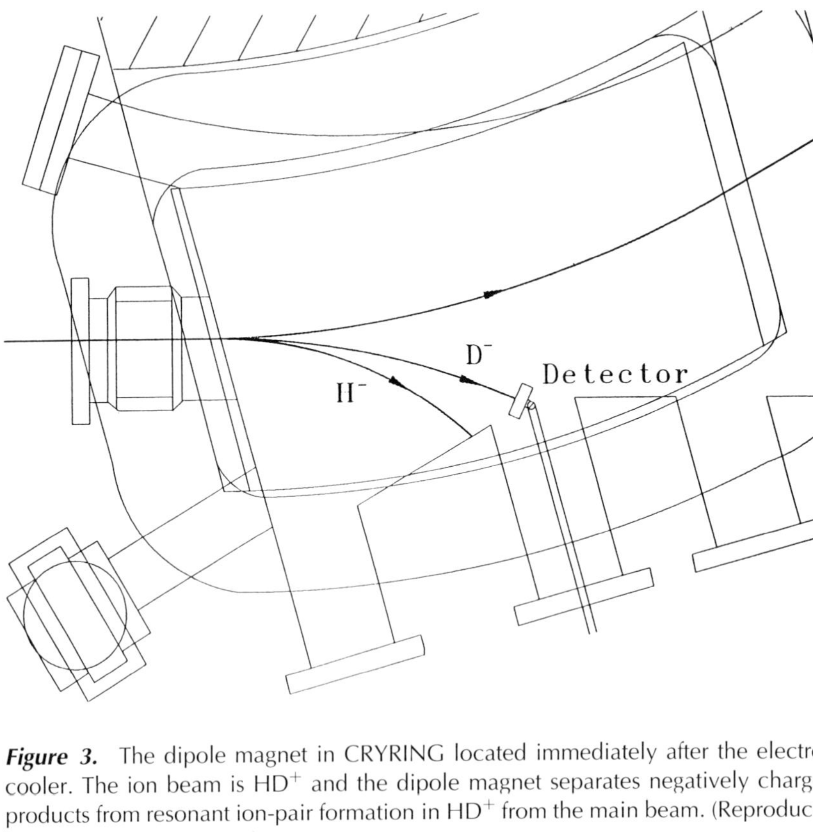

Figure 3. The dipole magnet in CRYRING located immediately after the electron cooler. The ion beam is HD^+ and the dipole magnet separates negatively charged products from resonant ion-pair formation in HD^+ from the main beam. (Reproduced with permission from Ref. [55].)

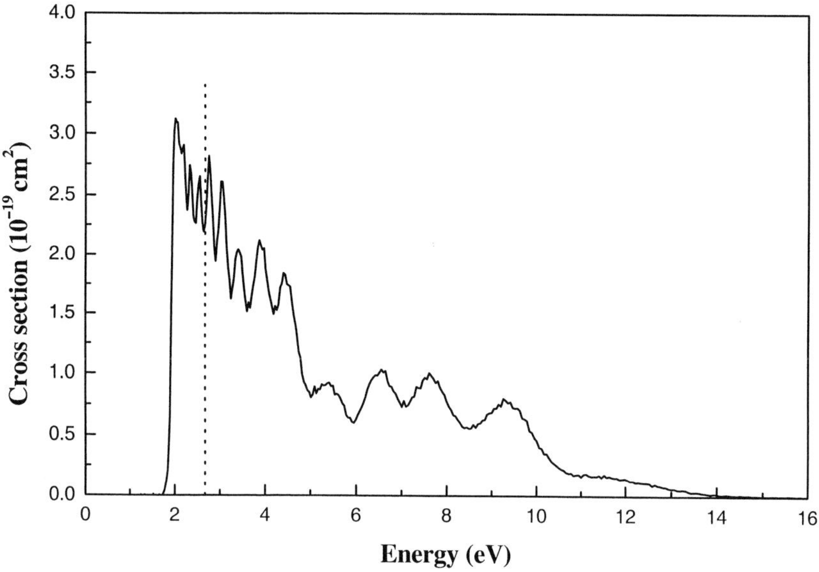

Figure 4. The experimental cross section for the $HD^+ + e^- \rightarrow H^+ + D^-$ reaction as a function of electron energy, as measured in CRYRING. The vertical dotted line shows the dissociation energy of the HD^+ ion. (Reproduced with permission from Ref. [55].)

electron cooler, where the electron beam is deflected in and out of the interaction region, collisions at higher energies will occur as compared with the situation for parallel beams in the interaction region. Thus, even when the collision energy in the interaction region is smaller than 1.92 eV, there will be a contribution at the edges of the cooler where the collision energies are larger than 1.92 eV. The data shown in Figure 4 have been corrected for this effect.

In the measured cross section for ion pair formation, 14 peaks are observed. To explain these peaks by electron capture into vibrationally excited Rydberg states, subsequently followed by predissociation, does not work because only five of the peaks are observed below the dissociation energy of HD^+ at 2.667 eV. For obvious reasons, this explanation breaks down for peaks observed to the right of the dotted vertical line in Figure 4. Instead, another explanation must be sought. Figure 5 illustrates in a simplified form how one can conceive of the origin of the peaks in the cross section. The dissociating wave can either diabatically follow the $(2p\sigma_u)^2\ {}^1\Sigma_g^+$ state to the ion-pair limit, or make a transition to the $(1s\sigma_g 3\sigma_g)$ $5\ {}^1\Sigma_g^+$ Rydberg state, propagate along this curve until large internuclear

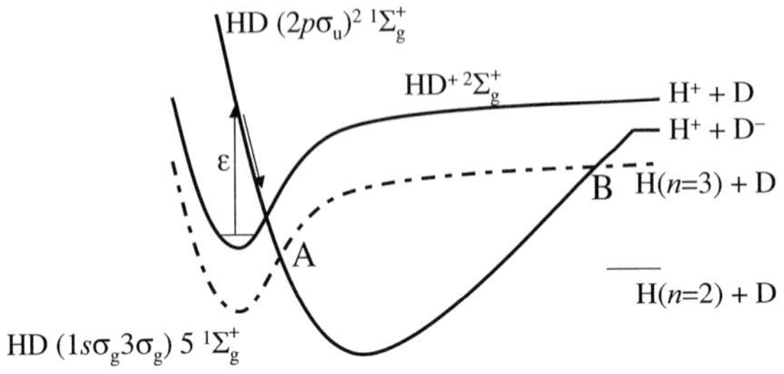

Figure 5. Simplified potential energy curves of HD^+ and HD drawn to illustrate the chemical "double slit" interference effect which occurs when the $H^+ + D^-$ limit can be reached in two different ways. After the electron has been captured into the $(2p\sigma_u)^2\ {}^1\Sigma_g^+$ resonant state, dissociation to $H^+ + D^-$ can either occur diabatically along the resonant state, or partly along the $(1s\sigma_g 3\sigma_g)$ 5 $^1\Sigma_g^+$ Rydberg state by making non-adiabatic transitions at points A and B. The possibility to reach $H^+ + D^-$ along two different decay paths give rise to interference, just as in a double slit experiment.

distance and then switch back to the $(2p\sigma_u)^2\ {}^1\Sigma_g^+$ state again before reaching the ion-pair limit. The competition between the pathways gives rise to a quantum interference which affects the total cross section. Just as for the photodissociation of H_2O,[56] the observed interference pattern provides a chemical analog of a double-slit experiment.

The arguments given above are simple and qualitative. In order to test the conjecture of quantum interference, two different types of calculations were performed.[55] A Landau-Zener model[57] was used to treat each individual crossing between the $(2p\sigma_u)^2\ {}^1\Sigma_g^+$ resonant state and the $(1s\sigma_g 2s\sigma_g)$, $(1s\sigma_g 3d\sigma_g)$, and $(1s\sigma_g 3s\sigma_g)$ Rydberg states. At electron energies above 6 eV, the $(2p\sigma_u)^2\ {}^1\Sigma_g^+$ state is no longer the only operative resonant state, and also higher excited resonant states, $(2p\sigma_u 2s\sigma_g)\ {}^1\Sigma_u^+$ and $(2p\sigma_u 3p\sigma_u)\ {}^1\Sigma_g^+$, were included in the calculations. Figures 6 and 7 show the result when the flux contributions are added incoherently and coherently, respectively. In the former case, the interference effect disappears and a smooth cross section is obtained. In the latter case, peaks are obtained that are in good agreement with the experimental result.

One may finally ask the question whether the opening of the ion-pair channel manifests in the dissociative recombination cross section leading to neutral products. The likelihood for this seems small at first sight since the DR cross section even at 1–2 eV, where it goes through a minimum, is an order of magnitude larger than the RIP cross section. Nevertheless, by carefully measuring the DR cross section between 1.0 eV and 2.5 eV at the TSR,[58,59] a small dip in the cross section was observed at the electron energy

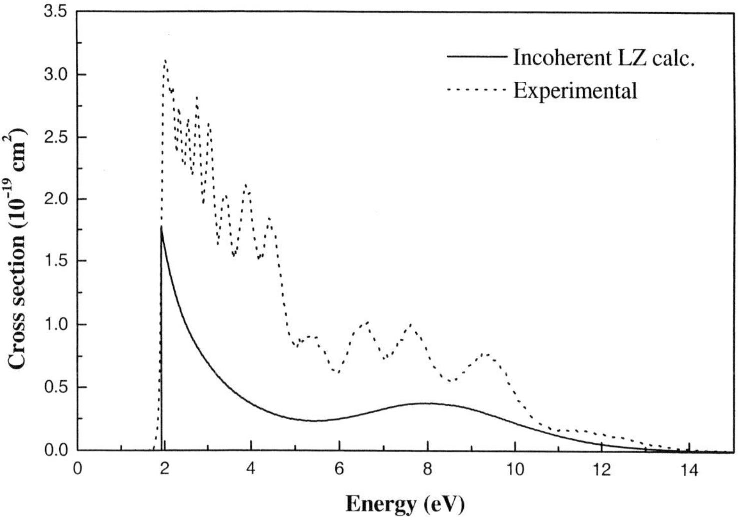

Figure 6. The cross section for D^- formation calculated using an incoherent sum of fluxes going to the ion-pair limit (solid line). The dotted curve represents the experimental cross section. (Reproduced with permission from Ref. [55].)

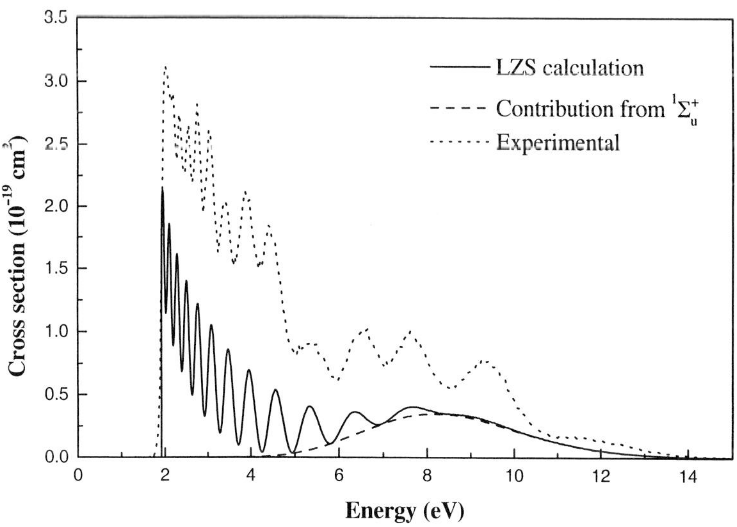

Figure 7. The cross section for D^- formation calculated taking the quantum interference into account (solid line). The dotted curve represents the experimental cross section. LZS stands for Landau-Zener-Stückelberg. (Reproduced with permission from Ref. [55].)

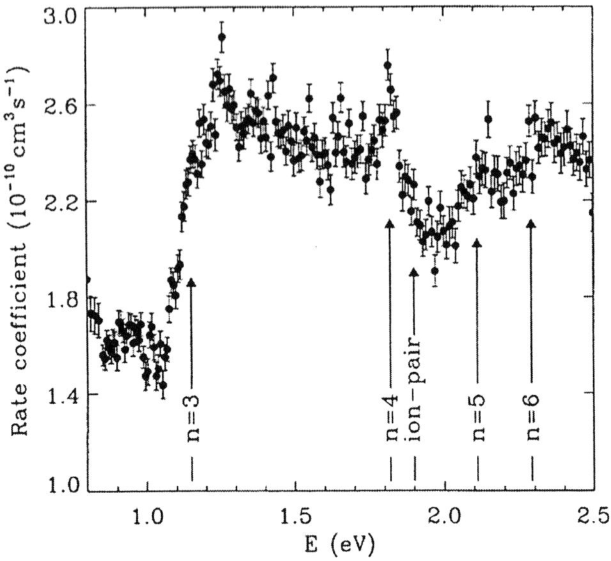

Figure 8. The rate coefficient for the dissociative recombination of HD^+ measured at the TSR (the cross section is obtained by division by the electron velocity). The arrows indicate the calculated position of the opening of new dissociation channels. (Reproduced with permission from Ref. [58].)

corresponding to the onset of RIP, as shown in Figure 8. It should be mentioned that the recording of the data in Figure 8 required a substantial investment in beam time (several days).

B. H_3^+

It was noted earlier that there is not yet a consensus regarding the thermal DR rate coefficient for H_3^+. The reader is referred to Ref. 28 for a detailed discussion of all experimental results to date. It does not seem possible at this point to reconcile the different experimental results given in Table 1. However, it is worth pointing out that the absolute DR cross section was measured recently[43] at TARN II and found to be in excellent agreement with the result from the CRYRING.[34,60,61]

Two other factors, however, combine to make a pressing need to reach a definitive experimental value for the thermal DR rate coefficient. The observation of H_3^+ in diffuse interstellar clouds has made it possible to determine the abundance of H_3^+.[30,31,62] The factor of 10 spread in the

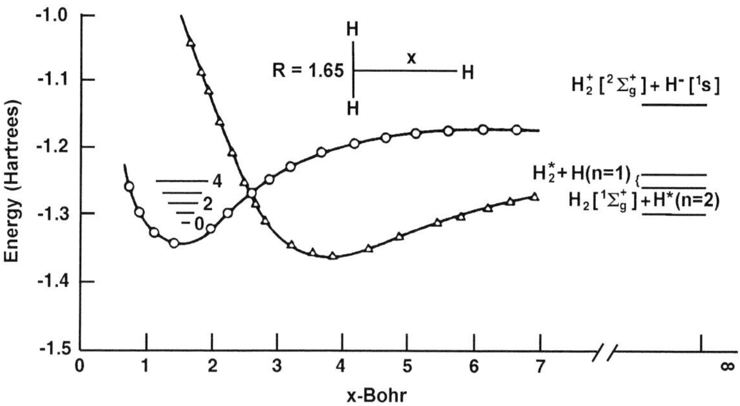

Figure 9. Potential energy curves for H_3 ($^2A_1$ resonant state: Δ—Δ—Δ—) and H_3^+ (X $^1A_1$: o—o—o—). (Adopted from Ref. [65].)

experimental results given in Table 1 directly translates into an uncertainty in size of the interstellar cloud by the same factor. Attempts have been made[63] to construct a cloud model that accounts for the observed abundance using the DR rate coefficient measured at the CRYRING, but it does not seem that this model is generally accepted.

The second factor is that, for the first time, theoretical calculations of the low-energy DR cross section have been made.[64] Figure 9 shows potential energy curves of H_3^+ and H_3 that illustrates why it was argued for a long time that H_3^+ must recombine very slowly unless substantially vibrationally excited. The argument (and the curves shown in Figure 9) dates back to a paper published in 1984[65] which established that recombination of H_3^+ ($\nu=0$) through the $^2A_1$ resonant state must be very small because the curve crossing occurs so far away from the right turning point of the $\nu=0$ state that the Franck-Condon factor is negligible. This was the prevailing theoretical standpoint until it was realized that HeH^+, which completely lacks a curve crossing in its electronic ground state, recombines dissociatively through states situated below the ion potential.[66–68] The ion state, plus the free electron, couple to these states by non-adiabatic coupling (rather than electron coupling which prevails in the crossing case). The theoretical results are confirmed by single-pass and storage ring merged-beams experiments.[68–70] Figure 10 shows an energy level diagram of H_3^+ and various dissociation limits of H_3^+ and H_3. By making an analog to HeH^+, dissociative recombination of H_3^+ would occur through the $^2E'$ state. In the first step of the theoretical work, predissociation of bound Rydberg states converging to the X $^1A_1$ state of H_3^+ was studied. The reason for this, is that there exists a considerable amount of experimental data for the

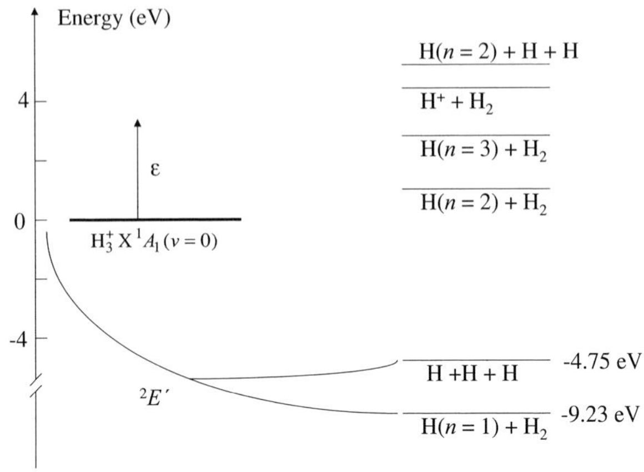

Figure 10. Schematic energy level diagram of H_3^+ and the lowest asymptotic limits of H_3. (Based on Figure 1 in Ref. [72].)

predissociation of these Rydberg states, obtained by a fast ion-beam technique.[71–75] By accurately calculating the non-adiabatic coupling elements between the degenerate $^2E'$ ground state of H_3 and the $2s(^2A_1')$, $3s(^2A_1')$, and $3p(^2A_1')$ Rydberg states, it was possible to theoretically reproduce the measured predissociation rates.[76] This is encouraging progress because it shows that the decay mechanism for electronic states of H_3 populated below the X $^1A_1$ state of H_3^+ is very similar to the one that drives dissociative recombination of HeH. Also autoionizing resonances have been treated by theoretical calculations and found to be in good agreement with experiment.[77]

Theoretical calculations[64,78,79] of the dissociative recombination of H_3^+ using the same approach as for the predissociation problem give cross sections that are smaller than all experimental results given in Table 1. However, these are only restricted, two-dimensional theoretical results, and cannot be used to draw definitive conclusions. Three-dimensional calculations, including all relevant electronic states, will be needed in the future. What seems clear at this point is that direct dissociative recombination, H_3^+ (X $^1A_1$) + $e^- \rightarrow$ H_3 (X $^2E'$) $\rightarrow$ H + H + H or H + H_2, is very slow.[64] For HeH^+ it was found that the indirect mechanism, which involves vibrationally excited Rydberg states as intermediates, enhances the cross section.[66] According to the preliminary calculations, the effect of the indirect mechanism is even more pronounced for H_3^+. Model calculations have shown that channel mixing effects can enhance the DR cross section substantially when the indirect mechanism is taken into account.[80] More theoretical work is needed before firm conclusions can be drawn.

Dissociative recombination of H_3^+ ($\nu = 0$) with zero eV electrons can only lead to the three channels given below:

$$H_3^+(\nu = 0) + e^-(\varepsilon = 0 \text{ eV}) \rightarrow H(1s) + H(1s) + H(1s) \quad (11a)$$

$$\rightarrow H_2(X^1\Sigma_g^+ \nu) + H(1s) \quad (11b)$$

$$\rightarrow H_3^* \quad (11c)$$

The branching ratios for these channels were measured in the CRYRING and found to be 75% for channel 11a, 25% for 11b and 0% for 11c.[81] It was further apparent in the CRYRING experiment that when the electron energy was increased so that the $^2A_1$ state (shown in Figure 9) became important, the branching ratio for channel 11b increased. This is in accord with the observation at TARN II that channel 11b was important when H_3^+ was vibrationally excited.[82] Thus, one can access the $^2A_1$ resonant state either by increasing the vibrational excitation or by increasing the electron energy. In a recent experiment at the TSR, it was found that the two-body channel 11b gives rise to a broad distribution of vibrational levels in H_2, and that three-body channel 11a has a predominance for nearly linear momentum geometries.[83] Dissociative recombination of H_3^+ is indeed a fascinating problem which is far from being solved.

IV. DISSOCIATIVE RECOMBINATION OF O_2^+ AND THE GREEN AIRGLOW

Several diatomic molecular ions have been investigated with respect to the final atomic state distribution in dissociative recombination. The reader is referred to Ref. 12 for a fairly complete discussion of those diatomic ions that have been studied in ion storage rings. In this section, we deal exclusively with O_2^+.

The night sky is illuminated by sources located outside our atmosphere, but also from photochemical processes which have their origin in the atmosphere. These processes give rise to a faint glow to which the name airglow is given. The airglow is orders of magnitude stronger during the daytime but is too feeble to be detectable by eye because of the scattered sunlight. The green airglow at 557.7 nm arises from the forbidden $^1S \rightarrow {}^1D$ transition in atomic oxygen, and in the F-region of the Earth's ionosphere, between 200 and 300 km, dissociative recombination of O_2^+ is believed to be the dominant source of $O(^1S)$. The $O(^1D)$ gives rise to the red airglow at

630.0 nm. Dissociative recombination of O_2^+ gives rise to the following metastable atoms[7]

$$O_2^+(X^2\Pi_g) + e^- \rightarrow O_2(^3\Sigma_u^-) \rightarrow O(^3P) + O(^1D) + 5.0 \text{ eV} \quad (12a)$$

$$\rightarrow O_2(^1\Delta_u) \rightarrow 2O(^1D) + 3.0 \text{ eV} \quad (12b)$$

$$\rightarrow O_2(^1\Sigma_u^+) \rightarrow O(^1D) + O(^1S) + 0.8 \text{ eV} \quad (12c)$$

Figure 11 illustrates the potential curves of O_2 involved in processes 12a to 12c and where they cross the electronic ground state of O_2^+. Whereas the $^3\Sigma_u^-$ and $^1\Delta_u$ states cross the X $^2\Pi_g$ state of O_2^+ close to the turning points of the zeroth vibrational level of the X $^2\Pi_g$ state, thus causing them to have a large vibrational overlap with this level, the $^1\Sigma_u^+$ state crosses the ground state somewhere between the first and second vibrational level. This means that O_2^+ ($v=0$) is not likely to recombine through the $^1\Sigma_u^+$ state, and hence is not expected to produce any substantial amounts of oxygen atoms in the ^{1}S state. Quantitative calculations found the quantum yield, f_S, for reaction 12c to be 0.0016 at $T_e = 300$ K when only the zeroth vibrational level is populated.[85] It was also shown that the quantum yields for $v=1$ and 2 are factors 23 and 78 times, respectively, larger than the quantum

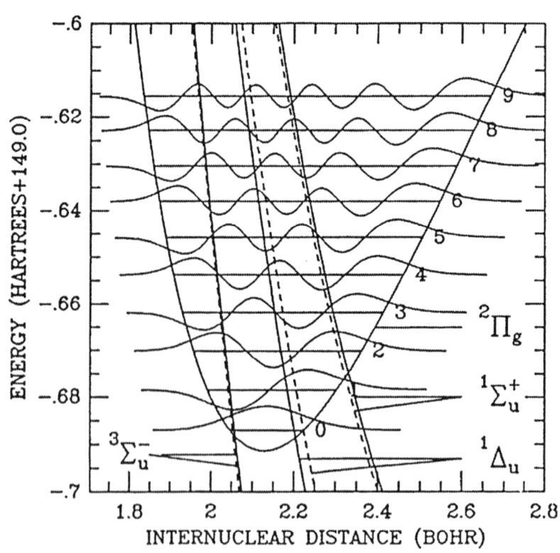

Figure 11. Potential energy curves of O_2^+ and O_2 (with wavefunctions and positions of lowest vibrational levels). The potential curves for the $^3\Sigma_u^-$, $^1\Delta_u$, and $^1\Sigma_u^+$ states were calculated ab initio to first (dashed line) and second (solid line) order. (Reproduced with permission from Ref. [84].)

yield for $\nu = 0$. This strong dependence on the vibrational quantum number caused some interpretation problems of data from rocket and ground based observations of the Earth's ionosphere.[86–94] The observations gave higher quantum yields, in the range 0.01 to 0.23, than the calculated one for $\nu = 0$. But was this due to the presence of vibrationally excited O_2^+ in the ionosphere, or was the theoretical calculation producing too small a quantum yield for O_2^+ ($\nu = 0$)? One problem in the interpretation of the rocket data is that the in-situ O_2^+ is formed in an unknown vibrational distribution, and experiences collisional vibrational deactivation by

$$O_2^+(\nu') + O \rightarrow O_2(\nu < \nu') + O \tag{13}$$

and other processes, none of which are well known. Also laboratory experiments[95–98] have had considerable difficulties in dealing with, in particular, the preparation of O_2^+ in its zeroth vibrational level. For example, it was found in a flowing afterglow experiment[98] that the low theoretical quantum yield[85] was difficult to reconcile with the experimental data, even though it was not possible to obtain data that were not influenced by O_2^+ in its $\nu = 1$ and 2 levels. In order to test whether it was a limitation to include only the direct process 12c in the calculation of the quantum yield,[85] MQDT calculations including also the indirect mechanism, which involves vibrationally excited Rydberg states as intermediates, were carried out.[99] This had the effect of decreasing the quantum yield to 0.0012 for O_2^+ ($\nu = 0$), and consequently did provide a route to reconcile the experimental and observational data.

The first direct evidence for the quantum yield being larger than previously predicted by theory[85,99] was obtained at the storage ring ASTRID.[100] A two-dimensional imaging detector was used to measure the distribution of projected distances for dissociative recombination of O_2^+ ($\nu = 0 - 5$). The broad distribution of vibrational levels in the stored ion beam caused contributions from different levels to overlap, but by means of a peak-fitting routine, it was possible to determine the quantum yield to be 0.05 ± 0.02 at an electron energy of nominally 0 eV. At the same time, new calculations[101] were carried out which explained why the quantum yield was higher than had been predicted by earlier calculations.[85,99] Figure 12 shows how the new mechanism is envisioned.[101] The electron is captured into the $^3\Sigma_u^-$ state, which according to reaction 12a leads to the formation of $O(^3P) + O(^1D)$. But the two oxygen atoms travel only briefly on the potential curve of the $^3\Sigma_u^-$ state, and then they switch to a vibrationally bound (but autoionizing) Rydberg state of spin-orbit mixed $^3\Sigma_u^-/^1\Sigma_u^+$ character. The Rydberg state is effectively predissociated by the $^1\Sigma_u^+$ state, which dissociates to $O(^1D) + O(^1S)$ in accordance with 12c. According to the MQDT calculations,[101] this redirection of

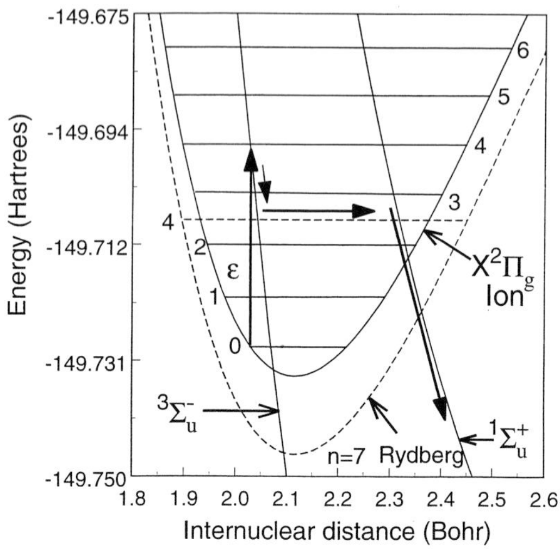

Figure 12. Potential energy curves for O_2^+ and O_2. The $^1\Sigma_u^+$ and $^3\Sigma_u^-$ states are repulsive states of O_2. The vertical, upward pointing, arrow shows direct capture into the $^3\Sigma_u^-$ state. The Rydberg state is of mixed $^1\Sigma_u^+/^3\Sigma_u^-$ character. The transition from the $^3\Sigma_u^-$ state to the $^1\Sigma_u^+$ state occurs through the $^1\Sigma_u^+/^3\Sigma_u^-$ Rydberg state. The $^1\Sigma_u^+$ state dissociates to $O(^1S)+O(^1D)$. (Reproduced with permission from Ref. [101].)

dissociation flux from the $^3\Sigma_u^-$ state to the $^1\Sigma_u^+$ state is sufficient to increase the $O(^1S)$ quantum yield from 0.0012 to 0.020 at an electron temperature of 300 K.

Also, according to the calculations, the amount of $^3\Sigma_u^-$(Rydberg) $\sim$ $^1\Sigma_u^+$(Rydberg) mixing varies with the principal quantum number n. There is also an increase in the predissociation width with decreasing n. Rydberg states of different n values are involved at different electron energies ε, which implies that one would anticipate that the partial cross section for $O(^1S)+O(^1D)$ production should vary as a function of electron energy. This hypothesis was not tested at ASTRID[100] since the experiment was only carried out at $\varepsilon = 0$ eV. Three improvements of the CRYRING and the detector system have made it possible to make a detailed study of the $O(^1S)$ quantum yield as a function of electron energy:

- The electron cooler was supplied with a supraconducting magnet, which made it possible to expand the electron beam a factor of 100.[102] This led to a transverse electron temperature only slightly above 1 meV.
- A high-pressure (>0.1 torr) hollow cathode ion source was used to produce vibrationally cold O_2^+ (being a homonuclear molecule it does

not relax vibrationally in the storage ring; even the addition of a small dipole moment by replacing one of the ^{16}O to form $^{18}O^{16}O$ is not sufficient to obtain vibrationally relaxed ions[100]).

- The three-dimensional imaging detector was improved in order to reduce the noise, see Figure 13.

Figure 14 shows experimental distance spectra taken at electron energies 0 meV, 5.5 meV, and 11 meV.[103] The timing information has been used so that primarily oxygen atoms deflected perpendicular with respect to the electron and ion beam axis are shown, i.e., atoms for which the difference in arrival time on the detector is small (<800 ps). The peak corresponding to $O(^1S)+O(^1D)$ undergoes a drastic change over a very small interval of electron energies. At 11 meV the peak has disappeared. The result over a broader range of electron energies is shown in Figure 15 together with the thermally averaged yield. Also shown in Figure 15 are results from MQDT calculations.[103] The agreement between experiment and theory is quite good at low electron energies, whereas at higher energies (>0.1 eV) only one data point of low accuracy is available. The reason for the few

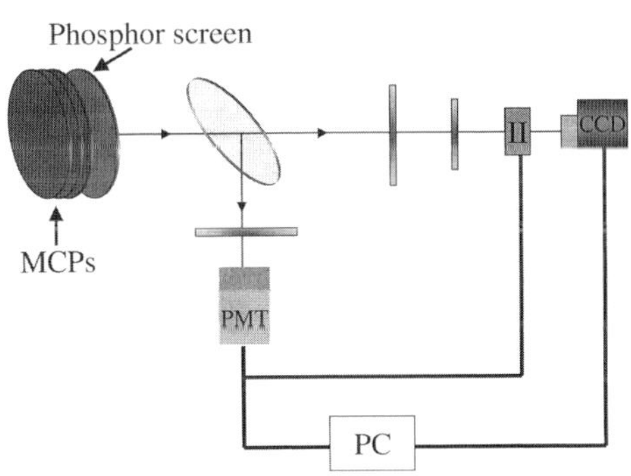

Figure 13. Three-dimensional imaging detector used for the O_2^+ experiment in the CRYRING. Fast, neutral particles impinge on a multichannel plate (MCP) and create a burst of electrons. The burst is amplified in two stages by two additional MCPs, and gives rise to a flash on a phosphor screen. The light from the phosphor screen is focused on a CCD camera, and on a photomultiplier (PMT) consisting of 16 primary dynodes. The photomultiplier gives information on the difference in arrival time between the two oxygen atoms. The photomultiplier also serves to give a signal to the image intensifier (II), which is only turned on when two different primary dynodes have been illuminated. This mode gives rise to a strong reduction of the background because the reaction $O_2^+ +$ residual gas $\rightarrow O^+ + O$ is not recorded unless two uncorrelated background events happen to occur within a time window of less than 30 ns. The information from the PMT and the CCD camera are combined in a computer (PC).

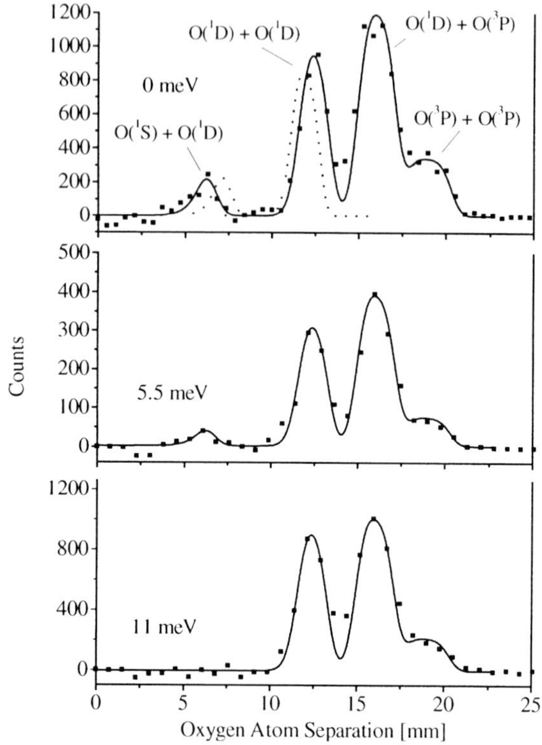

Figure 14. Experimental distance spectra. Solid lines are synthetic spectra, determined by the energetics of O_2^+ and O_2 and apparatus parameters. Three examples of distance spectra are given, at 0 meV (upper), 5.5 meV (middle), and 11 meV (lower). The peaks shown with a dotted line in the upper spectrum indicate expected positions if $O_2^+(\nu = 1)$ were present in the beam (peak at 7.5 mm) and if recombination of O_2^+ $(\nu = 0)$ would give rise to $O(^3P) + O(^1S)$ (peak at 12 mm). (Reproduced with permission from Ref. [103].)

data points at higher electron energies is of course related to the decrease in the DR cross section, which makes data taking at higher electron energies much more difficult. What is shown in Figure 15 is probably the first observation of resonances in an atomic product channel in dissociative recombination. The origin of the resonant behavior is the spin-orbit driven mechanism involving Rydberg states[101] and hence has a very different origin from the resonances observed in resonant ion pair formation in HD^+ (see Figure 5).

Finally one can note that the quantum yield for $O(^1D)$ was found to be 1.15 ± 0.05 and essentially independent of the electron energy.[103,104]

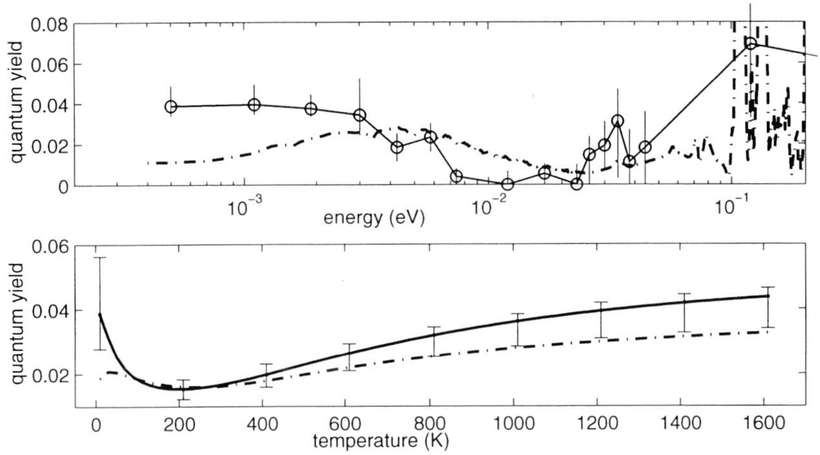

Figure 15. The quantum yield for $O(^1S)+O(^1D)$ as a function of electron energy and electron temperature. Experimental yields are shown as solid lines whereas the MQDT results are given as dot–dashed lines. Upper part: Yields (f_S) as a function of electron energy. The theoretical yields have been convoluted with the experimental resolution of about 1.3 meV. Lower part: Thermally averaged yields. (Reproduced with permission from Ref. [103].)

This raises the question whether one can use the relative strengths of the green and red airglows as a measure of the electron temperature. Before this question can be answered, however, the discrepancy[103] between green airglow data[89] from rocket measurement as a function of altitude, and the quantum yield derived for $O_2^+(\nu=0)$ from the CRYRING experiment[103] and from MQDT calculations[101,103] must be resolved. A possible source of this discrepancy could be the presence of vibrationally excited O_2^+ in the ionosphere. Theoretical work to explore the effects of vibrational excitations and isotopic substitutions has been started.[105]

V. BRANCHING RATIOS IN DISSOCIATIVE RECOMBINATION OF H_3O^+ AND OTHER POLYATOMIC IONS

Astrochemistry is a young branch of chemistry which has evolved during the last 30 years owing to the discoveries by radioastronomy of a large number of discrete spectral lines that derives from molecular transitions.[106] The synthesis of complex molecules is thought to occur via chains of ion–molecule reactions followed by dissociative recombination.[107] For example, the synthesis of water could occur through

$$H_3^+ + O \rightarrow OH^+ + H_2 \quad (13a)$$

$$OH^+ + H_2 \rightarrow H_2O^+ + H \quad (13b)$$

$$H_2O^+ + H_2 \rightarrow H_3O^+ + H \quad (13c)$$

followed by the dissociative recombination:

$$H_3O^+ + e^- \rightarrow H_2O + H \quad (14a)$$

However, there are also other energetically allowed decay channels when H_3O^+ recombines with electrons, such as

$$H_3O^+ + e^- \rightarrow OH + H + H \quad (14b)$$

$$\rightarrow OH + H_2 \quad (14c)$$

$$\rightarrow O + H + H_2 \quad (14d)$$

The branching ratios for the four decay channels 14a to 14d are by no means evident and cannot be deduced by chemical intuition, and various attempts to theoretically predict branching ratios have not always been successful.[108,109]

Progress has been made in the experimental determination of branching ratios owing to the application of ion storage rings[12] and to the development of the flowing afterglow technique.[110] In the latter technique, methods have been developed which allow the detection of the hydrogen and oxygen atoms and the OH radical. The flowing afterglow technique does not normally allow a determination of all neutral product channels.

The ion storage ring technique for determining branching ratios exploits the high energy available in the rings, and builds on an entirely different approach from the flowing afterglow technique. The original idea[111] behind the technique goes back to long before ion storage rings were used in atomic physics, and it has also been used in single-pass experiments.[112,113] Powerful variations of the technique were developed at CRYRING,[81,114,115] although for most parts the technique used in storage rings today is essentially as described in Ref. 112.

Figure 16 shows a pulse height spectrum recorded in an ion-implanted-silicon surface barrier detector mounted in the zero degree direction of the electron cooler in CRYRING.[116] A stored beam of 4.4 MeV D_3O^+ ions interacts with a beam of velocity matched electrons, thus the collision energy

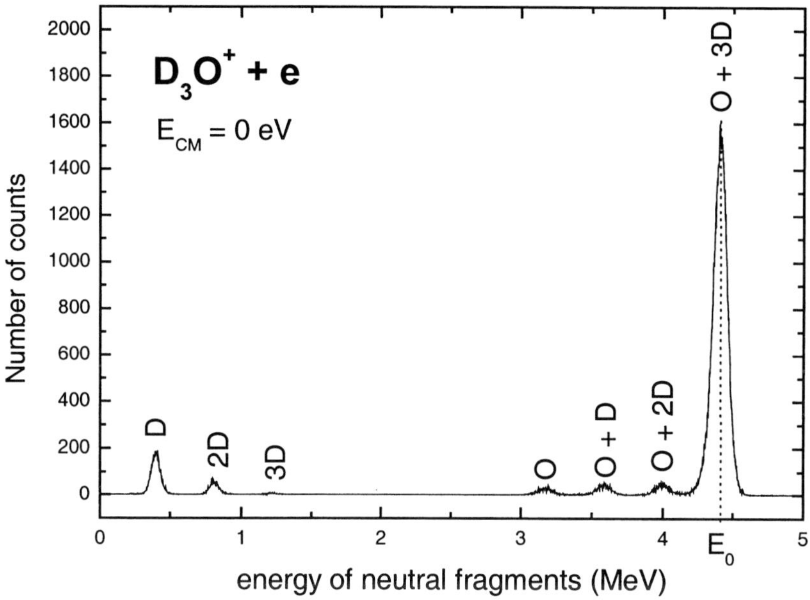

Figure 16. Pulse height spectrum recorded at the CRYRING with a 4.4 MeV stored beam of D_3O^+ ions and the electron cooler set to the velocity-matching condition. The peak at 4.4 MeV, corresponding to particles with a combined mass of 22 amu, originates from dissociative recombination, whereas the six other peaks are due to collisions of D_3O^+ with rest gas molecules. (Reproduced with permission from Ref. [116].)

is centered at 0 eV. Dissociative recombination of D_3O^+ gives rise to a peak at full beam energy, i.e., at 4.4 MeV. This peak is composed of the deuterated equivalent of the product channels 14a–14d, and there is no way that one can determine the branching ratios from this peak alone. There are also six other peaks, which all derive from collisions between the stored D_3O^+ ions and rest gas molecules present in the electron cooler section. These rest gas collisions give rise to a broad set of neutral and charged particles, and since the charged particles are deflected off the zero-degree direction, only the neutral particles arrive at the detector. In the next step, a 50-μm-thick stainless steel grid is inserted in front of the detector. The grid contains a large number of small holes, each 70 μm in diameter. The grid transmits neutral particles with a probability T, and as a consequence, the full beam energy peak will decrease at the expense of the other peaks in the pulse height spectrum, as shown in Figure 17.

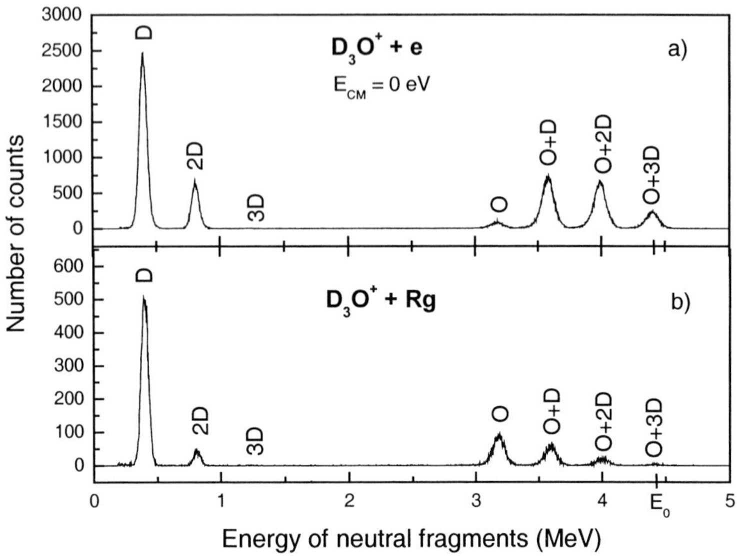

Figure 17. Pulse height spectra when a 50-μm-thick stainless steel grid has been inserted in front of the ion-implanted-silicon detector. (a) Electron beam on and set to 0 eV collision energy. (b) Collisions of 4.4 MeV D_3O^+ ions with rest gas molecules in the electron cooler section. Note that the peak at 4.4 MeV, corresponding to mass 22 amu, is almost absent, the reason being that the cross section for electron capture from the rest gas molecules is small at 4.4 MeV. If the beam energy was decreased, this peak would increase. (Reproduced with permission from Ref. [116].)

In the H_3O^+/D_3O^+ experiments in CRYRING,[116] also used to exemplify the grid technique, the results were:

$$
\begin{aligned}
&n_{14a}(H_2O + H) = 0.18 \pm 0.05 && n_{14a}(D_2O + D) = 0.17 \pm 0.05 \\
&n_{14b}(OH + H + H) = 0.67 \pm 0.06 && n_{14b}(OD + D + D) = 0.70 \pm 0.06 \\
&n_{14c}(OH + H_2) = 0.11 \pm 0.05 && n_{14c}(OD + D_2) = 0.13 \pm 0.03 \\
&n_{14d}(O + H_2 + H) = 0.04 \pm 0.06 && n_{14d}(O + D_2 + D) = 0.00 \pm 0.04
\end{aligned}
\tag{15}
$$

where the branching ratios $n_{14a\text{–}d}$ have been normalized so that their sum is unity. Within the error bars, there is no isotope effect in the branching ratios. Experiments at ASTRID[117,118] gave a water channel 14a which was 0.33, primarily at the expense of the three-body channel 14b, which was 0.48. However, more recent work from ASTRID gives better agreement with the CRYRING results.[119] The discrepancy with the flowing afterglow result[120] is more serious. In contrast to the storage ring experiments,[116–119] the flowing afterglow experiment gives a substantial production of oxygen atoms, i.e., reaction 14d is found to have a branching ratio of 0.30.[120] The water channel 14a was found to be only 0.05, i.e., almost negligible. The comparison of the storage ring[116–119] and flowing afterglow[120] results are important because this is so far the only time a flowing afterglow experiment has given the full set of branching ratios. Possible sources of error in the afterglow experiment is discussed in Ref. 121, but the conclusion of this discussion is that it seems difficult to identify processes that could lead to an apparent too high yield of O. In the storage ring experiments, the background must be measured accurately and the grid transmission must be known. The experiment itself is fairly straightforward and it is difficult to see where such a major discrepancy as found for H_3O^+ could come from. An interesting observation was made in the ASTRID experiment.[116] The detector was not sufficiently large to collect all particles, and for the channel 14a a fraction of the H atoms missed the detector. Thus, even without the grid it was possible to conclude that the water channel 14a must be at least 10% from the size of the peak at 18/19 of the full beam energy, which corresponds to H_2O hitting the detector while the H atom in reaction 14a is missing the detector.

Many more examples could be discussed, but since this is not intended to be a comprehensive review, the reader is referred to Ref. [12] for a more extensive discussion of branching ratios. The general conclusion from the ring experiments carried out so far is that:

- Dissociative recombination of polyatomic ions leads to more fragmentation than previously believed.
- Dissociative recombination of triatomic ions gives branching ratios for the three-body channel which consistently is about 0.6–0.7.[12]
- The theory for DR branching ratios has not yet been developed to a point where reliable predictions are possible.
- The dynamics of the fragmentation processes is essentially unknown.

This last point was addressed in recent experiments at the CRYRING[122,123] and TSR[83] with H_2O^+ and H_3^+, respectively. In these experiments, the grid technique cannot be used; instead, the dynamics information is obtained by the imaging technique. In the case of H_2O^+, an additional procedure was needed in order to identify the O atom on the imaging detector. A 5-mm-diameter Al disc, 2.5 μm thick, was placed just in front of the imaging detector's first MCP. At 250 keV/amu beam energy, the range of protons in Al is 2.2 μm whereas O atoms have a range of 3.2 μm. Thus, for a 4.5 MeV (250 keV/amu) beam of H_2O^+, a hit behind the Al foil must by necessity be due to an O atom since the H atoms are stopped by the Al foil. The results are best reproduced if it is assumed that the kinetic energy is randomly shared between the hydrogen atoms. Another interesting observation is that the two hydrogen atoms come out at an angle much smaller than the bond angle in H_2O^+ and H_2O. It was speculated at a recent Faraday discussion meeting that the dissociation proceeds via Rydberg states associated with the second excited superbent B $^2B_2$ state of H_2O^+.[124] The imaging data also made it possible to determine that the O atoms are formed in the ratio $O(^3P):O(^1D) = (3.5 \pm 1.0):1$.

VI. CONCLUSIONS

In this article, a few selected examples have been chosen in order to illustrate the progress that has been made in the study of dissociative recombination and similar processes by means of the ion-storage-ring-based merged-beams technique. Great progress has been made since the first experiments started in 1992. The progress has also stimulated theoretical developments, and comparisons of experiment and theory plays an important role. It is also clear that other techniques, such as the flowing afterglow technique, have undergone a development in recent years, maybe partly stimulated by the storage ring experiments.

It was mentioned in Section III that H_2^+ cools vibrationally in storage rings on time scales much shorter than the radiative vibrational lifetimes. The effect was shown to be correlated with the interaction of the stored ion beam with the electron beam,[51] and it was concluded that since higher vibrational levels recombine faster than lower levels, there will be an increase of the relative population of lower vibrational levels as a function of time. More recent work has shown that superelastic collisions

$$H_2^+(\nu) + e^-(\varepsilon) \rightarrow H_2^+(\nu' < \nu)e^-(\varepsilon + \Delta\varepsilon) \tag{16}$$

where $\Delta\varepsilon > 0$, leads to vibrational cooling.[125,126] It remains to be investigated whether this vibrational cooling mechanism also works for other homonuclear molecules.

One of the great challenges for the future will be to investigate the dynamics of dissociative recombination of polyatomic ions. Such experiments have already started. Another challenge is to study heavier systems. There is no clear upper limit to how complex systems can be studied in ion storage rings with velocity-matched electron beams. Recent experiments at the CRYRING with cluster ions such as $H_7O_3^+$ and $H_9O_4^+$ show that a broad range of ions are amenable.

ACKNOWLEDGMENTS

Support by the Swedish Natural Science Research Council (NFR) is gratefully acknowledged. The author would like to thank colleagues who provided preprints of their work, J. Semaniak and R. Thomas for valuable comments on the draft manuscript, and A. Derkatch for help with some of the figures.

REFERENCES

[1] Larsson, M. *Rep. Prog. Phys.* **1995**, *58*, 1267.
[2] Larsson, M. *Annu. Rev. Phys. Chem.* **1997**, *48*, 151.
[3] Biondi, M.A.; Brown, S.C. *Phys. Rev.* **1949**, *76*, 1697.
[4] Bates, D.R. *Phys. Rev.* **1950**, *77*, 718.
[5] Biondi, M.A.; Bardsley, J.N. *Adv. At. Mol. Phys.* **1970**, *6*, 1.
[6] Mitchell, J.B.A. *Phys. Rep.* **1990**, *186*, 215.
[7] Bates, D.R. *Adv. At. Mol. Opt. Phys.* **1994**, *34*, 427.
[8] Mitchell, J.B.A.; Guberman, S.L. (Eds.) *Dissociative Recombination: Theory, Experiment and Applications*; World Scientific: Singapore, 1989.
[9] Rowe, B.R.; Mitchell, J.B.A.; Canosa, A. (Eds.) *Dissociative Recombination: Theory, Experiment, and Applications*; NATO ASI Series B: Physics Vol. 313; Plenum Press: New York, 1993.
[10] Zajfman, D.; Mitchell, J.B.A.; Schwalm, D.; Rowe, B.R. (Eds.) *Dissociative Recombination: Theory, Experiment and Applications III*; World Scientific: Singapore, 1996.
[11] Larsson, M.; Mitchell, J.B.A.; Schneider, I.F. (Eds.) *Dissociative Recombination: Theory, Experiment and Applications IV*; World Scientific: Singapore, 2000.
[12] Larsson, M. In *Adv. Ser. Phys. Chem.: Photoionization and Photodetachment*; Ng, C.-Y., Ed.; World Scientific: Singapore, 2000, p 693.
[13] Larsson, M. In *The Physics of Electronic and Atomic Collisions, XXI International Conference*; AIP Conference Proceedings 500, American Institute of Physics: New York, 2000, p 44.
[14] Dunn, G.H.; Djuric, N. In *Novel Aspects of Electron-Molecule Scattering*; Becker, K., Ed.; World Scientific: Singapore, 1998, p 241.
[15] Andersen, L.H.; Andersen, T.; Hvelplund, P. *Adv. At. Mol. Opt. Phys.* **1998**, *38*, 155.
[16] Phaneuf, R.A.; Havener, C.C.; Dunn, G.H.; Müller, A. *Rep. Prog. Phys.* **1999**, *62*, 1143.
[17] Müller, A.; Wolf, A. In *Accelerator-Based Atomic Physics Techniques and Applications*; Shafroth, S.; Austin, J., Eds.; AIP: New York, 1997, p 147.
[18] Auerbach, D.J.; Cacak, R.; Caudano, R.; Gaily, T.D.; Keyser, C.J.; McGowan, J.Wm.; Mitchell, J.B.A.; Wilk, S.F.J. *J. Phys. B* **1977**, *10*, 3797.
[19] Schneider, I.F.; Dulieu, O.; Giusti-Suzor, A. *J. Phys. B* **1991**, *24*, L289.
[20] Van der Donk, P.; Yousif, F.B.; Mitchell, J.B.A.; Hickman, A.P. *Phys. Rev. Lett.* **1991**, *67*, 42.
[21] Schneider, I.F.; Dulieu, O.; Giusti-Suzor, A. *Phys. Rev. Lett.* **1992**, *68*, 2251.
[22] Van der Donk, P.; Yousif, F.B.; Mitchell, J.B.A.; Hickman, A.P. *Phys. Rev. Lett.* **1992**, *68*, 2252.
[23] Takagi, H. *J. Phys. B* **1993**, *26*, 4815.
[24] Canosa, A.; Gomel, J.C.; Rowe, B.R.; Mitchell, J.B.A.; Queffelec, J.L. *J. Chem. Phys.* **1992**, *97*, 1028.
[25] Larsson, M. *Annu. Rev. Phys. Chem.*, **1997**, *48*, 151.

[26] Dalgarno, *Adv. At. Mol. Opt. Phys.*, **1994**, *32*, 57.

[27] Gougousi, T.; Johnsen, R.; Golde, M.F. *Int. J. Mass. Spectrom. Ion. Proc.*, **1993**, *129*, 163.

[28] Larsson, M. *Phil. Trans. R. Soc. Lond. A*, **2000**, *358*, 2433.

[29] Geballe, T.R.; Oka, T. *Nature*, **1996**, *384*, 334.

[30] McCall, B.J.; Geballe, T.; Hinkle, K.H.; Oka, T. *Science*, **1998**, *279*, 1910.

[31] McCall, B.J.; Oka, T. *Science*, **2000**, *287*, 5460.

[32] Lee, H.-H.; Roueff, E.; Pineau des Forêts, G.; Shalabiea, O.M.; Terzieva, R.; Herbst, E. *Astron. Astrophys.*, **1998**, *334*, 1047, and references therein.

[33] Amano, T. *J. Chem. Phys.*, **1990**, *92*, 6492.

[34] Sundström, G.; Mowat, J.R.; Danared, H.; Datz, S.; Broström, L.; Filevich, A.; Källberg, A.; Mannervik, S.; Rensfelt, K.-G.; Sigray, P.; af Ugglas, M.; Larsson, M. *Science*, **1994**, *263*, 785.

[35] Laubé, S.; Le Padellec, A.; Sidko, O.; Rebrion-Rowe, C.; Mitchell, J.B.A.; Rowe, B.R. *J. Phys. B*, **1998**, *31*, 2111.

[36] Smith, D.; Spanel, P. *Int. J. Mass Spectrom. Ion Proc.*, **1993**, *149/150*, 131.

[37] Yousif, F.B.; Rogelstad, M.; Mitchell, J.B.A. In *Atomic and Molecular Physics: 4th US/Mexico Symposium*, Alvarez, I.; Cisneros, S.; Morgan, T.J., Eds.; World Scientific: Singapore, 1995, p 343.

[38] Forck, P.; Grieser, M.; Habs, D.; Lampert, A.; Repnow, P.; Schwalm, D.; Wolf, A.; Zajfman, D. *Phys. Rev. Lett.*, **1993**, *70*, 426.

[39] Strömholm, C.; Schneider, I.F.; Sundström, G.; Carata, L.; Danared, H.; Datz, S.; Dulieu, O.; Källberg, A.; af Ugglas, M.; Urbain, X.; Zengin, V.; Suzor-Weiner, A.; Larsson, M. *Phys. Rev. A*, **1995**, *52*, R4320.

[40] Tanabe, T.; Katayama, I.; Kamegaya, H.; Chida, K.; Arakaki, Y.; Watanabe, M.; Yoshizawa, M.; Saito, M.; Haruyama, Y.; Hosono, K.; Hatanaka, K.; Honma, T.; Noda, K.; Ohtani, S.; Takagi, H. *Phys. Rev. Lett.*, **1995**, *75*, 1066.

[41] Schneider, I.F.; Strömholm, C.; Caråta, L.; Urbain, X.; Larsson, M.; Suzor-Weiner, A. *J. Phys. B*, **1997**, *30*, 2687.

[42] Danared, H.; Andler, G.; Bagge, L.; Herrlander, C.J.; Hilke, J.; Jeansson, J.; Källberg, A.; Nilsson, A.; Paál, A.; Rensfelt, K.-G.; Rosengård, U.; Starker, J.; af Ugglas, M. *Phys. Rev. Lett.*, **1994**, *72*, 3775.

[43] Tanabe, T.; Chida, K.; Watanabe, T.; Arakaki, Y.; Takagi, H.; Katayama, I.; Haruyma, Y.; Saito, M.; Nomura, I.; Honma, T.; Noda, K.; Hoson, K. In *Dissociative Recombination: Theory, Experiment and Applications IV*, Larsson, M.; Mitchell, J.B.A.; Schneider, I.F., Eds.; World Scientific: Singapore, 2000, p 170.

[44] Al-Khalili, A.; van der Zande, W.J.; Rosén, S.; Danared, H.; Derkatch, A.M.; Källberg, A.; Le Padellec, A.; Neau, A.; Semaniak, J.; Thomas, R.; af Ugglas, M.; Vikor, L.; Zong, W.; Larsson, M.; Urbain, X.; Jensen, M.J.; Bilodeau, R.C.; Heber, O.; Pedersen, H.B.; Safvan, C.P.; Andersen, L.H.; Lange, M.; Levin, J.; Gwinner, G.; Knoll, L.; Kreckel, H.; Scheffel, M.; Schwalm, D.; Wolf, A.; Zajfman, D. (in preparation).

[45] Vager, Z.; Naaman, R.; Kanter, E.P. *Science*, **1989**, *244*, 426.

[46] Amitay, Z.; Baer, A.; Dahan, M.; Knoll, L.; Lange, M.; Levin, J.; Schneider, I.F.; Schwalm, D.; Suzor-Weiner, A.; Vager, Z.; Wester, R.; Wolf, A.; Zajfman, D. *Science*, **1998**, *281*, 75.

[47] Amitay, Z.; Baer, A.; Dahan, N.; Levin, J.; Vager, Z.; Zajfman, D.; Knoll, L.; Lange, M.; Schwalm, D.; Wester, R.; Wolf, A.; Schneider, I.F.; Suzor-Weiner, A. *Phys. Rev. A*, **1999**, *60*, 3769.

[48] Amitay, Z.; Zajfman, D. *Rev. Sci. Instrum.*, **1997**, *68*, 1387.

[49] Rosén, S.; Peverall, R.; ter Horst, J.; Sundström, G.; Semaniak, J.; Sundqvist, O.; Larsson, M.; de Wilde, M.; van der Zande, W.J. *Hyperfine Interact.*, **1998**, *115*, 201.

[50] Zajfman, D.; Amitay, Z.; Broude, C.; Forck, P.; Seidel, B.; Grieser, M.; Habs, D.; Schwalm, D.; Wolf, A. *Phys. Rev. Lett.*, **1995**, *75*, 814.
[51] van der Zande, W.J.; Semaniak, J.; Zengin, V.; Sundström, G.; Rosén, S.; Strömholm, C.; Datz, S.; Danared, H.; Larsson, M. *Phys. Rev. A*, **1996**, *54*, 5010.
[52] Andersen, L.H.; Johnson, P.J.; Kella, D.; Pedersen, H.B.; Vejby-Christensen, L. *Phys. Rev. A*, **1997**, *55*, 2799.
[53] Zajfman, D.; Amitay, Z.; Lange, M.; Hechtfischer, U.; Knoll, L.; Schwalm, D.; Wester, R.; Wolf, A.; Urbain, X. *Phys. Rev. Lett.*, **1997**, *79*, 1829.
[54] Zong, W.; Dunn, G.H.; Djuric, N.; Larsson, M.; Greene, C.H.; Al-Khalili, A.; Neau, A.; Derkatch, A.; Vikor, L.; Shi, W.; Le Padellec, A.; Rosén, S.; Danared, H.; af Ugglas, M. *Phys. Rev. Lett.*, **1999**, *83*, 951.
[55] Larson, Å.; Djuric, N.; Zong, W.; Greene, C.H.; Orel, A.E.; Al-Khalili, A.; Derkatch, A.M.; Le Padellec, A.; Neau, A.; Rosén, S.; Shi, W.; Vikor, L.; Danared, H.; af Ugglas, M.; Larsson, M.; Dunn, G.H. *Phys. Rev. A*, **2000**, *62*, 042707.
[56] Dixon, R.N.; Hwang, D.W.; Yang, X.F.; Harich, S.; Lin, J.J.; Yang, X. *Science*, **1999**, *285*, 1249.
[57] Zener, C. *Proc. Roy. Soc. (London) A*, **1932**, *137*, 696.
[58] Zajfman, D.; Amitay, Z.; Krohn, S.; Wolf, A.; Schwalm, D.; Wester, R.; Knoll, K.; Lange, M.; Levin, J.; Urbain, X. In *Dissociative Recombination: Theory, Experiment and Applications IV*, Larsson, M.; Mitchell, J.B.A.; Schneider, I.F., Eds.; World Scientific: Singapore, 2000, p 146.
[59] Lange, M.; Levin, J.; Gwinner, G.; Hechtfischer, U.; Knoll, L.; Schwalm, D.; Wester, R.; Wolf, A.; Urbain, X.; Zajfman, D. *Phys. Rev. Lett.*, **1999**, *83*, 4979.
[60] Larsson, M.; Danared, H.; Mowat, J.R.; Sigray, P.; Sundström, G.; Broström, L.; Filevich, A.; Källberg, A.; Mannervik, S.; Rensfelt, K.-G.; Datz, S. *Phys. Rev. Lett.*, **1993**, *70*, 430.
[61] Le Padellec, A.; Larsson, M.; Danared, H.; Larson, Å.; Peterson, J.R.; Rosén, S.; Semaniak, J.; Strömholm, C. *Phys. Scripta*, **1998**, *57*, 215.
[62] Geballe, T.R.; McCall, B.J.; Hinkle, K.H.; Oka, T. *Astrophys. J.*, **1999**, *510*, 251.
[63] Cecchi-Pestellini, C.; Dalgarno, A. *Mon. Not. R. Astron. Soc.*, **2000**, *313*, L6.
[64] Schneider, I.F.; Larsson, M.; Orel, A.E.; Suzor-Weiner, A. In *Dissociative Recombination: Theory, Experiment and Applications IV*, Larsson, M.; Mitchell, J.B.A.; Schneider, I.F., Eds.; World Scientific: Singapore, 2000, p 131.
[65] Michels, H.H.; Hobs, R.H. *Astrophys. J.*, **1984**, *286*, L27.
[66] Guberman, S.L. *Phys. Rev. A*, **1994**, *49*, R4277.
[67] Saroal, B.K.; Tennyson, J.; Morgan, L.A. *J. Phys. B*, **1994**, *27*, 5943.
[68] Tanabe, T.; Katayama, I.; Ono, S.; Chida, K.; Watanabe, T.; Arakaki, Y.; Haruyama, Y.; Saito, M.; Odagiri, T.; Hosono, K.; Noda, K.; Honma, T.; Takagi, H. *J. Phys. B*, **1998**, *31*, L297.
[69] Yousif, F.B.; Mitchell, J.B.A. *Phys. Rev. A*, **1989**, *40*, 4318.
[70] Strömholm, C.; Semaniak, J.; Rosén, S.; Danared, H.; Datz, S.; van der Zande, W.; Larsson, M. *Phys. Rev. A*, **1996**, *54*, 3086.
[71] Cosby, P.C.; Helm, H. *Phys. Rev. Lett.*, **1988**, *61*, 298.
[72] Peterson, J.R.; Devynck, P.; Hertzler, Ch.; Graham, W.G.; *J. Chem. Phys.*, **1992**, *96*, 8128.
[73] Müller, U.; Majer, U.; Reichle, R.; Braun, M. *J. Chem. Phys.*, **1997**, *106*, 7958.
[74] Müller, U.; Cosby, P.C. *Phys. Rev. A*, **1999**, *59*, 3632.
[75] Müller, U.; Eckert, T.; Braun, M.; Helm, H. *Phys. Rev. Lett.*, **1999**, *83*, 2718.
[76] Schneider, I.F.; Orel, A.E. *J. Chem. Phys.*, **1999**, *111*, 5873.
[77] Mistrik, I.; Reichle, R.; Müller, U.; Helm, H.; Jungen, M.; Stephens, J.A. *Phys. Rev. A*, **2000**, *61*, 033410.
[78] Orel, A.E.; Schneider, I.F.; Suzor-Weiner, A. *Phil. Trans. R. Soc. Lond. A*, **2000**, *358*, 2445.
[79] Schneider, I.F.; Orel, A.E.; Suzor-Weiner, A. *Phys. Rev. Lett.*, **2000**, *85*, 3735.

[80] Schneider, I.F.; Orel, A.E.; Suzor-Weiner, A. In *Dissociative Recombination: Theory, Experiment and Applications IV*, Larsson, M.; Mitchell, J.B.A.; Schneider, I.F. Eds.; World Scientific: Singapore, 2000, p 295.
[81] Datz, S.; Sundström, G.; Biedermann, Ch.; Broström, L.; Danared, H.; Mannervik, S.; Mowat, J.R.; Larsson, M. *Phys. Rev. Lett.*, **1995**, *74*, 896.
[82] Tanabe, T.; Katayama, I.; Kamegaya, H.; Chida, K.; Arakaki, Y.; Watanabe, T.; Yoshizawa, M.; Saito, M.; Haruyama, Y.; Hosono, K.; Hatanaka, K.; Honma, T.; Noda, K.; Ohtani, S. In *Dissociative Recombination: Theory, Experiment and Applications III*, Zajfman, D.; Mitchell, J.B.A.; Schwalm, D.; Rowe, B.R. Eds.; World Scientific: Singapore, 1996, p 84.
[83] Strasser, D.; Lammich, L.; Krohn, S.; Lange, M.; Kreckel, H.; Levin, J.; Schwalm, D.; Vager, Z.; Wester, R.; Wolf, A.; Zajfman, D. *Phys. Rev. Lett.*, **2001**, *86*, 779.
[84] Guberman, S.L. In *Dissociative Recombination: Theory, Experiment and Applications*; Mitchell, J.B.A.; Guberman, S.L., Eds.; World Scientific: Singapore, 1989, p 45.
[85] Guberman, S.L. *Nature*, **1987**, *327*, 408.
[86] Bates, D.R. In *Applied Collision Physics*; Massey, H.S.W.; Bates, D.R. Eds.; Academic Press, New York, 1982, p 149.
[87] Frederick, J.E.; Rusch, D.W.; Victor, G.A.; Sharp, W.E.; Hays, P.B.; Brinton, H.C. *J. Geophys. Res.*, **1976**, *81*, 3923.
[88] Hays, P.B.; Sharp, W.E.; *J. Geophys. Res.*, **1973**, *78*, 1153.
[89] Takashi, H.; Clemesha, B.R.; Batista, P.P.; Sahai, Y.; Abdu, M.A.; Muralikrishna, P. *Planet. Space Sci.*, **1990**, *38*, 547.
[90] Bates, D.R. *Planet. Space Sci.*, **1990**, *38*, 889.
[91] Abreau, V.J.; Solomon, S.C.; Sharp, W.E.; Hays, P.B. *J. Geophys. Res.*, **1983**, *88*, 4140.
[92] Sobral, J.H.A.; Takashi, H.; Abdu, M.A.; Muralikrishna, P.; Sahai, Y.; Zamlutti, C.J. *Planet. Space Sci.*, **1992**, *40*, 607.
[93] Bates, D.R. *Planet. Space Sci.*, **1992**, *40*, 893.
[94] Yee, J.-H.; Abreu, V.J.; Colwell, W.B. In *Dissociative Recombination: Theory, Experiment and Applications*; Mitchell, J.B.A.; Guberman, S.L., Eds.; World Scientific: Singapore, 1989, p 286.
[95] Zipf, E.C. *Bull. Am. Phys. Soc.*, **1970**, *15*, 418.
[96] Zipf, E.C. *J. Geophys. Res.*, **1980**, *85*, 4232.
[97] Zipf, E.C. *Plant. Space Sci.*, **1988**, *36*, 621.
[98] Queffelec, J.L.; Rowe, B.R.; Vallée, F.; Gomet, J.C.; Morlais, M. *J. Chem. Phys.*, **1989**, *91*, 5335.
[99] Guberman, S.L.; Giusti-Suzor, A. *J. Chem. Phys.*, **1991**, *95*, 2602.
[100] Kella, D.; Vejby-Christensen, L.; Johnson, P.J.; Pedersen, H.B.; Andersen, L.H. *Science*, **1997**, *276*, 1530; *ibid.*, **1997**, *277*, 167.
[101] Guberman, S.L. *Science*, **1997**, *278*, 1276.
[102] Danared, H. *Hyperfine Int.*, **1998**, *115*, 61.
[103] Peverall, R.; Rosén, S.; Larsson, M.; Peterson, J.R.; Bobbenkamp, R.; Guberman, S.L.; Danared, H.; af Ugglas, M.; Al-Khalili, A.; Maurellis, A.N.; van der Zande, W.J. *Geophys. Res. Lett.*, **2000**, *27*, 481.
[104] Peverall, R.; Rosén, S.; Peterson, J.R.; Larsson, M.; Al-Khalili, A.; Vikor, L.; Semaniak, J.; Bobbenkamp, R.; Le Padellec, A.; Maurellis, A.N.; van der Zande, W.J. *J. Chem. Phys.*, **2001**, *114*, 6679.
[105] Guberman, S.L. In *Dissociative Recombination: Theory, Experiment and Applications IV*; Larsson, M.; Mitchell, J.B.A.; Schneider, I.F., Eds.; World Scientific: Singapore, 2000, p 111.

[106] Wilson, T.L.; Johnston, K.J. (Eds.) *The Structure and Content of Molecular Clouds*; Lecture Notes in Physics; Springer-Verlag: New York, 1994; Thaddeus, P. *IAU Symp.* **1997**, *170*, 3.

[107] Herbst, E. In *Advances in Gas-Phase Ion Chemistry*; Adams, N.G.; Babcock, L.M., Eds.; JAI Press: Greenwich, CT, Vol. 3, p 1.

[108] Herbst, E. *Astrophys. J.*, **1978**, *222*, 508.

[109] Bates, D.R. *Astrophys. J.*, **1986**, *306*, L45.

[110] Adams, N.G. In *Advances in Gas-Phase Ion Chemistry*; Adams, N.G.; Babcock, L.M., Eds.; JAI Press: Greenwich, CT, 1992, Vol. 1, p 271.

[111] Berkner, K.H.; Morgan, T.J.; Pyle, R.V.; Stearns, J.W. In *The Physics of Electronic and Atomic Collisions, 7th Int. Conf.*; Amsterdam: The Netherlands, 1971, p 422.

[112] Forand, J.L.; Mitchell, J.B.A.; McGowan, J.Wm. *J. Phys. E*, **1985**, *18*, 623.

[113] Mitchell, J.B.A.; Forand, J.L.; Ng, C.T.; Levac, D.P.; Mitchell, R.E.; Mul, P.M.; Clayes, W.; Sen, A.; McGowan, J.Wm. *Phys. Rev. Lett.*, **1983**, *51*, 885.

[114] Datz, S.; Larsson, M.; Strömholm, C.; Sundström, G.; Zengin, V.; Danared, H.; Källberg, A.; af Ugglas, M. *Phys. Rev. A*, **1995**, *52*, 2901.

[115] Larsson, M.; Lepp, S.; Dalgarno, A.; Strömholm, C.; Sundström, G.; Zengin, V.; Danared, H.; Källberg, A.; af Ugglas, M.; Datz, S. *Astron. Astrophys.*, **1996**, *309*, L1.

[116] Neau, A.; Al-Khalili, A.; Rosén, S.; Le Padellec, A.; Derkatch, A.M.; Shi, W.; Vikor, L.; Semaniak, J.; Thomas, R.; Någård, M.; Andersson, K.; Danared, H.; af Ugglas, M. *J. Chem. Phys.*, **2000**, *113*, 1762.

[117] Andersen, L.H.; Heber, O.; Kella, D.; Pedersen, H.B.; Vejby-Christensen, L.; Zajfman, D. *Phys. Rev. Lett*., **1996**, *77*, 4891.

[118] Vejby-Christensen, L.; Andersen, L.H.; Heber, O.; Kella, D.; Pedersen, H.B.; Schmidt, H.; Zajfman, D. *Astrophys. J.*, **1997**, *483*, 531.

[119] Jensen, M.J.; Bilodeau, R.C.; Safvan, C.P.; Seiersen, K.; Andersen, L.H.; Pedersen, H.B.; Heber, O. *Astrophys. J.*, **2000**, *543*, 764.

[120] Williams, T.L.; Adams, N.G.; Babcock, L.M.; Herd, C.R.; Geoghegan, M. *Mon. Not. R. Astron. Soc.*, **1996**, *282*, 413.

[121] Adams, N.G.; Babcock, L.M. In *Dissociative Recombination: Theory, Experiment and Applications IV*; Larsson, M.; Mitchell, J.B.A.; Schneider, I.F., Eds.; World Scientific: Singapore, 2000, p 190.

[122] Datz, S.; Rosén, S.; Thomas, R.; Al-Khalili, A.; Derkatch, A.; Shi, W.; Vikor, L.; Larsson, M.; van der Zande, W. In *Dissociative Recombination: Theory, Experiment and Applications IV*; Larsson, M.; Mitchell, J.B.A.; Schneider, I.F., Eds.; World Scientific: Singapore, 2000, p 210.

[123] Datz, S.; Thomas, R.; Rosén, S.; Larsson, M.; Derkatch, A.; Hellberg, F.; van der Zande, W.J. *Phys. Rev. Lett.*, **2000**, *85*, 5555.

[124] Rosén, S.; Derkatch, A.; Semaniak, J.; Neau, A.; Al-Khalili, A.; Le Padellec, A.; Vikor, L.; Thomas, R.; Danared, H.; af Ugglas, M.; Larsson, M. *Faraday Discuss.*, **2000**, *115*, 295.

[125] Tanabe, T.; Takagi, H.; Katayama, I.; Chida, K.; Watanabe, T.; Arakaki, Y.; Haruyama, Y.; Saito, M.; Nomura, I.; Honma, T.; Noda, K.; Hosono, K. *Phys. Rev. Lett.*, **1999**, *83*, 2163.

[126] Krohn, S.; Amitay, Z.; Baer, A.; Zajfman, D.; Lange, M.; Knoll, L.; Levin, J.; Schwalm, D.; Wester, R.; Wolf, A. *Phys. Rev. A*, **2000**, *62*, 032713.

NEUTRAL PRODUCTS FROM GAS PHASE REARRANGEMENTS OF SIMPLE CARBOCATIONS

Thomas Hellman Morton

OUTLINE

Advances in Gas-Phase Ion Chemistry
Volume 4, pages 213–256.

ABSTRACT

Analyzing the neutral products from ionic reactions in the gas phase provides information that cannot be gained by mass spectrometric methods alone. Neutrals have been recovered using three general techniques for generating ions in sufficient quantities: nuclear decay of multiply tritiated precursors, γ-radiolysis studies, and electron bombardment flow (EBFlow) experiments. Analyses of the uncharged reaction products of ion–molecule reactions are most effectively interpreted in conjunction with parallel mass spectrometric investigations. Taken together, these combined studies demonstrate the propensity of gaseous cations to undergo similar sorts of isomerizations as have been reported in condensed media. The absence of solvent and counterions makes it possible to produce ions in the gas phase that cannot be formed in solution. Despite the difference in reaction medium, the same two general categories of rearrangement—ring closure/ring opening and atom/group transfer—account for the variety of ion structures that give rise to the observed neutral products.

I. INTRODUCTION

Cationic rearrangements play a major role in organic and biological chemistry. For example, they account for the enormous variety of carbon skeletons found in natural products. Those rearrangements do not typically take place in bulk solution. Instead they occur within enzyme active sites. The number of enzymes is at least as great as the number of structures they produce, yet it has long been recognized that nearly all cationic rearrangement pathways fall into two categories: ring closure/ring opening and atom/group transfer.[1–3] Both move the electric charge onto a new center.

How can such a limited repertoire of mechanisms lead to the vast range of naturally occurring molecular structures? Biologists are endeavoring to answer that question by genetic manipulation,[4] but the outlines of the answer are already apparent: controlling the microscopic environment of the cation by steric hindrance, positioning of counterions, orientation of permanent dipoles and proximity of induced dipoles determine which of many possible avenues a cation follows. Guiding the rate of capture of the ion and the site at which it occurs must also play an important role in stopping the reaction at just the right point. Implicit in this picture is the notion that most carbocations enjoy many thermally accessible reaction pathways, which are governed by conformation and electric fields on the molecular scale. Understanding what cations do in the absence of solvent and counterions provides a guide to the consequences of these external effects.

From a historical perspective, their relevance to natural products explains why cationic rearrangements have attracted chemists' attention for more than a century. Recent years have seen a focus on increasingly simple structures, with the development of strategies to examine isolated cations. Out of such experiments, a clearer picture is emerging of their intrinsic reactivity, which offers insight as to how environment affects mechanism. NMR studies in non-nucleophilic solvents, and mass spectrometric studies in the gas phase, have greatly expanded our knowledge, but there are many categories of simple ions whose chemistry cannot be explored by these techniques. Some species are inaccessible for NMR studies because their precursors form different positively charged species under highly acidic conditions, or else the cations rearrange to more stable structures too rapidly. Mass spectrometry by itself typically gives limited information regarding ion structures. A variety of approaches have interrogated the neutral products that result from cationic rearrangements, with a view towards examining isomer and (in the case of labeled ions) isotopomeric distributions. The two categories of rearrangement have been explored in this fashion. Sections III–IV of this chapter will focus on carbocations with an even number of electrons. Where appropriate, previously unpublished computational results using density functional calculations (B3LYP/6-31G** with zero point energy correction[5]) from the author's laboratory supplement discussions of experimental results. First, however, it will be useful to clarify some general aspects of branching between competing pathways.

II. THERMODYNAMIC VERSUS KINETIC CONTROL

The first question to be posed in discussing rearrangement chemistry concerns whether a given step exhibits thermodynamic or kinetic control, i.e. whether the distribution of products reflects equilibrium proportions. This issue arises at several stages in a reaction sequence: the production of ions, their isomerizations, and the processes by which ions become neutralized. Ions often form (e.g. by bond cleavage) under conditions that reflect thermodynamic control, while their ultimate neutralization (e.g. gas phase proton transfer) tends to operate with kinetic control.

A. Bond Cleavage: Homolysis, Heterolysis, and Mesolysis

Direct ionization of neutral precursors typically produces ions with an odd number of electrons. Even-electron ions form subsequently, either by bond cleavage or via ion–molecule reactions. With regard to single bond cleavages, the terms homolysis and heterolysis are well understood. When the electron pair in a covalent bond splits between two fragments, the cleavage is a homolysis. When one fragment retains both electrons, the cleavage is a heterolysis. Application of this terminology to ions is, in many cases, straightforward. If the electric charge in the starting material can be viewed as localized on one side of the breaking bond, the position of the charge after cleavage determines whether homolysis or heterolysis has taken place. When the charge stays on the same side, the bond fission is a homolysis. When the charge migrates to the other side, the bond fission is a heterolysis.

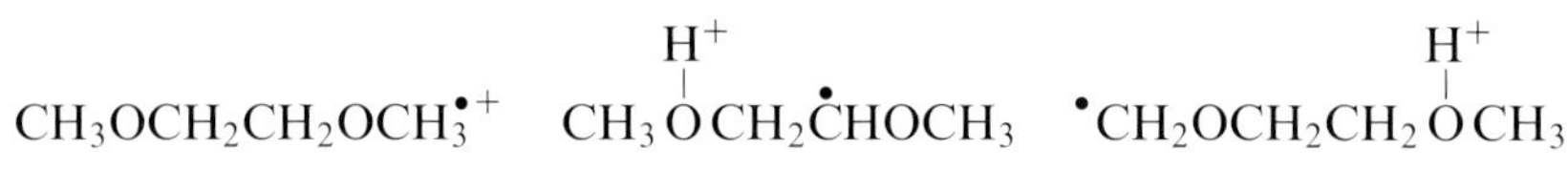

A **B** **C**

A simple example illustrates this distinction for the set of radical cations **A**–**C**. When prepared at low ionizing energies, the molecular ion of 1,2-dimethoxyethane (**A**) decomposes as though it rearranges to a mixture of the two distonic ions **B** and **C** drawn above.[6] In both distonic ions, the positive charge is localized on the protonated methoxy group. Ion **B** loses neutral methanol to yield ionized $CH_2{=}CHOCH_3$. In other words, the charge moves from methanol to the other side of the breaking bond, and the cleavage is a heterolysis.

Ion **C** loses neutral formaldehyde to form the distonic radical ion $^\bullet CH_2CH_2O(H)CH_3^+$. In this case, the positive charge is retained by the fragment that originally carried it, and the cleavage is a homolysis. The cleavage of the carbon–carbon bond of **A** (which represents the base peak in the 70 eV mass spectrum) into $CH_3O{=}CH_2^+$ and $CH_3OCH_2\bullet$ cannot be be classified as homolysis or heterolysis, because the charge is completely delocalized between the two oxygens in the molecular ion. This is an example of what Maslak has termed a mesolysis,[7] which he defines as the cleavage of a radical ion into a radical and an ion. While examples of homolysis and heterolysis have been included under this rubric,[7,8] as well, it would seem to make more sense (at least in the context of gas phase ion chemistry) to restrict mesolysis to describe only those cleavages that cannot be categorized as either homolysis or heterolysis.

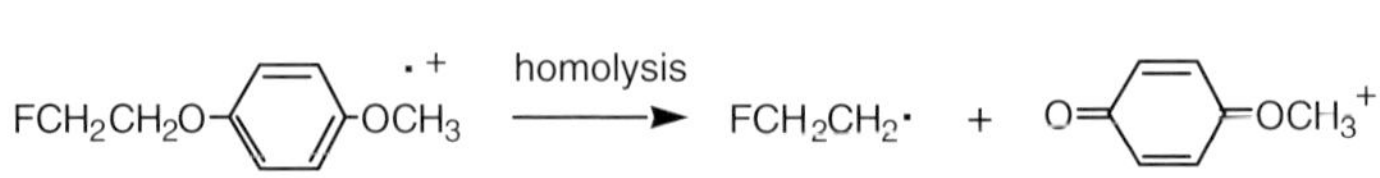

(1)

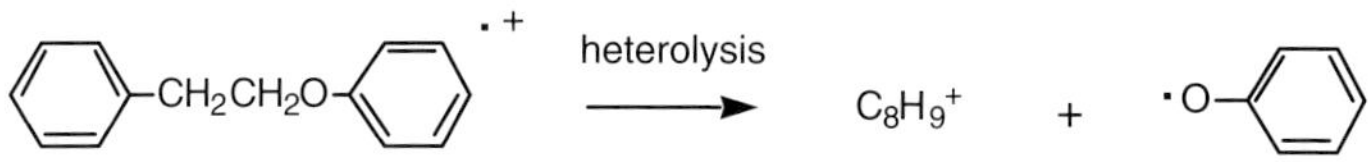

(2)

Now consider instances of homolysis and heterolysis that produce even-electron ions from odd-electron precursors. The cleavages of the two ionized molecular ions in equations 1 and 2 correspond to the most abundant bond fissions in the mass spectra of the corresponding aryl ethers. Each one yields a neutral radical and a cation. Since the charge can be viewed as being localized in the aryloxy group in both starting materials, the fission in equation 1 represents a homolysis, while the one in equation 2 is a heterolysis. Bond cleavages of radical cations (be they homolytic, heterolytic, or mesolytic) tend to exhibit thermodynamic control, in the sense that the charge resides on the most easily ionized fragment. This is sometimes called the Audier–Stevenson rule.[9,10]

B. Reaction Path Degeneracy

When a reaction does not produce an equilibrium distribution of products, it is exhibiting some form of kinetic control. In one limiting case, kinetic control gives relative yields that are completely insensitive to thermodynamic considerations. Under such circumstances, the product ratio reflects the degeneracies of the competing reaction paths. In 1978, Coulson[11] and Pollak and Pechoukas[12] published back-to-back papers, in which they prove a simple algorithm for determining the reaction path degeneracy of a single-step reaction. The algorithm can be formulated as follows:

(*i*) Find the symmetry numbers σ of the reactant and of the transition state;
(*ii*) Multiply σ by 1/2 if the corresponding species is not superimposable on its mirror image (i.e. is chiral, or, to use the terminology of Pollak and Pechoukas,[12] is "optically active") to get a weighted symmetry number;
(*iii*) Divide the weighted symmetry number of the reactant by that of the transition state to get the reaction path degeneracy for the forward direction.

The first step applies rigorously to overall molecular rotations. At temperatures where internal barriers can be neglected (typically <10 kT; e.g. for most single-bond torsions at room temperature), the algorithm can be extended to include internal rotations. A simple example illustrates the use of the algorithm: the unimolecular dissociation of ionized isopropyl

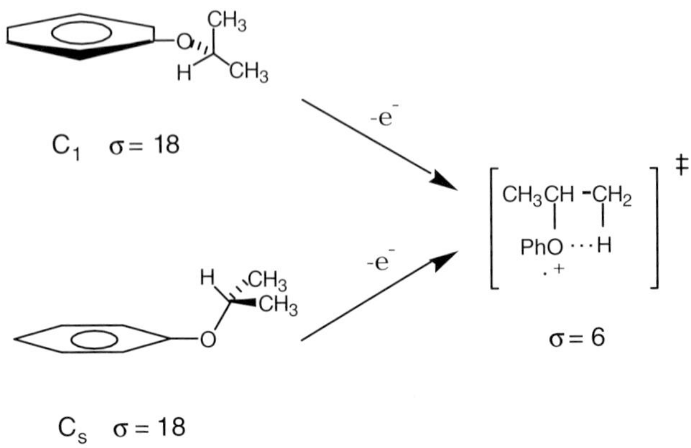

Scheme 1.

phenyl ether into ionized phenol and neutral propene. Ab initio calculations indicate that the neutral ether has two stable conformations, one with a plane of symmetry (C_s) and one without (C_1), as Scheme 1 illustrates. Laser REMPI spectroscopy of a supersonically cooled jet confirms the ab initio prediction that the C_1 is lower in energy.[13]

A transition state for concerted 4-center elimination from the molecule is depicted to the right in Scheme 1,[14] and we consider here the reaction path degeneracy for each conformation as it passes through this transition state. Internal rotation about the phenyl–oxygen bond is considered explicitly as a 2-fold free rotor in both conformations of the reactant as well as in the transition state. The C_1 and C_s reactants each possess, in addition, a pair of methyl rotors, for net symmetry numbers $\sigma = 18$. The transition state retains only one of these methyl rotors, for a net symmetry number $\sigma = 6$. The transition state and the C_1 reactant are both chiral, and the corresponding reaction path degeneracy is 3. The C_s reactant is achiral, so the reaction path degeneracy for this conformation is 6.

In other words, elimination from the more symmetrical structure is favored by naive statistics (even though this conformation happens to be energetically less accessible). The number of energetically equivalent pathways (as would be obtained by enumeration of equivalent hydrogens) is the same as the reaction path degeneracy: the methyl groups in the C_s conformation are rendered equivalent by the plane of symmetry, while in the C_1 conformation the methyl groups are distinguishable. While the reaction path degeneracy is not always necessarily the same as the number of chemically equivalent atoms,[12] it turns out that few of the examples where the two methods disagree are chemically realistic.

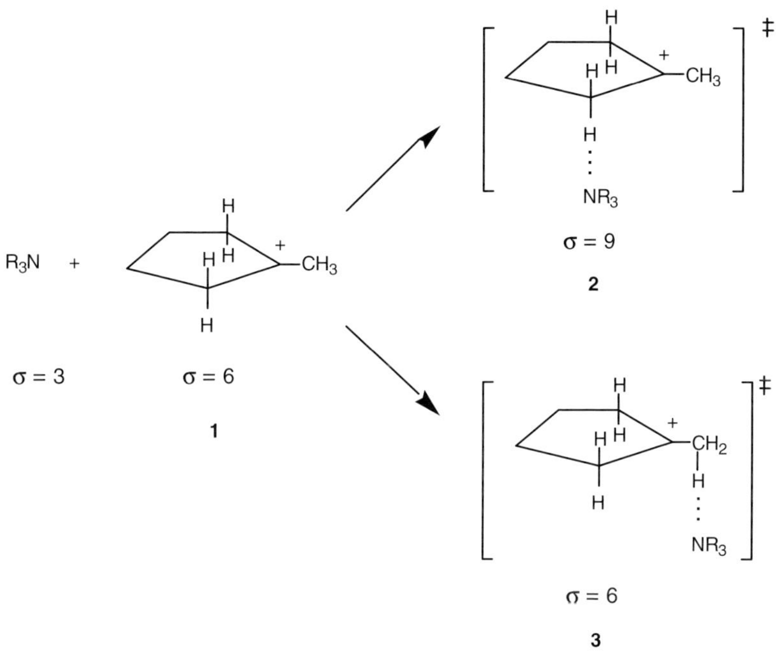

Scheme 2.

We now turn to the competition between two transition states with different symmetry numbers. Consider the reaction of the 1-methylcyclopentyl cation, **1**, with a symmetrically substituted amine R_3N, drawn in Scheme 2. Cation **1** is produced in both solution and the gas phase via unimolecular rearrangement of the cyclohexyl cation.[15] At room temperature, the gaseous cyclohexyl cation has a lifetime on the order of 1 ms before undergoing this exothermic isomerization.[16,17]

Two alternative types of intermediate (which we call the Eigen model and the Lewis model[18,19]), could describe the mechanism for the Brønsted acid–base reaction. The Eigen model supposes strong hydrogen-bonding between the acidic cation and the neutral base prior to proton transfer. The Lewis model supposes a covalent Lewis acid–base complex. In contrasting the predictions of these two models, we neglect internal rotation about N–R or within the R-groups, since that contribution does not change in going from reactant to transition state. Carbocation **1** has a planar skeleton and C_s symmetry, with $\sigma = 6$ since it possesses both an internal 2-fold rotor (the 5-member ring) and a methyl rotor, connected by an sp^2–sp^3 carbon–carbon single bond. The R_3N molecule has an overall C_3-axis of symmetry and $\sigma = 3$. Thus the net symmetry number of the reactants is 18.

The Eigen model requires two different hydrogen-bonded intermediates corresponding to the two alkene products that are formed. Each intermediate leads to a different transition state, structures **2** and **3** drawn in Scheme 2. The former, **2**, leads ultimately to 1-methylcyclopentene as the final product. This transition state has no overall molecular symmetry, but retains the methyl rotor as well as free rotation about the carbon–nitrogen single bond for a net symmetry number $\sigma = 9$. Because transition state **2** is chiral, the reaction path degeneracy is equal to $(18/\frac{9}{2}) = 4$.

The latter transition state, **3**, leads to methylenecyclopentane as the final product. This transition state retains C_s-symmetry, the 2-fold internal rotor, and 3-fold internal rotation about the carbon–nitrogen bond, for a net symmetry number $\sigma = 6$. Since neither the reactant nor this transition state is chiral, the reaction path degeneracy is $(18/6) = 3$.

Let us compare the ratio of reaction path degeneracies in Scheme 2 (4/3) to the ratio expected for net proton transfer via the Lewis model, which proceeds via unimolecular decomposition of a single intermediate, the quarternary ammonium ion **4** in Scheme 3. The reactive intermediate has C_s-symmetry and two internal 3-fold rotors, for $\sigma = 9$. Ion **4** can dissociate via two competing transition states, **5** and **6**, both of which

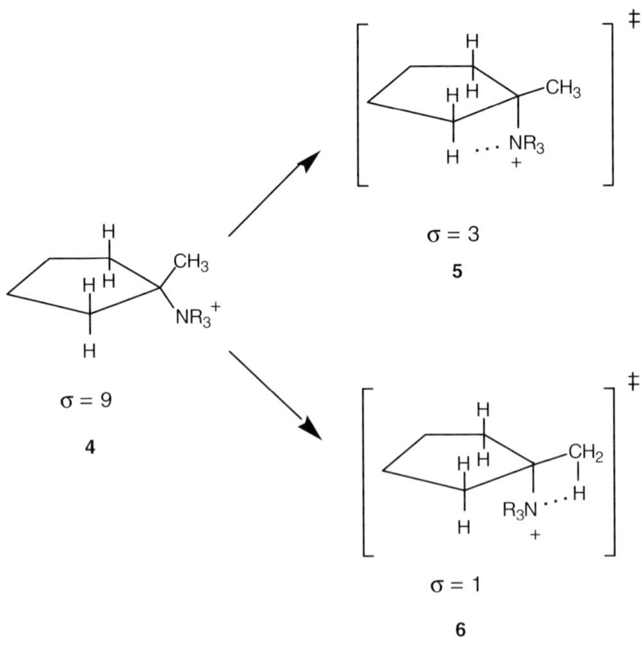

Scheme 3.

freeze internal rotation about the carbon–nitrogen single bond. Transition state **5** (which leads to 1-methylcyclopentene) retains its methyl rotor ($\sigma = 3$) but is chiral. Hence the reaction path degeneracy is 6. In terms of statistical factors this corresponds to the two chemically equivalent hydrogens *cis* to the nitrogen, multiplied by three equivalent orientations of the NR_3.

Transition state **6** (which leads to methylenecyclopentane) retains the plane of symmetry, but has no internal rotors. Therefore the reaction path degeneracy is 9. Thus, the predicted ratio of reaction path degeneracies is $6/9 = 2/3$, a factor of two different from that determined for Scheme 2. Repeated measurements of the 1-methylcyclopentene/methylenecyclopentane product ratio from the gas phase reaction of **1** with tertiary amines[15,19] (where energy barriers are negligible) give branching ratios within experimental error of 4/3 (e.g. 1.34 ± 0.02 for R = methyl[19]), which has been taken to argue in favor of the Eigen model rather than the Lewis model. With weaker bases, the product distribution shifts in the direction of the thermodynamically favored product.[18,19]

C. Distinguishing Transposition from Randomization

In principle, isotopic labeling ought to permit the neutral products of a cationic reaction to be deduced from mass spectrometric measurements alone. As an example, the elimination reaction drawn in Scheme 1 yields a mixture of *cis* and *trans* alkenes when the alkyl group has >3 carbons. If the reaction passes through transition state drawn in Scheme 1 (called a *syn*-transition state), the ratio of alkene isomers from the stereospecifically deuterated precursors in Scheme 4 should correspond to the ratio of $PhOH^{\bullet+}$ to $PhOD^{\bullet+}$ observed in the mass spectrum.[20]

Two caveats apply to such an inference. First, the deuterium label affects the outcome of the reaction, and its kinetic isotope effect must be taken into account. Second, many cations tend to scramble their atoms (although the radical ions in Scheme 4 do not do so). One must therefore survey additional isotopically labeled analogues in order to ensure that scrambling is not taking place.[21]

Deuterium scrambling can arise from different types of exchange, as Scheme 5 portrays for an idealized molecule containing a dideuterated and an undeuterated methylene group. Positional interchange transposes the labeled and unlabeled carbon atoms (in general, by passing through a structure where, in the absence of label, they would be chemically equivalent). This implies that carbon–deuterium bonds do not break, and the net result is illustrated at the top of Scheme 5. If a carbon happened to be isotopically labeled as well, the hydrogens originally attached to it would

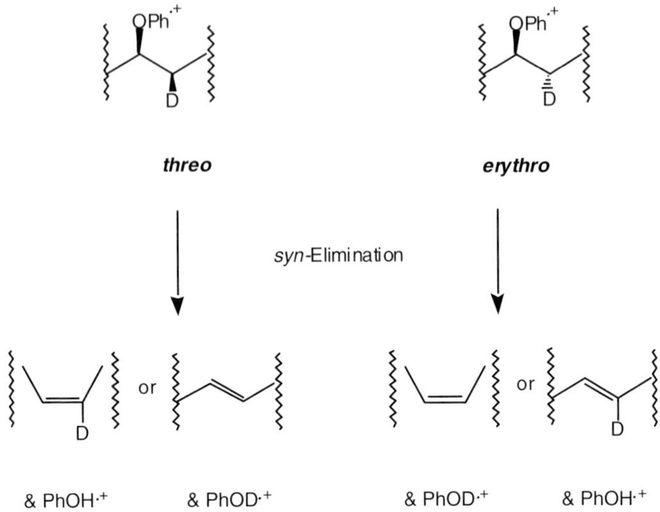

Scheme 4.

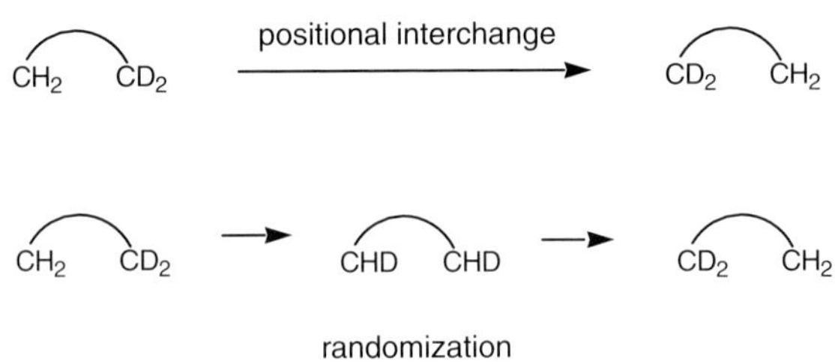

Scheme 5.

be observed not to have shifted. Randomization, by contrast, designates a process in which hydrogens move back and forth, leading to the outcome shown at the bottom of Scheme 5. The two types of exchange can be differentiated by the fact that species with CHD groups form during randomization, but not during positional interchange.

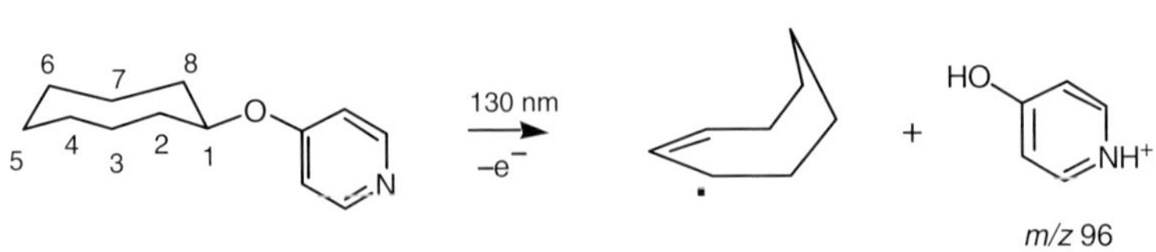

(3)

Mass spectrometry by itself cannot usually tell what positional isomers are present in a mixture. In favorable cases, however, it is possible to differentiate between positional interchange and randomization by analyzing product ion distributions. The double hydrogen transfer in equation 3 represents a case in point. Photoionization of cyclooctyl 4-pyridyl ether at 9.5 eV gives a high yield (approximately half of the total ionization) of protonated 4-hydroxypyridine (m/z 96), along with a neutral product that is presumed to be cyclooctenyl radical (for which the most favorable conformation is drawn). The calculated (B3LYP/6-31G**) appearance energy for this dissociation is 7.5 eV, far below that calculated for production of free cyclooctyl cation (AE = 9.3 eV). On one hand, deuteration of the cyclooctyl ring at positions 1 or 5 leads to a very low level of deuterium incorporation into the ion. On the other hand, perdeuteration at positions 2 and 8 or at positions 4 and 6 leads to a substantial increase in the deuterium content of the ion, and the d_0 (m/z 96) : d_1 (m/z 97) : d_2 (m/z 98) ion intensity ratios are nearly the same for both tetradeuterated starting materials.[22] Does this represent positional interchange or randomization?

Scheme 6 depicts a way to test whether randomization takes place, assuming that the double hydrogen transfer occurs in two steps. Suppose that randomization of m hydrogens with n deuteria takes place in the parent molecular ion, $M^{\bullet+}$, prior to hydrogen transfer. The first step can transfer either an H (to give intermediate **I**) or a D (to give intermediate **II**). If the ratio of **I** to **II** does not vary with time and if the second step obeys first-order kinetics, then the kinetic isotope effect k_H/k_D can be extracted from the experimental ion intensity ratios $[d_0/d_2]$ and $[d_1/d_2]$ by means of the expression in equation 4:

$$k_H/k_D = \frac{n-1}{2n}\left\{[d_1/d_2] + \left([d_1/d_2]^2 - \frac{4mn[d_0/d_2]}{(n-1)/(m-1)}\right)^{1/2}\right\} \tag{4}$$

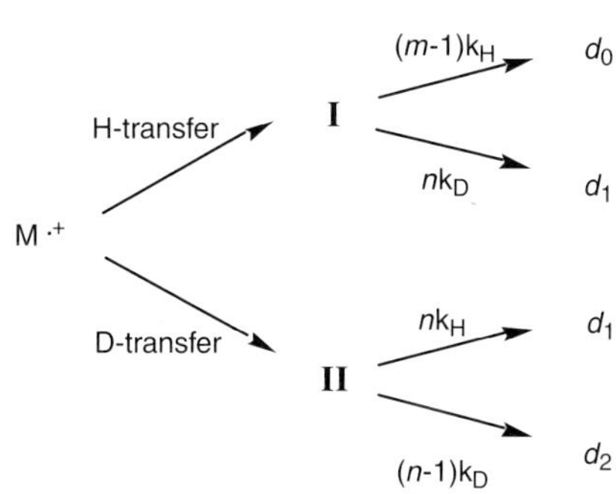

Scheme 6.

It turns out that all positive definite choices for m and n give imaginary values for k_H/k_D when the experimental values of the ion intensity ratios are substituted into the equation.[22] In other words, simple randomization cannot account for the fact that the two tetradeuterated isomers give the observed proportions of d_0, d_1, and d_2 fragment ions.

As is well known, kinetic arguments can be used to disprove mechanistic hypotheses (as in this instance), but never to prove them. The validity of first-order kinetics has been experimentally verified for monoenergetic ions,[23] but the general applicability to ions with a distribution of internal energies has been questioned.[24] Conditions under which first-order kinetics represent a usable approximation have been discussed,[25] and kinetic isotope effects (having positive definite values) have been extracted from experimental data for a number of unimolecular decompositions of ions produced by photoionization.[14,26–28] Our interpretation of the dissociation in Scheme 6 in terms of positional interchange is treated in the next section.

D. Formation of Ion–Neutral Complexes versus Simple Bond Fission

An ion–neutral complex is defined as an aggregate of an ion with one or more neutrals, in which at least one of the partners has a rotational degree of freedom about an axis perpendicular to the intermolecular direction. This definition is known as the "reorientation criterion",[30] and the characteristic degree of freedom within a complex has sometimes been called "planetary motion".[31] Ion–neutral complexes can form either as a result of a bimolecular reaction or from unimolecular dissociation. Theory suggests that some closed shell clusters, where one of the partners has tetrahedral symmetry (such as $P_4 \cdot Li^+$[31] or the proton-bound dimer of ammonia and fluoromethane,[32] which have been observed in the mass spectrometer), constitute stable ion–neutral complexes. Since fluorine does not tend to form strong hydrogen bonds, the ammonium ion within species such as $NH_4^+ \ldots FH$[33] or $NH_4^+ \ldots FCH_3$[32] at room temperature ought to enjoy virtually free rotation in three dimensions. Directed valence (such as operates in conventional hydrogen bonds) is therefore largely absent, and the bonding must be viewed as mostly ionic in character.

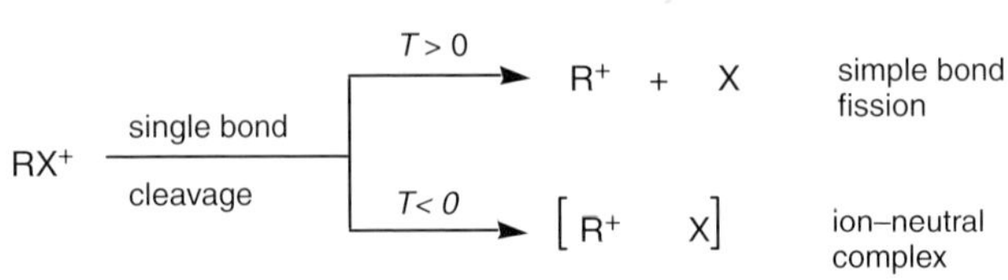

(5)

Short-lived ion–neutral complexes are created unimolecularly when bonds sever in parent ions and the fragments do not have enough kinetic energy to completely escape the electrostatic attraction between them. This can be thought of as a negative translational kinetic energy release ($T < 0$), as equation 5 represents schematically for cleavage of an R–X single bond. In a number of cases, simple bond fission (positive translational kinetic energy release, $T > 0$) and formation of ion–neutral complexes compete with one another. Transient ion–neutral complexes reveal their existence when they undergo an ion–molecule reaction between the partners faster than they collapse back to a covalently bound structure. Isotopic scrambling within the ionic partner frequently displays aspects characteristic of a free ion, but the lifetime of the complex is usually too short to permit skeletal rearrangement.[34] In the case of equation 3, photoionization was performed at a long enough wavelength that simple bond fission cannot compete effectively. Instead, cyclooctyl cations are formed in ion–neutral complexes with 4-pyridyloxy radicals.

Were it possible to look at the cyclooctenyl radicals from equation 3, the neutrals produced by the tetradeuterated precursors would be expected to contain sp^3 carbons with CH_2 or CD_2, but no CHD groups. It has not yet proven possible to examine those radicals directly, but this deduction arises from the arguments outlined in the previous section. The positional interchange agrees with what would be predicted for the [cyclooctyl cation $^{\bullet}OC_5H_4N$] ion–neutral complex to the left in Scheme 7, whose subsequent Brønsted acid–base reactions give the ion–neutral complexes containing the forms of cyclooctene that are drawn in brackets. The cyclooctyl cation is known (from solution phase NMR studies[35]) to have a bridged structure,

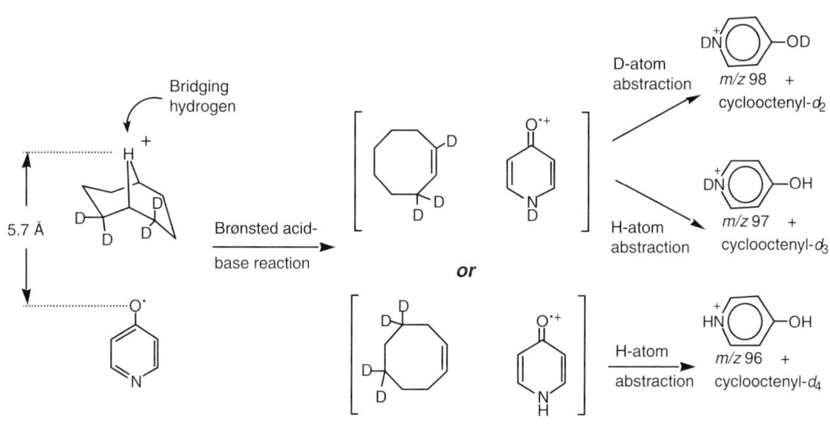

Scheme 7.

which confers a plane of symmetry upon the carbon skeleton that does not exist in monosubstituted cyclooctanes.

The observed proportions of d_0 (m/z 96), d_1 (m/z 97), and d_2 (m/z 98) ions result from a normal kinetic isotope effect on the order of 3.5 for the Brønsted acid–base reaction followed by a k_H/k_D of 2.8 for the subsequent atom abstraction. Cyclooctyl cations acquire symmetry due to the bridging hydrogen, as depicted, which accounts for the fact that both of the tetradeuterated precursors give the same product ion intensity ratios. After the Brønsted acid–base step, however, the cyclooctene-containing complexes are discrete and non-interconverting, which accounts for why the randomization model (equation 3) fails for this reaction.

Since an ion–neutral complex does not exhibit directed valence between the partners, it cannot be said to have an equilibrium structure. Some discussion of the interaction seems warranted at this point. The simplest physical picture for an ion–neutral complex portrays it in terms of a point charge and a point dipole. Density functional (DFT) calculations have tested this approximation for the [cyclooctyl cation $^{\bullet}OC_5H_4N$] complex. The permanent dipole moment of a 4-pyridyloxy radical in its lowest electronic state is calculated to be 1.1 Debye, and the calculated exact polarizability in the C–O bond direction is $\alpha_{zz} = 11.4$ Å^3 (at the B3LYP/6-31G** level). The geometry shown in Scheme 7 (which is constrained to have a plane of symmetry and corresponds to a center-of-charge to center-of-charge distance of 6.8 Å) was chosen at a separation where the partners lie outside of one another's steric radii. The classical point charge–point induced dipole attraction at that distance is 3.7 kJ/mol, while the point charge–permanent dipole attraction is 6.8 kJ/mol. The B3LYP/6-31G** calculated attraction (electronic energy difference including counterpoise corrections for basis set superposition error) is 22 kJ/mol. Rotating the 4-pyridyloxy radical so that its permanent dipole repels the positive charge (making nitrogen the atom closest to the cation and keeping the center-of-charge to center-of-charge distance at 6.8 Å) is calculated to cost 14 kJ/mol at the B3LYP/6-31G** level, almost exactly equal to the classical prediction. We conclude that a classical model does not give a bad picture for the charge–permanent dipole interaction, but that the other contributions are greatly underestimated by a point charge–point induced dipole.

If a [cyclooctyl cation $^{\bullet}OC_5H_4N$] complex is allowed to collapse (subject to the constraint of C_s symmetry) with the orientation shown in Scheme 7, the calculated distance between the bridging hydrogen and the oxygen shrinks to 4.5 Å and the bridging hydrogen–O–N angle is 172°. The classical point charge–point dipole attraction is 9.7 kJ/mol and the total B3LYP/6-31G** attractive potential is 40 kJ/mol. This structure has the highest energy of the four stable C_s geometries having a bridged cyclooctyl cation (all of which exhibit one imaginary vibrational frequency and will revert to the covalent radical cation if the symmetry constraint is removed). The zero point energy of the complex is 2 kJ/mol greater than that of the separated ion and neutral. Rotating the pyridyloxy ring 90° about the C–O bond (giving another structure with C_s symmetry) permits the oxygen to approach 0.1 Å closer to the cyclooctyl cation, and the bridging hydrogen–O–N angle decreases slightly to 167°. The DFT-calculated attraction increases by 1 kJ/mol. If, instead, the cyclooctyl ring is flipped over (still maintaining C_s symmetry) so that the bridging hydrogen is closest to the oxygen, the center-of-charge to center-of-charge distance between the two partners does not change, and the bridging hydrogen–O–N angle decreases to 164°. The DFT-calculated attraction increases to 46 kJ/mol. This structure has the lowest energy of the C_s geometries.

As Scheme 7 depicts, the two hydrogens that are transferred end up on opposite ends of the pyridyloxy. This signifies that the ring must have rotated 180° in the course of the ion decomposition. Thus the experiment demonstrates that the reorientation criterion[30] has been fulfilled. The majority of cyclooctyl cations are formed in ion–neutral complexes because the photon energy is not high enough to yield free cations efficiently. Other mass spectrometric studies show that bond cleavages often favor the formation of carbocations within ion–neutral complexes, even when the internal energy greatly exceeds the threshold for producing free cations.[26]

Ion–neutral complexes have been called "microscopic reaction 'vessels'" for gas-phase ion chemistry.[36] They intervene in both unimolecular and bimolecular reactions and behave like gas-phase analogues of cage effects in solution.[37,38] Unimolecular heterolyses operate as gas-phase solvolyses to form complexes,[38,39] where the ion–neutral complex plays the same role as does an ion pair in solution. The lifetimes of ions within complexes are brief, and rearrangements take place in free ions that occur too slowly to be detected from the neutral products of complexes.

III. PRODUCTS FROM TRITIUM DECAY

Nuclear decay transmutes one element into another. β-decay (the ejection of an electron from the nucleus) increases atomic number by 1 and imparts a positive charge to the remnant. This has proved a useful technique for creating gaseous cations under conditions where the neutral products of their ion–molecule reactions can be collected for analysis.[40–42] When the radionuclide is tritium, the net transformation is $^3H \rightarrow {}^3He^+$. When there is more than one tritium in the starting material, the cation remains radioactive after one of the nuclei decays. Its radioactive neutral products can then be identified.

$$ {}^{1}n_{0} \rightarrow {}^{1}\mathbf{p}_{1} + [W^{-}] \tag{6} $$

$$ [W^{-}] \rightarrow \bar{\nu}_{e} + \beta^{-} \tag{7} $$

β-decay is a three-body process, in which a neutron transforms to a proton. It can be viewed as a pair of successive two-body decays, summarized in equations 6 and 7. The first step (equation 6) is the decomposition to a pair of heavy particles, one of which is a virtual, negatively charged intermediate vector boson W^{-}. This virtual particle has a lifetime $<10^{-25}$ s; hence, its mass can be very much larger than that of the original neutron. Decay of the virtual particle into an electron (the β-particle) and an antineutrino (equation 7), as a second step, means that the angular distribution of the nuclear recoil is virtually independent of the direction of the β-particle. In other words, the momentum distribution of the antineutrino $\bar{\nu}_e$ is essentially isotropic (if one neglects spin correlation effects, which are small). The average recoil energy of an ^{3}He nucleus produced by β-decay of tritium is <1 eV, while the β-particle itself has an average kinetic energy on the order of 12 keV.[42]

$$ \text{R–T} \rightarrow \text{R–He}^{+} \rightarrow \text{R}^{+} + \text{He} \tag{8} $$

β-decay converts a tritiated organic molecule, R–T, into a cation, as equation 8 illustrates. Some (but not all) of the molecular fragments attached to a disintegrating atom become internally excited following nuclear decay. Three general mechanisms dominate: (***a***) an ejected particle can collide with the rest of the molecule, possibly forming a superexcited state; (***b***) the molecular fragment can be left behind in a deformed geometry, which relaxes to an equilibrium structure containing a high degree of vibrational excitation; or (***c***) the molecular fragment is not left as equation 8 portrays, but rather as a multiply charged cation or as two or more smaller fragments. Mass spectrometric studies of the ions produced by tritiated organic molecules suggest the last option can often be neglected.[41,42] The first mechanism can produce ions that contain >2 eV internal energy, but the second mechanism produces ions containing <2 eV of internal energy.

Consider excitation mechanism (***a***) above. The arrows in Figure 1 represent a subset of the trajectories of ejected particles with their origin at a tritium nucleus attached to an sp^3 carbon. From the vantage point at the tritium nucleus, the rest of the molecule subtends a region of space easily contained within a 120° cone, which is represented by dashed lines. The odds

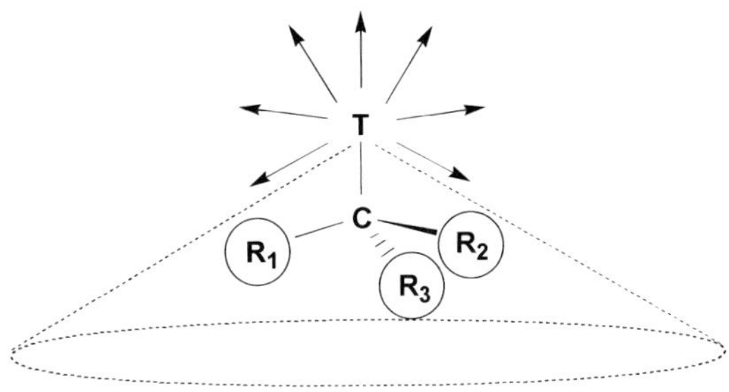

Figure 1. Schematic representation of the directions of β-emission and 3He loss that do not lead to collision of ejected particles with the remaining carbocation (represented by arrows), as compared with the volume of space (represented by the dashed 120° cone) where such collisions are possible.

against the β-particle exiting within the cone are 3 : 1. Therefore the odds against the ^{3}He atom exiting within the cone are also 3 : 1. We take the sum of the two probabilities as an upper bound for the net probability that one of the ejected particles collides with the carbocation from which it is departing. In other words, $\geq 50\%$ of the carbocations are formed without being struck by an ejected particle.

Now consider excitation pathway (***b***), vibrational excitation of carbocations that do not collide with a β-particle or a ^{3}He atom. Of the total kinetic energy liberated by the nuclear decay (18.61 keV), an average of one-third is carried off by the neutrino (which does not interact with ordinary matter). Conservation of momentum dictates that, of the remaining kinetic energy, $<0.02\%$ is imparted to the ^{3}He nucleus. The helium atom ought to carry off most of this recoil as translational kinetic energy when it detaches. The tritiated carbocation may be vibrationally excited, largely as a consequence of geometrical reorganization during departure of the helium.

β-emission takes place on the 10^{-18} s timescale. Therefore ^{3}He forms in association with the sp^3 carbon to which tritium was attached. At this carbon–helium distance, r_o, the helium atom is repelled by the cationized carbon (even if it remains tetrahedral), since the equilibrium C–He distance is much longer than a C–H bond. If a nascent carbocation undergoes no geometrical relaxation, SCF calculations (including BSSE) give a repulsive potential of 25–30 kJ mol^{-1} for expelling the helium atom. If the primary sp^2-cationic center is allowed to planarize, SCF calculations place the energy of the nascent cation on the order of 160 kJ mol^{-1} above the final state. No potential energy barrier intervenes between the nascent cation and the final geometry, which has an SCF electronic energy only 0.5 kJ below that of a free carbocation plus a helium atom. It is not clear whether the local minimum corresponds to a bound geometry, since the calculated zero point energy difference is comparable to the well depth.

Rehybridization should eject the helium atom as though from a slingshot, with coupling between the relaxation of the cation geometry and translation of the helium atom. It seems unlikely that all of the deformation energy is left in the carbocation when the ^3He departs. The atom is accelerated to a velocity v over a distance r by the repulsive potential. If nuclear decay were to create an initially stationary helium atom, most of the relaxation energy would be carried away by the ^3He as translational kinetic energy, T_{He}, as depicted by the lower curve in Figure 2. In that case, nearly all the potential energy would be converted to translational motion of the helium atom, and the atom could be accelerated to a final velocity as high as 13.5 $km\,s^{-1}$ relative to the cation. However, nuclear decay forms ^3He with a high initial velocity, v_o. If the maximum possible recoil energy were wholly converted into v_o of the helium atom (3.4 eV), the ^3He would have an initial velocity $v_o \approx 15$ $km\,s^{-1}$ in the center-of-mass frame. But because the neutrino typically carries off about one-third of the decay energy, the average tritium decay imparts an initial velocity on the order of $v_o = 10$ $km\,s^{-1}$. As v_o and v have comparable magnitudes,

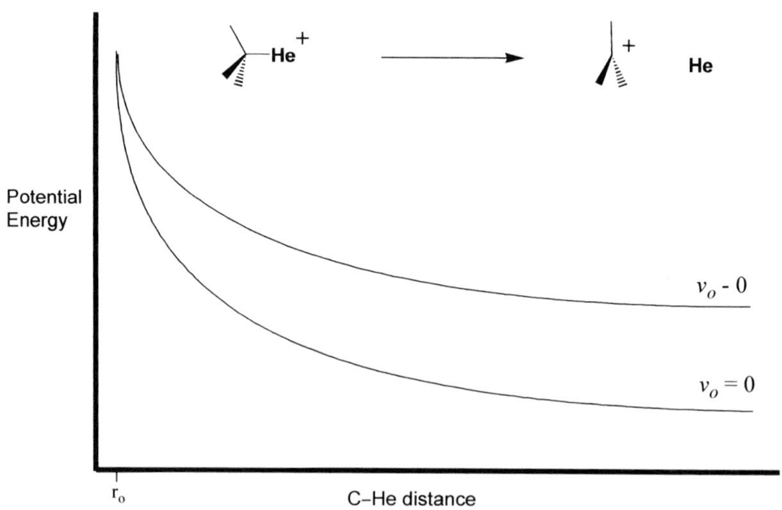

Figure 2. Schematic picture of the conversion of strain energy into kinetic energy as a function of distance for an ejected He atom starting at rest at position r_o (with velocity $v_o = 0$) versus one starting with initial velocity v_o.

the helium atom cannot be said to escape instantaneously from the carbocation.

$$T_{\mathrm{He}} = \int_{r_o}^{\infty} \frac{v}{v_o + v}\left(\frac{\mathrm{d}E}{\mathrm{d}r}\right)\mathrm{d}r \tag{9}$$

If v_o were very much larger than v, the helium atom would be gone before the cation had time to rearrange at all. Nuclear decay of tritium bound to an sp^3 carbon represents a case intermediate between $v_o = 0$ and $v_o \gg v$. The helium atom is moving fast to begin with, so its outward motion is not strongly coupled to the geometrical relaxation of the cation. However, the shape of the repulsive potential curve does not exhibit a strong dependence on the direction the helium is travelling (so long as it is not within the dashed cone drawn in Figure 1). The force exerted on the helium atom, $\mathrm{d}E/\mathrm{d}r$, should exhibit the same dependence on the C–^{3}He distance, $r_o + r$, regardless of which of the solid arrows in Figure 1 comes closest to the direction of v_o. We derive equation 9 as a classical approximation for the relaxation energy carried off as T_{He} when $v_o \neq 0$ (in the limit where $T_{\mathrm{He}} = \Delta E$ when $v_o = 0$). Evaluation of SCF potential energy curves gives a value of T_{He} on the order of 110 kJ mol^{-1} for a tritiated methyl group. In other words, the initial velocity v_o of the ^{3}He is fast enough to have gone before the cation has completely relaxed, carrying off only about 40% of the available relaxation energy. The partially relaxed cation structure that remains behind has a nearly planarized cation center.

Many experiments have been performed using tritium-containing ions derived from β-decay of multiply tritiated hydrocarbons.[12] Nuclear decay of tritiated methyl groups produces primary carbocations, which (unless they are conjugated with a double bond or cyclopropane ring) do not correspond to stable structures on their respective potential energy surfaces. Tritiated cations from *n*-alkanes rapidly transpose hydrogens to form secondary carbocations and, on a much slower timescale, undergo skeletal rearrangements to tertiary carbocations. The radiolabeled cations are captured by gaseous nucleophiles. The proportion of neutral products from secondary cations increases (and the proportion of tertiary cation products decreases) as the partial pressure of the nucleophile is raised.[42] A few percent of ostensible primary cation product has also been reported, probably from valence isomerization of primary cations to corner-protonated cyclopropanes.

Pertritiated methane (produced by the reaction of aluminum carbide with T_2O) yields tritiated methyl cations (symbolized as CT_3^+), whose gas phase chemistry has been extensively studied by Nefedov's group in Russia. One of their first major contributions was to provide evidence for methyl migration around protonated benzene rings, as summarized in Table 1.[43] In mixtures of xylenes (at their vapor pressure) and pertritiated methane,

Table 1. Radiochemical vields of Trimethylbenzenes from Reaction of Xylenes with CT_4. Numbers in Parentheses Indicate Statistical Proportions

		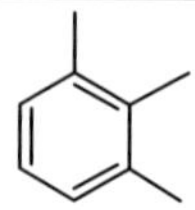	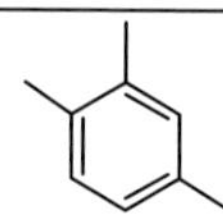	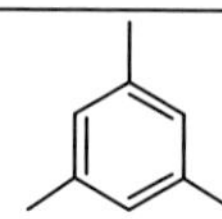
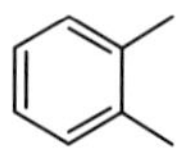	$\xrightarrow{CT_3^+}$	0.11 (1/2)	0.24 (1/2)	0.16 (0)
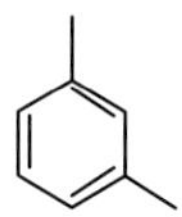	$\xrightarrow{CT_3^+}$	0.12 (1/4)	0.26 (1/2)	0.20 (1/4)
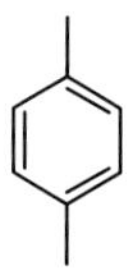	$\xrightarrow{CT_3^+}$	0.07 (0)	0.49 (1)	0.13 (0)

all three trimethylbenzenes are recovered as radiolabeled neutral products. In Table 1, the fractions in parentheses represent the statistical proportions of the methylation products that would be expected if methyl groups could not transpose (calculated using the algorithm outlined in Section II.B above). As is apparent from the experimental yields, the different xylene isomers afford different proportions of products, signifying that complete equilibration does not take place during the lifetime of the ion. These experimental results present one of the earliest demonstrations of orbital symmetry-allowed suprafacial sigmatropic shifts of methyl groups.

Another test of orbital symmetry comes from the reaction of methyl cations with ketones. ICR experiments have shown at least 10 different pathways for reaction of CD_3^+ with acetone. One of the minor pathways produces $C_3H_5^+$ cation.[44] Thermodynamically it is plausible that the neutral products are formaldehyde and molecular hydrogen, as equation 10 depicts for the orbital-symmetry allowed decomposition of the

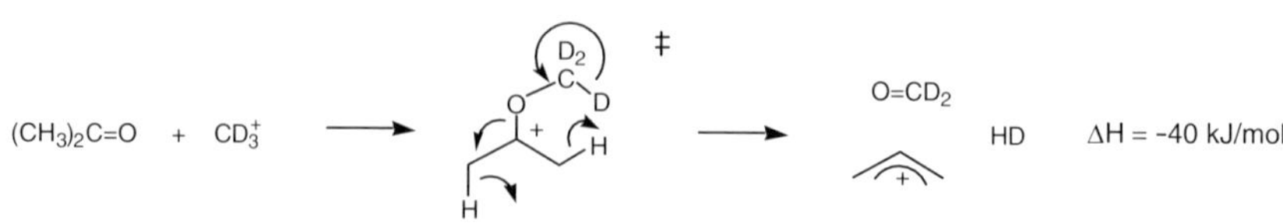

(10)

vibrationally excited adduct ion (the overall ion–molecule reaction would still be about 20 kJ/mol exothermic if the product ion has the $CH_3C{=}CH_2^+$ structure instead of the more stable allyl cation). However, a systematic study of the neutral products from CT_3^+ on a variety of gaseous ketones (analyzed by gas chromatography with a radioactivity detector) showed that the major tritiated neutral product is methanol rather than formaldehyde.[45]

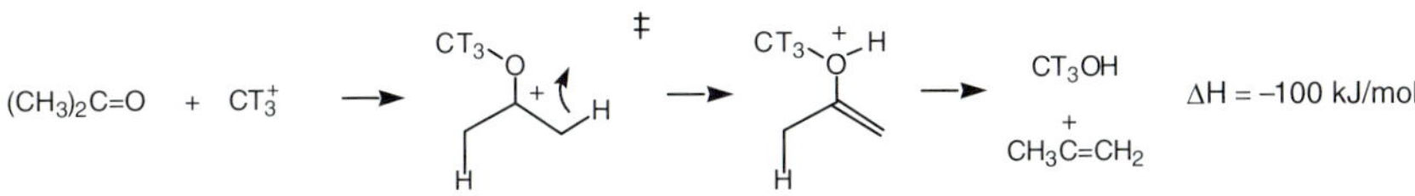

(11)

Assuming that the principal product is CT_3OH, Speranza has proposed the isomerization of the vibrationally excited adduct ion via a concerted, unimolecular 1,3-hydrogen shift, as equation 11 portrays.[42] Orbital symmetry forbids this as a suprafacial isomerization. The question has lately been posed, whether planar, even-electron gaseous ions of this sort rearrange antarafacially.[16] If so, the molecular skeleton ought to retain its plane of symmetry in the transition state, and the migrating hydrogen should pass through that plane, as drawn to the left in Scheme 8.

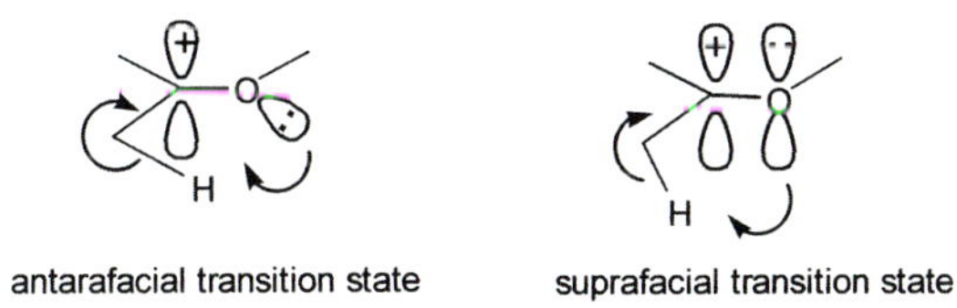

Scheme 8.

Oxygen has two lone pairs, and Scheme 8 contrasts two possible transition states. In the thermally forbidden suprafacial transition state (to the right), the itinerant hydrogen moves onto the lone pair that is parallel to the vacant p-orbital of the cation. The molecular skeleton can become nonplanar (e.g. by twisting about the sp^2 carbon–oxygen bond), and C=O double bond character can be sacrificed to increase the double bond character of the newly forming C=C double bond. By contrast, the antarafacial transition state transfers hydrogen to the other oxygen lone pair, which lies in the molecular plane. The C=C double bond develops in a highly twisted geometry. The question remains open whether the distortion required for the symmetry-allowed antarafacial transition state destabilizes it to such an extent that the suprafacial one is lower in energy.

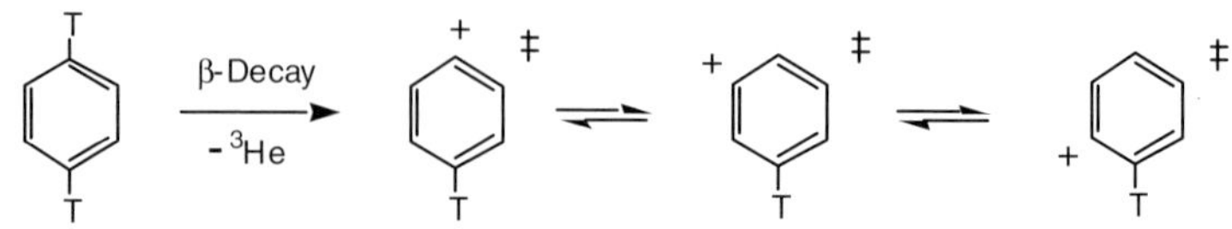

(12)

Researchers in Rome have examined many other multiply tritiated hydrocarbons. This body of work includes comprehensive studies of the phenyl cation produced by the β-decay in equation 12. The phenyl cation is difficult to prepare in condensed media but forms directly from β-decay of tritiated benzene. Its neutral reaction products in the gas phase have been carefully probed by a number of well-controlled experiments. An elegant series of investigations has demonstrated that intramolecular hydrogen migration within the vibrationally excited cation occurs on the 10–100 ns timescale, as equation 12 depicts (despite initial theoretical predictions that this isomerization should have a barrier too high to permit this).[42] Methylated phenyl cations rapidly transpose ring hydrogens, too, without detectable isomerization to the much more stable benzyl cation. The proportions of neutral methylanisoles summarized in equation 13 have been observed by capture of the cation with methanol and are probably not far from representing the equilibrium distribution of isomeric ions.[47] The bimolecular reactivity of other monosubstituted substituted phenyl cations $XC_6H_3T^+$ ($X=F$, Cl, Br, NO_2, OCH_3, and CN) has also been reported.[48]

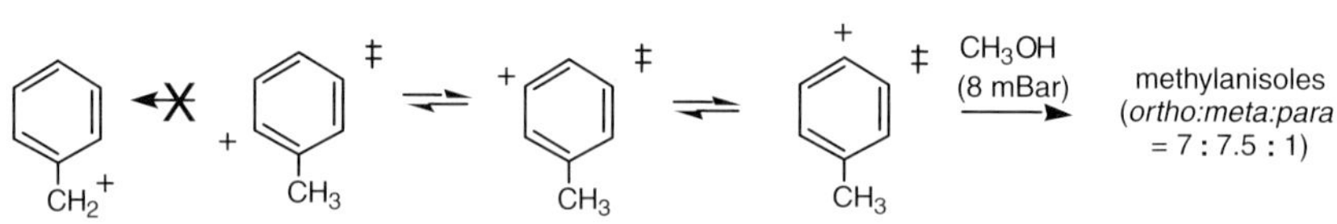

(13)

Capture of phenyl cations with methyl *n*-propyl ether gives products in which rearrangement has taken place.[49] Initial capture gives the oxonium ion drawn to the left in equation 14. Most of this ion eliminates propene, and deprotonation of the resulting ion affords anisole (PhOR, where $R=$methyl), which constitutes 75% of the radiochemical yield. Among the other products are isopropyl phenyl ether (roughly 3% of the radiochemical yield, about 3 times more abundant than *n*-propyl phenyl ether) and the isopropyl anisoles (also roughly 3% of the radiochemical yield).

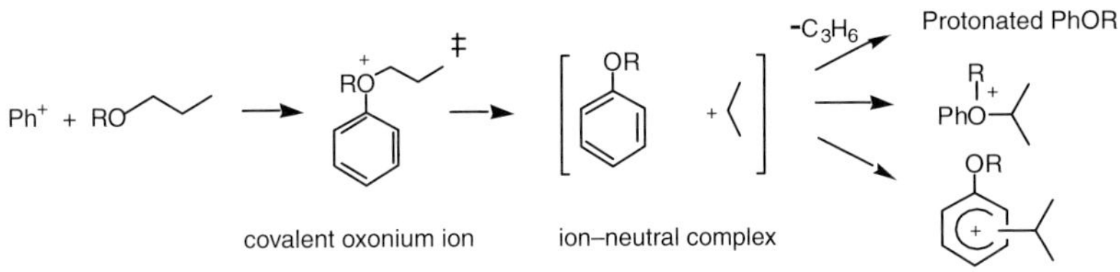

(14)

The most economical explanation supposes that the initially formed adduct of Ph^+ with an *n*-propoxy group, $PhO(R)CH_2CH_2CH_3^+$, behaves the same way as does protonated *n*-propyl phenyl ether (R=H), whose rearrangements have been examined using mass spectrometry of a number of specifically deuterated analogues.[50–53] The protonated ion decomposes via propene loss. The mass spectrometric studies also reveal that the hydrogens of the *n*-propyl group randomize and that they exchange with the hydrogen initially bound to oxygen (but not with the ring hydrogens). The intermediacy of the [PhOR $(CH_3)_2CH^+$] ion–neutral complexes accounts for the results. These complexes interconvert (but do not equilibrate completely) with [PhO(H)R^+ $CH_3CH{=}CH_2$] complexes. Kinetic analysis of the ion intensity ratios from protonated *n*-propyl phenyl ethers (analogous to that described in Section II.C above) shows that interconversion of the covalent oxonium ion with the latter complex cannot, by itself, account for the deuterium labeling data measured by mass spectrometry.[51]

Another mass spectrometric experiment from this laboratory (P.S. Mayer and T.H. Morton, unpublished results) confirms the pathway for decomposition of the covalent oxonium ion shown in Scheme 13 for R=C_2H_5. Chemical ionization of *n*-propyl phenyl ether with methane reagent gas at 0.0001 mbar produces a substantial peak corresponding to an adduct with ethyl cation. This adduct ion decomposes extensively via propene expulsion to give protonated $PhOC_2H_5$ (*m*/*z* 123). When deuterated *n*-propyl phenyl ethers are chemically ionized in this fashion, both *m*/*z* 123 and *m*/*z* 124 form. In the chemical ionization source $PhOCH_2CH_2CHD_2$ gives an *m*/*z* 123 : *m*/*z* 124 intensity ratio of 2.4 : 1; $PhOCD_2CH_2CH_3$ gives a ratio of 2.85 : 1; $PhOCH_2CH_2CD_3$ gives a ratio of 1.5 : 1; and $PhOCD_2CD_2CH_3$ gives a ratio of 1.15 : 1. Intermediacy of ion–neutral complexes such as drawn in Scheme 13 predicts this sort of result, in which the transferred hydron can come from every position of the propyl chain.

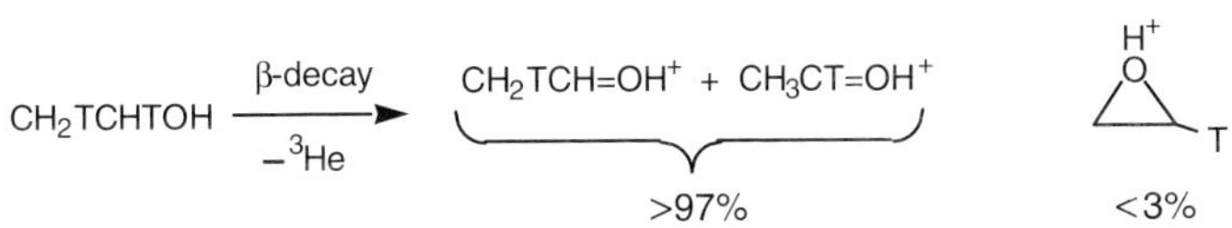

(15)

A few experiments have been performed on multiply tritiated molecules containing functional groups. Tritiated ethanol, in which the principal isotopomer is $CH_2TCHTOH$, has been permitted to undergo β-decay in the presence of 0.5 bar of gaseous trimethylamine. At that pressure the ion-molecule collision rate is faster than 10^{10} s^{-1}. The recovered radioactive neutral product is predominantly acetaldehyde, with a barely detectable yield of ethylene oxide (in a ratio $\geq 40:1$). As equation 15 summarizes, this implies that nuclear decay produces, for the most part, protonated acetaldehyde, regardless of which tritium expels a β-particle.[52]

Since the two tritium atoms are equally likely to decay, one would expect half of the initially formed ions to have the nascent primary cation structure $HOCHTCH_2^+$. Yet very little of the recovered neutral product comes from protonated ethylene oxide. Two possible explanations may be advanced to account for this. One is that hydride shift is much more favorable than oxygen bridging, even though oxygen bridging is calculated to have no energy barrier. The alternative explanation holds that protonated ethylene oxide does form, but with such high vibrational energy that it rearranges completely on the 10^{-10} s timescale before deprotonation by collision with trimethylamine.

One test of the latter hypothesis involves preparing a higher homologue, whose isomerization ought to yield characteristic products. To that end, the ditritiated *n*-propanol shown in Scheme 9 was synthesized and the neutral products from its β-decay analyzed. Decay of the tritium at position 3 yields tritiated ethylene by cleavage of a carbon–carbon bond. Both propylene oxide and propionaldehyde are recovered as tritiated products from decay at position 2 in the presence of 0.5 bar trimethylamine, but no acetone is seen.[55] Protonated acetone would be expected among the thermal rearrangement products of protonated propylene oxide. The absence of acetone among the neutral products indicates that protonated epoxides

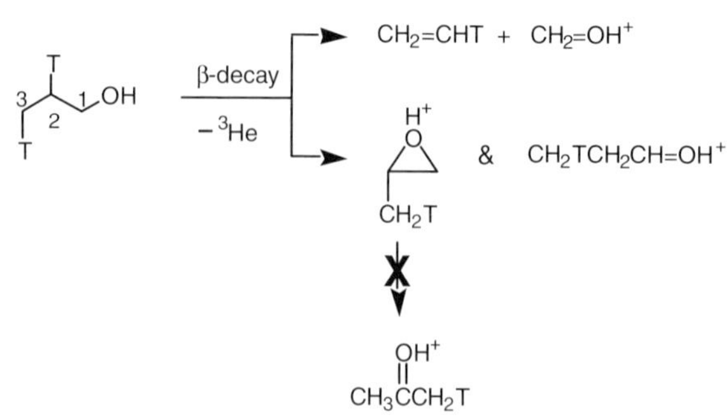

Scheme 9.

formed by oxygen bridging are stable on the 10^{-10} s timescale. Therefore the low yield of ethylene oxide from ditritiated ethanol must reflect a preference for hydride shift versus oxygen bridging. Mass spectrometric studies of unimolecularly generated [$CH_4D_2O^+$ $PhO^•$] ion–neutral complexes confirm this inference.[54]

IV. PRODUCTS FROM RADIOLYSIS EXPERIMENTS

Various forms of radiation have been used to produce ions in sufficient quantitites to yield neutral products for subsequent analysis. In principle, it should be possible to use intense beams of UV below ionization threshold for this purpose. To date, however, efforts to collect neutrals from resonant multiphoton ionization (REMPI) have not succeeded. In one experiment, 1 mbar of gaseous *n*-propyl phenyl ether was irradiated at room temperature with a 0.1 W beam of 266 nm ultraviolet (from an 800 Hz laser that gives 8 n pulses) concurrent with a 0.5 W beam at 532 nm.[56] The beams were intense enough not only to ionize the ether in the mass spectrometer, but also to excite it so that it expels propene.[26] After several hours of irradiation <10% of the starting material remained. Production of carbon monoxide and acetylene (decomposition products of the phenoxy group) could be detected by infrared absorption spectroscopy, but the yield of neutral propene (as measured by NMR spectroscopy) was infinitesimal.

This outcome illustrates one of the obstacles confronting the analysis of neutral products. Ion detection has such high sensitivity that REMPI can be effective in a mass spectrometric experiment even if it produces <10 ions per laser pulse. The techniques for examining neutrals (e.g. GC-MS or NMR) provide structural information that ion studies cannot accomplish, but they require much greater amounts of product. In order to observe uncharged dissociation products directly, REMPI would have to produce $>10^8$ ions per laser pulse.

Ionizing radiation (γ-radiolysis[57–59], vacuum ultraviolet,[60] electron impact[18,61]) has been the method of choice for neutral product collection studies under a variety of experimental conditions. On one hand, irradiations with γ-rays or vacuum ultraviolet are typically performed on static samples at pressures >1 mbar. On the other hand, electron beams with energies <100 eV are hard to sustain at pressures >0.001 mbar. Electron impact experiments are run on flowing samples at low pressure, which reproduces the conditions found in many ordinary mass spectrometers. Neutral product studies have provided the best measure of the scope and limitations of the two categories of cation rearrangement—ring closure/ring opening and atom/group transfer—listed in the opening paragraph of this chapter. The first category comprises ring closure and its reverse reaction, ring opening.

A. Ring Closure/Ring Opening

Equation 16 illustrates a celebrated example where ring closure competes with vicinal hydride shift (a common form of atom transfer in cations, which does not take place in free radicals or anions). The gas phase reaction was explored by preparing the dimethylfluoronium ion, $(CH_3)_2F^+$, by γ-radiolysis of fluoromethane. Exothermic methylation of a sample of ^{13}C-β-phenethyl chloride (where the asterisk in equation 16 symbolizes the labeled position) in the gas mixture gives a vibrationally excited ion that loses chloromethane to form two isomeric ions, α-phenethyl cation and spirooctadienyl cation (sometimes called ethylenebenzenium). Nucleophilic attack by methanol in the reaction mixture yields $PhCH(CH_3)OCH_3$, whose isotopic label remains almost entirely at the methyl group. The recovered $PhCH_2CH_2OCH_3$ contains ^{13}C equally distributed between the two methylene positions.[62] The spirooctadienyl ion does not isomerize to α-phenethyl cation, even though DFT calculations predict the latter to be 55 kJ/mol more stable.

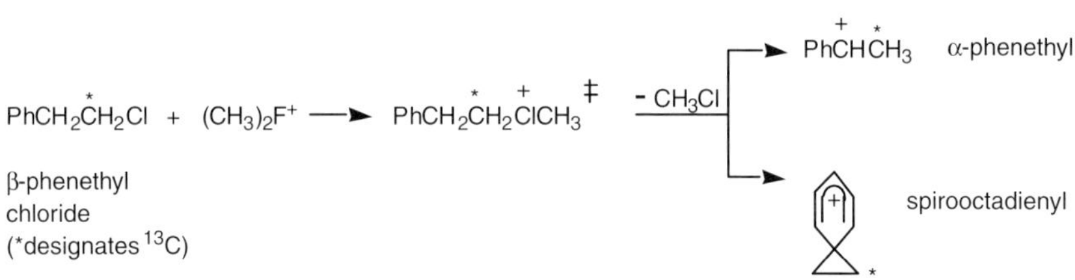

(16)

Nucleophilic capture of the spirooctadienyl cation opens the 3-member ring. This behavior characterizes many reactions of many other cyclopropane-containing carbocations, as well. γ-radiolysis of perdeuterated propane forms $C_3D_7^+$ ions, most of which either transfer D^+ or form isopropyl adducts. As the propane pressure is raised from 1000 mbar to 2000 mbar, however, the isopropyl/n-propyl adduct ratio falls from 30 : 1 to about 5.5 : 1.[63] This implies the formation of corner-protonated cyclopropane, which reacts with nucleophiles as though it were an n-propyl cation. With increased pressure, vibrationally excited protonated cyclopropane experiences more frequent nonreactive collisions, which deactivate it and slow down its rate of unimolecular isomerization to isopropyl cation.

Cationic cyclization and its reverse have particular relevance, since ring closure and ring opening have been invoked to account for a large number of biosynthetic pathways.[1,2] The gaseous cyclopropylcarbinyl–cyclobutyl–homoallyl system represents an extreme instance of a rapid, reversible isomerization of this sort, which has been studied over a period of 20 years both by β-decay of tritiated cyclobutane[64] and by radiolytic methods.[65] This interconversion is so fast that the three $C_4H_7^+$ structures have been discussed as resonance forms of a nonclassical ion. In any event, their energies are so close and the barriers among them (if they exist at all) are so low that medium and counterion effects very likely influence whether $C_4H_7^+$ is an equilibrating system or not.

Carbocations that contain small rings display a range of unimolecular reactivity, in the sense that some do not rearrange to more stable isomers (such as the spirooctadienyl cation in equation 16); some isomerize at measurable rates (such as corner-protonated cyclopropanes); and some exhibit very rapid interconversion (such as cyclopropylcarbinyl–cyclobutyl–homoallyl). What happens with larger rings? β-decay experiments on tritiated cyclopentane in the presence of unlabeled cyclopentane show a measurable yield of radioactive 1-pentene (and complete absence of 2-pentene), suggesting that cyclopentyl cation can undergo ring opening.[66] Since $CH_2{=}CHCH_2CH_2CH_2^+$ does not correspond to a stable structure on the $C_5H_9^+$ potential energy surface, the sequence summarized in equation 17 seems likely to be taking place. DFT calculations portray the 4-penten-2-yl cation as a resonance hybrid with the 2-methylcyclopropylcarbinyl cation, as drawn. The second step of equation 17, quenching $CH_2{=}CHCH_2CHCH_3^+$ by hydride transfer from neutral cyclopentane, is endothermic and is probably driven by residual vibration excitation in the pentenyl cation.

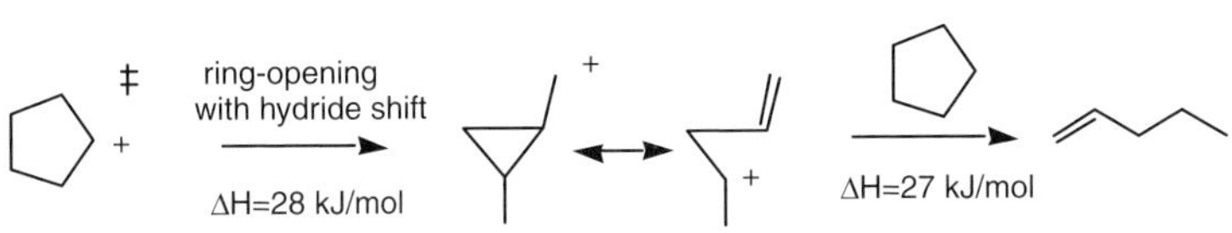

(17)

Ring closure of nascent $CH_2{=}CHCH_2CH_2CH_2^+$ within ion–neutral complexes has been studied using a specially designed Electron Bombardment Flow (EBFlow) reactor, schematically drawn in Figure 3. This apparatus has the advantage that the conditions under which ions are formed and react (70 eV electron impact; pressure ≤ 0.001 mbar) closely parallel those in mass spectrometer sources. The neutral product yields are routinely interpreted with reference to the ionic products observed by the mass spectrometry. Hypotheses based on EBFlow results for ion–neutral complexes are further tested by comparison with mass spectrometry.

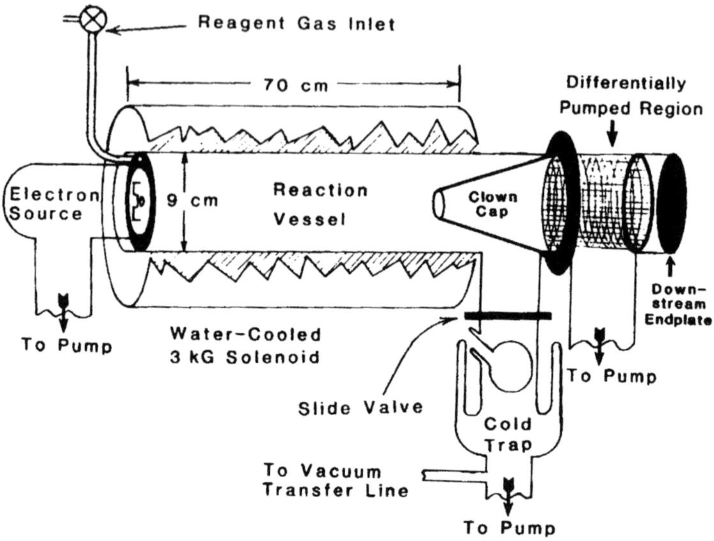

Figure 3. Schematic of an Electron Bombardment Flow (EBFlow) reactor. The magnetic field maintains the electron beam along the axis of the solenoid. The beam's negative charge prevents cations from migrating to the walls, and all charged species are dumped into the differentially pumped region downstream of the clown cap, where they are pumped away and do not contaminate the neutrals collected in the cold trap, which come from the EBFlow reaction vessel.

The cylindrical EBFlow reaction vessel is modeled on a conventional electron ionization (EI) source, but with the path of electron beam enlarged by a factor of 1000. An external solenoid electromagnet keeps the electrons on the cylinder axis, and the space-charge of the beam (along with the magnetic field) prevents collisions from driving ions to the walls. At the far end of the vessel, electrons and ions exit to a differentially pumped chamber via a conical Faraday plate ("clown cap"), while neutrals find their way into a liquid nitrogen-cooled trap, where they are collected.[18] Keeping ions off the walls is important, since surface neutralization can produce the same sorts of products as homogeneous gas-phase reactions (this has previously been discussed with regard to an earlier EBFlow design[15]). Efficiency of neutral product collection has been estimated to be on the order of 85%.[18] EBFlow yields are normalized relative to the beam current and expressed in units of μmole per ampere-second (i.e. per coulomb) of incident electrons, μmol/A-s. A yield of 1 molecule per electron corresponds to 10.36 μmol/A-s. Given typical cross sections for 70 eV electron ionization, a pathlength on the order of 1 metre, and a reactant pressure of 0.005 mbar, theoretical yields should be on the order of 2–3 μmol/A-s.[18]

The electron beam produces free radicals as well as cations. Neutral radicals tend not to isomerize (or, if they do, they rearrange by pathways different from those of cations). As a result, the neutral products from cationic rearrangements can generally be distinguished from the ones that arise via neutral free radicals. To judge from the relative abundances of neutral products recovered from the EBFlow, 70 eV electron impact produces more cations than radical pairs. Only very light hydrocarbons (e.g. ethylene and acetylene) are formed in yields comparable to the yields of cationic products.

$$\mathrm{PhOCH_2CH_2R}^{\cdot+} \xrightarrow{\text{alkyl group rearrangement}} \left[\mathrm{PhO}^{\cdot}\quad \mathrm{CH_3\overset{+}{C}HR}\right] \xrightarrow{-\,\mathrm{PhOH}^{\cdot+}} \mathrm{CH_2{=}CHR} \qquad (18)$$

One major focus of study has been the neutral products from 70 eV electron bombardment of primary phenoxyalkanes of the general formula $PhOCH_2CH_2R$. Unlike secondary alkyl phenyl ethers (whose radical cations expel alkenes via *syn*-elimination, as exemplified by Schemes 1 and 4), primary alkyl phenyl ether molecular ions decompose by formation of ion–neutral complexes via bond heterolysis accompanied by alkyl group rearrangements, such as equation 18 illustrates[13,26–29,61,67]. The corresponding neutral products are typically alkenes, which have been collected in the EBFlow and subsequently analyzed by chromatographic or spectroscopic means.[18,34,61,68–73] In all cases, vicinal hydride shift (as in equation 18) takes place. With alkyl groups having more than two methylenes, hydride shifts can continue down the chain. The question arises as to what other reactions compete with equation 18.

EBFlow radiolyses of 1-phenoxyalkenes of the general formula $PhOCH_2(CH_2)_nCH{=}CH_2$ give measurable yields of cycloalkenes, signaling that the ring closure represented in Scheme 10 competes with hydride shift. Internal nucleophilic displacement of phenoxy radical by the double bond leads to ion–neutral complexes containing cycloalkyl cations. Bond homolysis is ruled out by the fact that 4-penten-1-yl radicals do not cyclize,[74] as well as by the virtual absence of methylenecyclopentane among

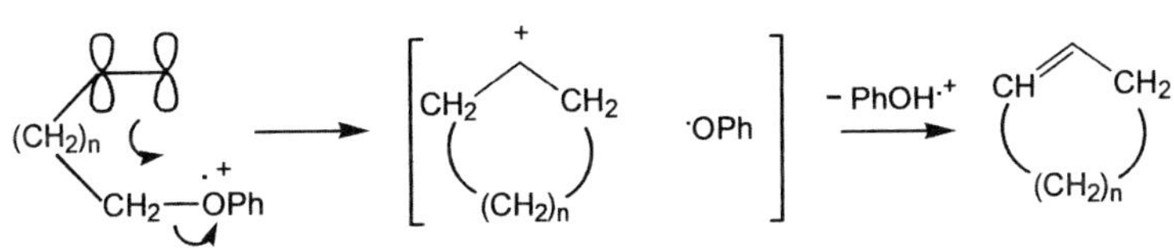

Scheme 10.

the products for $n=3$ (which would be the cycloalkene expected to result from intramolecular cyclization of 5-hexen-1-yl radicals[74]). Previously published GC traces display the product distributions for $n=2$ and 3.[18] For $n=2$ the yield of cyclopentene from Scheme 10 is 0.2 μmol/A-s, while the yield of linear pentadienes from hydride shift is about 10 times greater. The 1,3-pentadienes (which must have come from 1,2-hydride shifts) constitute >60% of the C_5H_8 neutral products, 1.5 μmol/A-s in a *trans*/*cis* ratio of 2.4 : 1. Skeletal rearrangement to the most stable allylic cation gives isoprene as a neutral product in a low yield (0.03 μmol/A-s), which probably comes about in the small amount of free $C_5H_9^+$ produced by 70 eV electron impact on $PhOCH_2CH_2CH_2CH{=}CH_2$. As mentioned above, mass spectrometric studies complement EBFlow experiments. The proportion of $PhOD^{\bullet +}$ seen in the 70 eV mass spectrum of $PhOCH_2CH_2CH_2CH{=}CD_2$ is in accordance with the yield of cyclopentene in the EBFlow.

Cyclization to form 6-membered rings competes more favorably with hydride shift. Linear hexadienes with terminal CH_2-groups (expected products of equation 18) are recovered in a yield of 0.6 μmol/A-s for the case $n=3$, while cyclohexene constitutes a yield of 0.35 μmol/A-s; 1-methylcyclopentene 0.08 μmol/A-s; and the other methylcyclopentenes 0.06 μmol/A-s. Products characteristic of free $C_6H_{11}^+$ cations that rearrange to the most stable allylic structures—2,4-hexadienes and branched C_6H_{10} dienes—are recovered in a yield of 0.15 μmol/A-s. As in the $n=2$ case, the products from cations produced in ion–neutral complexes greatly exceed those from free cations produced by simple bond fission, just as the 70 eV mass spectrum would predict.

EBFlow experiments also provide evidence for ring closure to form a 7-membered ring, but products from other cyclizations are more abundant. Unfortunately, there are so many C_7H_{12} isomers that it is difficult to separate them all by gas chromatography. Cycloheptene stands well apart, though, and is recovered in a yield of 0.02 μmol/A-s, as compared to a yield of 0.05 μmol/A-s for 1-methylcyclohexene and 0.05 μmol/A-s for other methylcyclohexenes and ethylcyclopentenes (which presumably arise via ring closures that take place after the hydride shift). Linear dienes from hydride shifts are recovered in a yield of 0.25 μmol/A-s.

Comparing cyclization yields with those of linear dienes with terminal CH_2-groups gives a measure of the dependence of Scheme 10 on chainlength. Closure to a 6-membered ring ($n=3$) goes 0.6 times as fast as vicinal hydride shift, while closure to a 5-membered ring ($n=2$) goes 0.1 times as fast and closure to a 7-membered ring goes only 0.07 times as fast. This outcome can be compared with the low probability of forming an epoxide in equation 15. However, other cationic cyclizations give a higher probability of 3-member ring formation than does equation 15.

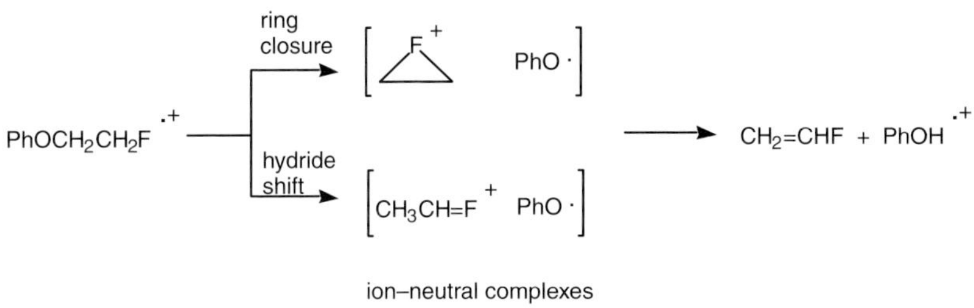

(19)

Cyclic, 3-member halonium ions are well-known intermediates for chlorine, bromine, and iodine. The question as to whether cyclic fluoronium ions might exist as stable entities was decided by an EBFlow experiment that generated them in ion–neutral complexes. As in the case of ring closure to the spirooctadienyl cation (equation 16), ring closure takes place in competition with hydride shift. Ionized β-fluorophenetole dissociates via two competing pathways, as equation 19 represents, which would be indistinguishable in the absence of isotopic labeling. Both involve heterolysis of the sp^3-carbon oxygen. The upper pathway drawn has the fluorine bridging between two carbons, yielding a symmetrical epifluoronium ion. The hydride shift in the lower pathway forms the most stable of the possible $C_2H_5F^+$ structures. The two competing mechanisms operate in about a 1 : 2 ratio (hydride shift favored), as revealed by ^{19}F NMR of the neutral fluoroethylenes from $PhOCH_2CD_2F$ and $PhOCD_2CH_2F$.[70] The upper pathway yields $CH_2{=}CDF$ and $CD_2{=}CHF$ regardless of which isomer is ionized. The lower pathway yields different products from the isomeric dideuterated starting materials, as Scheme 11 summarizes.

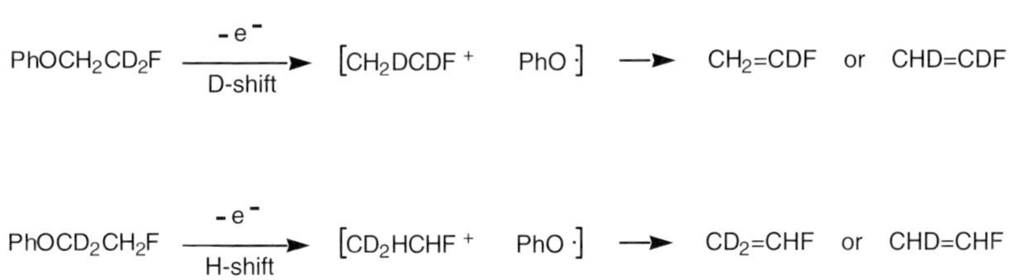

Scheme 11.

Consider three possibilities. If only hydride shift occurs, just the products shown in Scheme 11 will be detected. If, by contrast, bridging is followed by hydride shift, the EBFlow will recover all four possible deuterated fluoroethylenes from either starting material (although not necessarily in identical proportions). Finally, if (as in equation 16) bridging and hydride shift represent separate pathways that produce noninterconverting intermediates, each starting material will afford only three products (neglecting *cis-trans* isomerism). This last outcome has been observed. On the one hand, no CHD=CDF is recovered from $PhOCD_2CH_2F$, and no CHD=CHF is recovered from $PhOCH_2CD_2F$. On the other hand, the observation of CH_2=CDF from $PhOCD_2CH_2F$ and CD_2=CHF from $PhOCH_2CD_2F$ provides the first unequivocal evidence for an epifluoronium ion.[70]

B. Atom/Group Transfer

The evidence for cyclization presented in the previous section raises a series of questions. From a topological standpoint, any transfer of an atom/group from one position to another either passes through a cyclic structure or else involves a bond cleavage followed by recombination. How can one tell the difference? If an atom/group transfer involves a cyclic structure, does it operate via reversible ring closure/ring opening sequences? Or does the cyclic structure simply represent a transition state? Questions of this nature have intrigued mechanistic chemists for many years. Before attempting to address them, a survey of vicinal atom/group transfer will illustrate the scope of this process beyond the hydride transfer represented by equation 18.

Scheme 12 summarizes a γ-radiolysis performed on a mixture of benzene-d_6, CF_4, and ethylene. Ionization of CF_4 yields CF_3^+, which has been called an "ionic Lewis superacid."[75–77] Electrophilic attack on benzene-d_6 gives the conjugate acid of perdeuterated trifluorotoluene, $CF_3C_6D_6^+$, as the first step of Scheme 12. Trifluorotoluene is more basic than ethylene; hence, the ion–molecule reaction of $CF_3C_6D_6^+$ with ethylene cannot give a Brønsted acid–base reaction. The only thermochemically accessible pathway to ethyl trifluorotoluenes is via ion–neutral complexes containing the ethyl cation

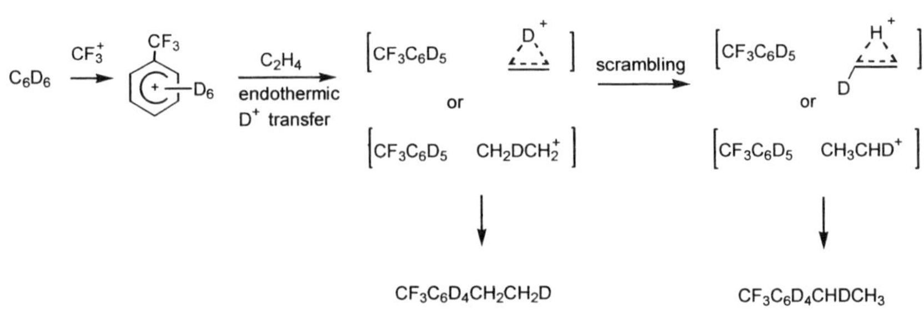

Scheme 12.

and neutral trifluorotoluene. Within the ion–neutral complexes, a small fraction (in the range of 7–9%, depending on the temperature of the reaction mixture) of the ethyl cations appear to scramble the label.[78]

Theory suggests that the ethyl cation has a nonclassical equilibrium geometry with a proton sitting in the middle of a double bond (the dashed-line structures in Scheme 12), but the interpretation of the results does not depend on whether the cation prefers a classical or nonclassical geometry. Scheme 12 represents both alternatives. Recovery of products with undeuterated methyl groups means that classical structures have been accessed, either as a transition state in passing from a D^+-bridged to a H^+-bridged geometry or as the preferred structure in the complex, which undergoes a vicinal hydride shift. The competition between scrambling and formation of the final products means that if the rate of scrambling is known, then the average lifetime of the ion–neutral complex can be deduced.[78]

Vicinal shifts occur not only in cations, but also in neutral reactants that are in the process of acquiring a positive charge. The hydride abstraction by *tert*-butyl cation depicted in Scheme 13 has not been observed by ICR[79] nor in radiolysis experiments at pressures <200 mbar.[57] But γ-radiolysis at higher pressures permits noncovalent clustering to form encounter complexes, in which the ion and neutral have time to adopt the right geometry for methyl migration to occur simultaneously with bimolecular hydride abstraction, as the curved arrows show. The transition state is thought to require *anti*-periplanar disposition of the reacting bonds. The entropic barrier for this concerted reaction apparently requires thermal activation of the encounter complex, even at the expense of collisionally dissipating the energy liberated by clustering. In the presence of added methanol the final product is the methyl ether drawn at the bottom of Scheme 13.[79]

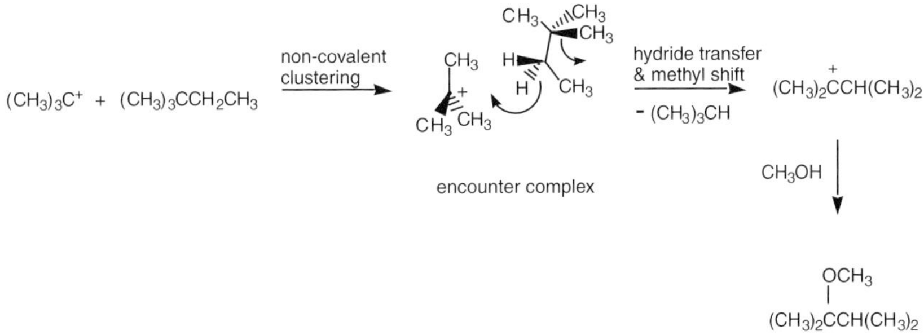

Scheme 13.

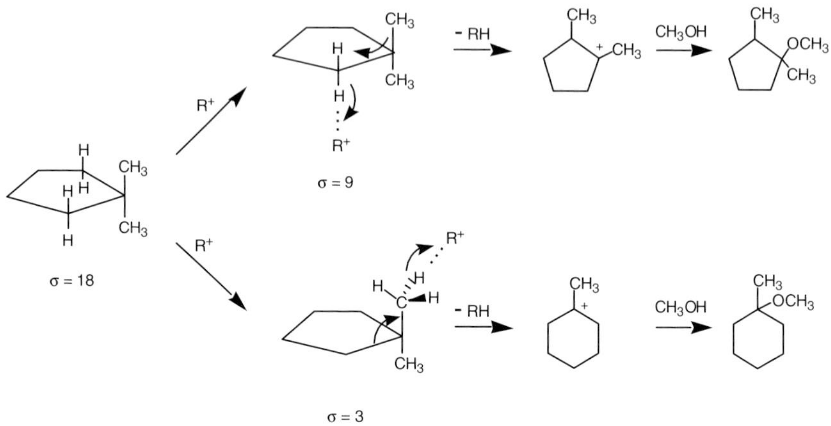

Scheme 14.

Neutral product distributions also measure the competition between the carbon–carbon bond shifts when a hydride is abstracted from 1,1-dimethylcyclopentane, as represented in Scheme 14. A radiolytically generated alkyl cation (R^+) can abstract from the ring (upper pathway) or from a methyl (lower pathway). As the symmetry numbers indicate, the lower pathway is favored on the basis of reaction path degeneracy. Also, according to DFT calculations, the methylcyclohexyl cation in the lower pathway is 1 kJ/mol more stable than the, dimethylcyclopentyl cation in the upper pathway. Nevertheless, in the presence of added methanol the corresponding methyl ethers are recovered in a 92 : 8 ratio in favor of the upper pathway when $R^+ = (CH_3)_3C^+$.[80] The *cis* and *trans* isomers from the upper pathway are recovered in virtually equal quantities, which tends to argue against the products coming from a terbody complex of R^+, methanol, and 1,1-dimethylcyclopentane. The lower pathway becomes more prominent as R^+ becomes more electrophilic, as would be expected from the reactivity–selectivity principle: the proportion of methylcyclohexyl ether roughly doubles when R^+ is the isopropyl cation.

Shifts of different atoms/groups to the same cationic center provide a measure of relative propensities to migrate. Three ion–neutral complexes form from the molecular ion drawn in Scheme 15. The EBFlow experiment required two isotopic labels (the asterisk designates ^{13}C) in order to tease apart the different pathways (including about 10% of the molecular ions that eliminate $PhOD^{\bullet+}$ via *syn*-elimination to yield undeuterated 2-fluoropropene). Product analysis was rendered more difficult because the two 1-fluoropropyl cations $CH_3CH_2CHF^+$ and $CH_3CHCH_2F^+$ have nearly the

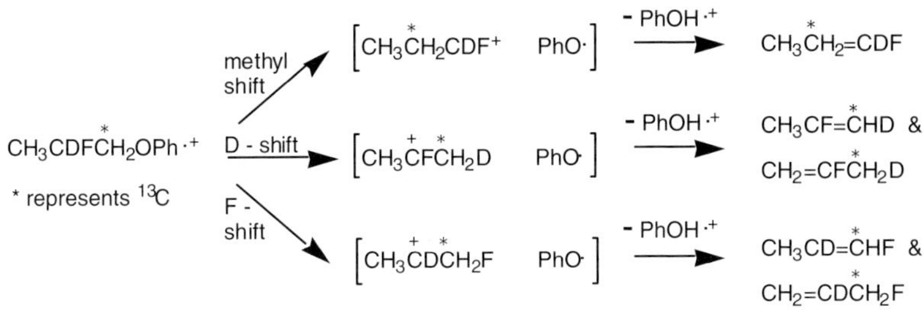

Scheme 15.

same stability and interconvert rapidly, scrambling (but not completely equilibrating) the deuterium label. ^{19}F NMR resolves five different isotopomeric/isomeric *Z*-1-fluoropropenes (as well as five *E*-1-fluoropropenes), permitting a quantitative assessment of migratory aptitudes—methyl shift : D-shift : F-shift = 6 : 11 : 2.[69]

$$CF_3^+ + CH_3CH_2CH{=}O \xrightarrow{-CF_2=O} CH_3CH_2CHF^+ \longleftrightarrow CH_3CH_2CH{=}F^+ \qquad (20)$$

The predominance of methyl and D-shift reflects the fact that a fluorine substituent attached to an sp^2-cation center exerts a stabilizing influence due to delocalization of a fluorine lone pair onto the electron-deficient carbon (as exemplified by the resonance structure at the right in equation 20). This effect is well attested and has been used to guide the regiochemistry of cationic organic syntheses.[81,82] The 1-fluoroisopropyl cation from F-shift in Scheme 15 does not enjoy this effect, since fluorine is attached to an sp^3-carbon instead. Although it is a secondary carbocation, the electronegativity of the vicinal fluorine destabilizes it, even as fluorine stabilizes the primary carbocation produced by methyl shift. For these reasons $CH_3CH_2CHF^+$ and $CH_3CHCH_2F^+$ have nearly the same stability.

$$CF_3^+ + CH_3CH_2\overset{*}{C}H{=}O \xrightarrow{-CF_2=O} \overset{*}{C}H_3CFCH_3^+ \ \& \ CH_3CF\overset{*}{C}H_3^+ \qquad (21)$$

$$80 : 20$$

Free fluoroalkyl cations are produced by metathesis of perfluoroalkyl cations with simple carbonyl compounds.[83,84] Bimolecular exchange of F^+ for O between CF_3^+ and acetone forms $(CH_3)_2CF^+$, which is the most stable structure on the $C_3H_6F^+$ potential energy surface.[69] A variety of gas phase reactions create that ion. Neutral products can be recovered in the EBFlow from its gas phase ion–molecule reactions: $CH_3CF{=}CH_2$ from proton transfer to neutral acetone, for example, or $(CH_3)_2CF_2$ produced by fluoride abstraction from *tert*-butyl fluoride.[85] $CH_3CFCD_3^+$ affords neutral products that have not transposed label between the two methyls (i.e. no products with partially deuterated methyl groups are seen).[86] Even when the ion has enough internal energy to expel HF unimolecularly, the neutral products from the resulting 2-propenyl ion exhibit no evidence for having scrambled hydrogens prior to its neutralization.[87]

When propionaldehyde reacts with CF_3^+ in the EBFlow, a small amount of $CH_3CH{=}CHF$ is recovered from the metathesis in equation 20 (followed by proton transfer to neutral propionaldehyde).[69] The principal C_3H_5F isomer, though, is $CH_3CF{=}CH_2$, signalling that $(CH_3)_2CF^+$ has been formed. It is not altogether clear if all of that ion comes from isomerization of initially formed $CH_3CH_2CHF^+$, but the results outlined below show that at least some of it does. An EBFlow experiment with ^{13}C-labeled propionaldehyde, summarized in equation 21, yields $CH_3CF{=}CH_2$ with ^{13}C in all positions (but not a random distribution). About one-fifth of the product has the label in the center carbon position, which must have arisen via unimolecular methyl shift within $CH_3CH_2CHF^+$.[88] The majority of ions, however, have come from a process that transposes fluorine.

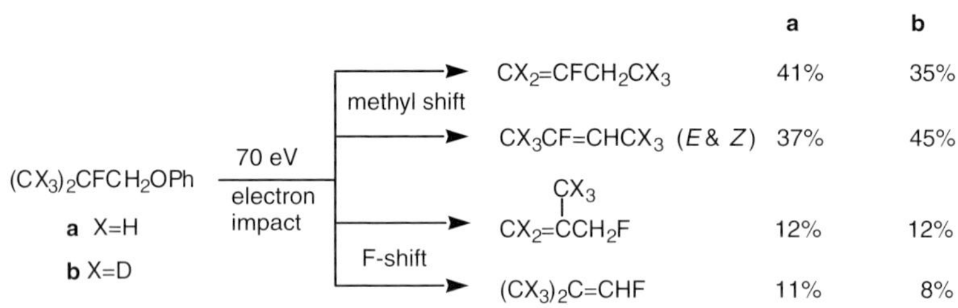

(22)

The first question posed at the beginning of this section inquires whether it is possible to distinguish atom transfer via a cyclic transition state from dissociation–recombination. An EBFlow study of unimolecular competition between methyl and fluorine shift, shown in equation 22, has addressed this issue.[73] The rearrangements occur in the course of forming ion–neutral complexes, and the product distributions change in going from unlabeled (**a**) to deuterated (**b**) starting material. A ^{19}F NMR measurement of the fluoroalkene distribution exhibits a degree of experimental error such that the ratio of F-shift products to methyl shift products cannot be said to change significantly with deuteration of the methyls, but statistically significant normal kinetic isotope effects are exerted on the distribution of positional isomers. The kinetic isotope effect on the ratios of the methyl shift products corresponds to $k_H/k_D = 1.4$ (with $k_H/k_D = 0.9$ on the *E*/*Z* 2-fluorobutene product ratio), and for the F-shift products $k_H/k_D = 1.6$. These are consistent with the $PhOH^{\bullet+}$: $PhOD^{\bullet+}$ ion intensity ratio observed in the 70 eV mass spectrum of the deuterated ether. They represent the combined primary and α-secondary isotope effects on the Brønsted acid–base reaction between the cation and the phenoxy radical in the complexes and are of the same magnitude as the isotope effects within the complexes shown in Scheme 15[69] as well as for deprotonation of cations within [*tert*-amyl cation PhO•] complexes from neopentyl phenyl ether.[34] They are smaller than the primary kinetic isotope effects reported for bimolecular hydride transfer.[80]

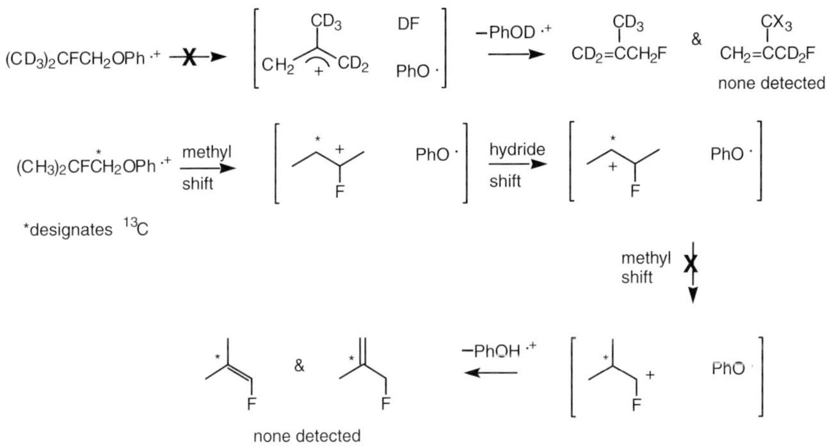

Scheme 16.

After correcting for reaction path degeneracies, methyl shift in equation 22 is approximately twice as likely as F-shift, compared with being three times as likely in the reaction summarized in Scheme 15. The absence of a significant isotope effect on methyl versus F-shift argues against a dissociation–recombination mechanism for F-shift. The deuterium labeling pattern in the neutral products rules out elimination–readdition rigorously. Had the transposition of fluorine occurred via the terbody complex drawn at the top of Scheme 16, neutral methallyl fluoride would have been recovered in which deuterium is attached to the same carbon as fluorine, but that product is not seen. The ^{13}C-labeling experiment summarized at the bottom of Scheme 16 confirms this and also demonstrates that a genuine fluorine shift has taken place, instead of the carbon skeleton rearrangement shown, which would not have attached fluorine to the tagged carbon. Both of the branched products from the ^{13}C-labeled precursor display NMR spin–spin couplings proving that fluorine is directly connected to the tagged carbon, while the linear fluorobutenes show only two-bond couplings between ^{19}F and ^{13}C.[72]

The second question raised at the beginning of this section asks which atom/group transfers proceed via reversible ring closure to a stable intermediate. Some cationic ring closures are irreversible (at least, on the timescale of the reactions that produce the recovered neutral products); for instance, equations 16 and 19. While SCF calculations predict that the epifluoronium ion in equation 19 lies in a potential energy well, they indicate that unsymmetrical alkyl substitution of the 3-membered ring destabilizes the cyclic structure so that it becomes a transition state.[72] If that is correct, the F-shifts in Scheme 15 and equation 22 pass through cyclic transition states and do not operate by reversible ring closure-ring opening.

F-shift exhibits two limiting cases: in one instance a ring closes, but does not open again until a base pulls a proton off from the epifluoronium ion; in other instances, the ring does not correspond to a minimum on the potential energy surface. In a number of different systems a variety of options are equally plausible, and experiment draws a distinction that theory cannot. For example, the results summarized in Scheme 9 argue that hydride shift from one carbon to the other in equation 15 does not proceed via ring closure to protonated ethylene oxide followed by ring opening concomitant with a hydrogen transposition.

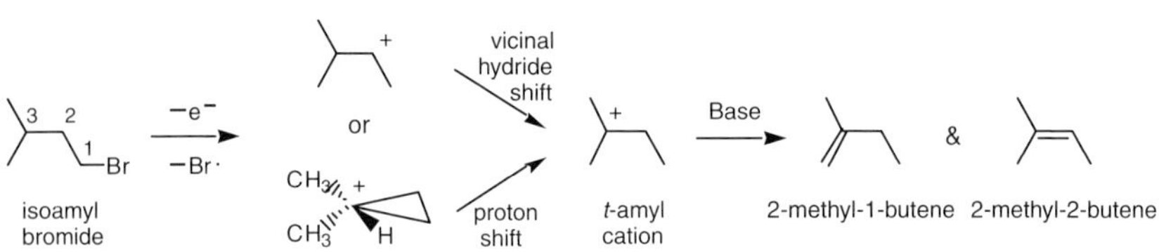

Scheme 17.

A number of cases remain to be resolved where theory gives an equivocal answer, such as in the formation of *t*-amyl cation from ionization of isoamyl bromide. An EBFlow experiment provides evidence for the net reaction illustrated in Scheme 17: 70 eV radiolysis of 0.0005 mbar isoamyl bromide plus 0.0003 mbar triethylamine yields 2-methyl-1-butene as the principal C_5H_{10} product (2.3 μmol/A-s) with a slightly lower yield of 2-methyl-2-butene.[15] Rearrangement takes place in the course of loss of a bromine atom from the molecular ion. At least two different mechanisms can describe what takes place. Vicinal hydride shift could occur to make a secondary isoamyl cation, $(CH_3)_2CHCHCH_3^+$, which might undergo another vicinal hydride shift to give *t*-amyl cation. Alternatively, a corner-protonated cyclopropane might form, which could undergo a corner-to-corner proton shift and open up to make the *t*-amyl cation. DFT calculations predict both intermediates to be stable, with the secondary cation 30 kJ/mol more stable than the corner-protonated cyclopropane. Successive vicinal hydride shifts predict a different isotopic substitution pattern on the *t*-amyl than does the net 1,3-transfer implied by the protonated cyclopropane intermediate, but the barrier for unimolecular randomization of all the hydrogens within *t*-amyl cation is so low as to render an isotopic labeling experiment problematic.[68] In any event, specific deuteration of isoamyl bromide at carbon 3 did not succeed in distinguishing the two pathways, since label scrambling within the $C_5H_9D^{\bullet+}$ parent ions foiled any attempt to use GC-MS to determine the position of the isotope in recovered 2-methyl-1-butene.

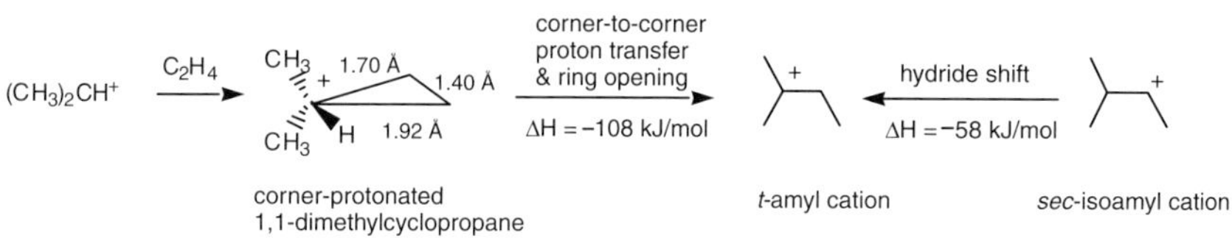

Scheme 18.

A different type of experiment demonstrates the viability of an intermediate protonated cyclopropane. Isopropyl cation (*i*Pr$^+$) was generated in the EBFlow by 70 eV electron bombardment of di-*n*-propyl ether (0.0004 mbar) in the presence of ethylene (0.0013 mbar). This reaction yields 0.09 μmol/A-s of 2-methyl-1-butene and 0.08 μmol/A-s of 2-methyl-2-butene.[15] Ion–molecule association reactions are ordinarily extremely inefficient at these low pressures because, in the absence of a third body collision, the cluster ion has enough internal energy to dissociate back to the reactants. But, as Scheme 18 depicts, *i*Pr$^+$ associates with ethylene to make *t*-amyl cation. It seems unlikely that this reaction passes through the *sec*-isoamyl cation, even though DFT calculations (summarized in Scheme 18) predict it to be 50 kJ/mol more stable than the corner-protonated cyclopropane. Formation of an intermediate *sec*-isoamyl cation would require migration of one of the hydrogens originally attached to ethylene.

DFT calculations show that the corner-protonated cyclopropane has the structure shown in Scheme 18. Its geometry that resembles an *i*Pr$^+$ sitting on top of an ethylene, but with the charge-bearing carbon pyramidalized. Rearrangement to *t*-amyl cation is highly exothermic. RRKM estimates of the rate of corner-to-corner proton transfer suggest that isomerization occurs rapidly enough to compete with dissociation.[68] Given that the experimental value for the net exothermicity of *i*Pr$^+$ + C_2H_4 → *t*-amyl cation has a value of $\Delta H = -140$ kJ/mol, rearrangement puts $C_5H_{11}^+$ into such a deep well that it can last for milliseconds without dissociating. But, in the absence of rearrangement, RRKM calculations predict a lifetime for the corner-protonated cyclopropane $<10^{-8}$ s (to be compared with experimental estimates on the order of a fraction of a nanosecond for the lifetimes of ion–neutral complexes[36,78]). If these estimates are correct, corner-to-corner proton transfer concomitant with ring opening must be at least as fast as RRKM calculations predict.

According to the rules of orbital symmetry, vicinal hydride shift in cations is a thermally allowed suprafacial 1,2-sigmatropic shift. Migration of a more distant hydrogen can occur by successive 1,2-shifts,[89] via a bridged structure (such as the cyclooctyl cation in Scheme 7[35]), or by corner-to-corner transfer in a protonated cyclopropane (which give net 1,3-shifts). An EBFlow experiment has been able to measure the competition between 1,2- and 1,3-shift in the 1-fluoropropyl system.

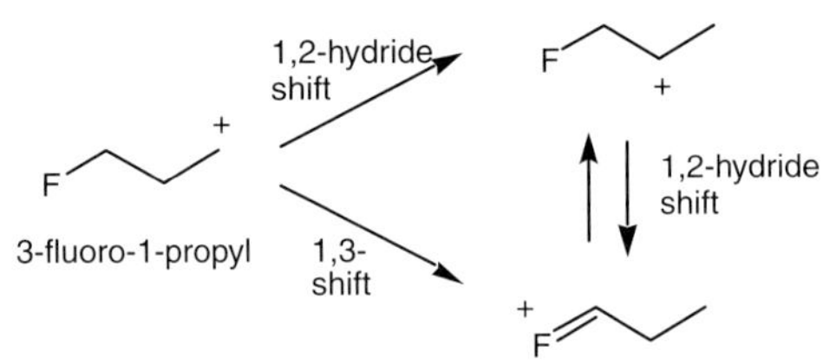

Scheme 19.

The 3-fluoro-1-propyl cation is not a stable species. It rearranges, as Scheme 19 portrays, to a mixture of two cations that have nearly the same stability (cf. Scheme 15 above). This isomerization can occur via two pathways—a 1,3-shift or successive 1,2-shifts. Experimentally, these pathways have been distinguished by looking at the EBFlow of $PhOCH_2CH_2CD_2F$, which forms ion–neutral complexes by C–O bond heterolysis (cf. equation 18). A 1,2-shift transfers hydrogen to form a methyl group. A 1,3-shift transfers deuterium directly to form a CH_2D group. Subsequent 1,2-shifts (which interconvert the two stable isomers drawn to the right in Scheme 19) do not affect the labeling pattern of the methyl. The ^{19}F NMR of the recovered neutral 1-fluoropropenes in Figure 4 shows that products with undeuterated methyl groups are 15 times more abundant than products with CH_2D-groups, which is taken to represent the relative rates of 1,2- versus 1,3-shift.[71]

A striking feature of this EBFlow radiolysis experiment is the low yield of 2-fluoropropene, all of which can be accounted for by the small amount of free fluoropropyl ions generated by 70 eV electron impact. This result implies that the skeletal rearrangement of linear to branched fluoropropyl cations (which gives rise to the neutral products in equation 21, for example) occurs too slowly to take place in the brief lifetime of ion–neutral complexes. In this regard the ion–neutral complexes formed by ionizing

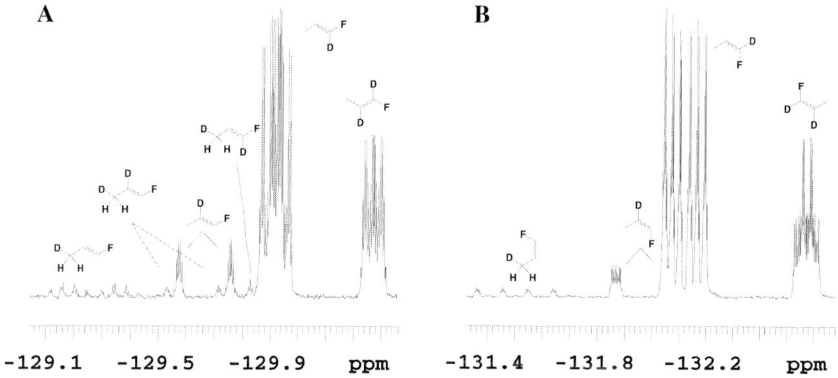

Figure 4. ^{19}F NMR spectrum of 1-fluoropropenes (**A** *trans*; **B** *cis*) from 70 eV EBFlow radiolysis of $PhOCH_2CH_2CD_2F$.

3-fluoropropyl phenyl ether closely parallel the behavior of complexes from *n*-butyl phenyl ether,[61] just as equation 22 parallels ionized neopentyl phenyl ether.[34] These outcomes support the view that, in a carbocation, a single fluorine substituent has many of the same properties as a methyl group.

V. CONCLUSION

Cationic rearrangements display considerable variety. Investigations of the simplest molecular systems have proven difficult in condensed phases but have been amenable to systematic study in the gas phase. Neutral product studies exhibit the scope and limitations of two general categories of isomerization—ring closure/ring opening and atom/group transfer—and do not suggest that any additional categories need be added to the repertoire. In terms of unimolecular reaction rates, three general classes of reactions operate: rearrangements that take place concomitantly with bond heterolyses, those that occur rapidly after heterolysis and whose neutral products can be detected from ion–neutral complexes, and those that happen in free cations but which are too slow to be observed to any great extent within complexes. Radiolytic and mass spectrometric studies (in which bonds heterolyze following ionization) find their complement in tritium decay studies (in which helium departs as a leaving group so rapidly that heterolysis is virtually simultaneous with ionization, and rearrangement occurs later).

At present, theory has developed a persuasive ability to predict cation stabilities. Current understanding of relative reaction rates, however, has not advanced as far. A number of processes for which computation predicts no energy barrier (such as the ring closure in equation 19 or the collapse of ion–neutral complexes to give rearranged, covalently bound ions[28]) occur slowly relative to reactions (such as hydride shift or proton transfer) that would appear to have low (but nonzero) potential energy barriers. Neutral product distributions give a view of the dynamics of competing reactions that is difficult to acquire by any other means. The long-term objectives of such experiments include an understanding of competing pathways that would permit an organization of reactions within enzyme active sites along the lines of the three classes listed above.

ACKNOWLEDGMENT

This work was supported by NSF grant CHE 9983610.

REFERENCES

[1] Coates, R.M. *Prog. Chem. Org. Nat. Prod.* **1976**, *33*, 74–230.

[2] Gibbs, R.A. In *Comprehensive Biological Catalysis: a Mechanistic Reference*; Vol 1 Sinnott, M., Ed.), Academic Press, London, 1998, p 31–118.

[3] Bohlmann, J.; Meyer-Gauen, G.; Croteau, R. *Proc. Nat. Acad. USA* **1998**, *95*, 4126–4133.

[4] Hoshino, T.; Abe, T.; Kouda, M. *JCS Chem. Commun.* **2000**, 441–442; Dang, T.; Prestwich, G.D. *Chemistry& Biology* **2000**, *7*, 643–649; Fernandez, S.M.S.; Kellogg, B.A.; Poulter, C.D. *Biochemistry* **2000**, *39*, 15316–15321; Wendt. K.U.; Schulz, G.E.; Corey, E.J.; Liu, D.R. *Angew. Chem. Int. Ed.* **2000**, *39*, 2812–2833.

[5] Kohn, W.; Becke, A.D.; Parr, R.G. *J. Phys. Chem.* **1996**, *100*, 12974–12980.

[6] Morton,T.H.; Beauchamp, J.L. *J. Am. Chem. Soc.* **1975**, *97*, 2355–2362; **1977**, *99*, 1288.

[7] Maslak, P.; McGuinn, J.M. *JCS Chem. Commun.* **1999**, 2467–2468.

[8] Baciocchi, E; Bietti, M.; Lanzalunga, O. *Acc. Chem. Res.* **2000**, *33*, 243–251.

[9] Stevenson, D.P. *Disc. Faraday Soc.* **1951**, *10*, 35–45.

[10] Audier, H.E. *Org. Mass Spectrom.* **1969**, *2*, 283–298.

[11] Coulson, D.R. *J. Am. Chem. Soc.* **1978**, *100*, 2992–2996.

[12] Pollak, E.; Pechoukas, P. *J. Am. Chem. Soc.* **1978**, *100*, 2984–2991.

[13] Weddle, G.H.; Dunbar, R.C.; Song, K.; Morton, T.H. *J. Am. Chem. Soc.* **1995**, *117*, 2573–2580.

[14] Traeger, J.C.; Luna, A.; Tortajada, J.C.; Morton, T.H. *J. Phys. Chem. A* **1999**, *103*, 2348–2358.

[15] Marinelli, W.J.; Morton, T.H. *J. Am. Chem. Soc.* **1978**, *100*, 3536–3539; **1979**, *101*, 1908.

[16] Attiná, M.; Cacace, F.; Cipollini, R.; Speranza, M. *J. Am. Chem. Soc.* **1985**, *107*, 4824–4828.

[17] Attiná, M.; Cacace, F.; DiMarzio, A. *J. Am. Chem. Soc.* **1989**, *111*, 6004–6008.

[18] Morton, T.H. In *Techniques for the Study of Ion-Molecule Reactions*; Farrar, J.M.; Saunders, W.H. Jr., Ed., *Techniques of Chemistry XX*, Wiley-Interscience, New York, 1988, p 119–164.
[19] Redman, E.W.; Morton, T.H. *J. Am. Chem. Soc.* **1986**, *108*, 5701–5708.
[20] Taphanel, M.H.; Morizur, J.P.; Leblanc, D.; Borchardt, D.; Morton, T.H. *Anal. Chem.* **1997**, *69*, 4191–4196.
[21] Morizur, J.P.; Taphanel, M.H.; Mayer, P. S.; Morton, T. H. *J. Org. Chem.* **2000**, *65*, 381–387.
[22] Biermann, H.W.; Freeman, W.P.; Morton, T.H. *J. Am. Chem. Soc.* **1982**, *104*, 2307–2308.
[23] Andlauer, B.; Ottinger, Ch. *J. Chem. Phys.* **1971**, *55*, 1471–1472.
[24] Fairweather, R.B.; McLafferty, F.W. *Org. Mass Spectrom.* **1970**, *4*, 221–224.
[25] Kondrat , R.W.; Morton, T.H. *Org. Mass Spectrom.* **1991**, *26*, 410–415.
[26] Chronister, E.L.; Morton, T.H. *J. Am. Chem. Soc.* **1990**, *112*, 133–139.
[27] Song, K.; van Eijk, A.; Shaler, T.A.; Morton, T.H. *J. Am. Chem. Soc.* **1994**, *116*, 4455–4460.
[28] Traeger, J.C.; Morton, T.H. *J. Am. Chem. Soc.* **1996**, *118*, 9661–9668.
[29] Kohler, B.E.; Morton, T.H.; Nguyen, V.; Shaler, T.A. *J. Phys. Chem. A* **1999**, *103*, 2302–2309
[30] Morton, T.H. *Org. Mass Spectrom.* **1992**, *27*, 353–368.
[31] Abboud, J.M.; Alkorta, I.; Dávalos, J.Z.; Gal, J.-F.; Herreros, M.; Maria, P.-C.; Mó, O.; Molina, M.T.; Notario, R.; Yañez, M. *J. Am. Chem. Soc.* **2000**, *122*, 4451–4454.
[32] Illies, A.J.; Morton, T.H. *Int. J. Mass Spectrom. Ion Processes* **1997**, *167/168*, 431–445.
[33] Midland, M.M.; Morton, T.H. *J. Am. Chem. Soc.* **1993**, *115*, 9596–9601.
[34] Morton, T.H. *J. Am. Chem. Soc.* **1980**, *102*, 1596–1602.
[35] Kirchen, R.P.; Sorenson, T.S. *J. Am. Chem. Soc.* **1979**, *101*, 3240–3243.
[36] Aschi, M.; Attiná, M.; Cacace, F.; D'Archangelo, G. *J. Am. Chem. Soc.* **1998**, *120*, 3982–3987.
[37] Audier, H.E.; Morton, T.H. *J. Am. Chem. Soc.* **1991**, *113*, 9001–9003.
[38] Morton, T.H. *Tetrahedron* **1982**, *38*, 3195–3243.
[39] McAdoo, D.J.; Morton, T.H. **1993**, *26*, 295–302.
[40] Cacace, F.; Speranza, M. In *Techniques for the Study of Ion–Molecule Reactions*; Farrar, J.M., Saunders, W.H., Jr., Eds.; *Techniques of Chemistry XX*; Wiley-Interscience, New York, 1988, p. 287–323.
[41] Cacace, F. *Science* **1990**, *250*, 392–399.
[42] Speranza, M. *Chem. Rev.* **1993**, *93*, 2833–2986.
[43] Nefedov, V.D.; Sinotova, E.N.; Akulov, G.P.; Korsakov, M.V. *J. Org. Chem. USSR* **1970**, *6*, 1220–1224.
[44] Smith, R.D.; Herold, D.A.; Elwood, T.A.; Futrell, J.H. *J. Am. Chem. Soc.* **1977**, *99*, 6042–6045.
[45] Nefedov, V.D.; Sinotova, E.N.; Bermudez, R.K. *J. Org. Chem. USSR* **1982**, *18*, 1642–1646.
[46] Hudson, C.E.; McAdoo, D.J. *J. Am. Soc. Mass Spectrom.* **1998**, *9*, 130–137.
[47] Cacace, F.; Ciranni, G.; Sparapani, C.; Speranza, M. *J. Am. Chem. Soc.* **1984**, *106*, 8046–8050.
[48] Filippi, A.; Lilla, G.; Occhiucci, G.; Sparapani, C.; Ursini, O.; Speranza, M. *J. Org. Chem.* **1995**, *60*, 1250–1264.
[49] Fornarini, S.; Speranza, M. *J. Chem. Soc. Perkin 2* **1984**, 171–177.
[50] Benoit, F.M.; Harrison, A.G. *Org. Mass Spectrom.* **1976**, *11*, 599–608.
[51] Bogdanov, B.; Matimba, H.E.K.; Ingemann, S.; Nibbering, N.M.M. *J. Am. Soc. Mass Spectrom.* **1996**, *7*, 639–652.
[52] Kondrat, R.W.; Morton, T.H. *J. Org. Chem.* **1991**, *56*, 952–957.

[53] Jacquet, J.-P.; Morton, T.H. *J. Mass Spectrom.* **1997**, *32*, 251–252.
[54] Nguyen, V.; Bennett, J.S.; Morton, T.H. *J. Am. Chem. Soc.* **1997**, *119*, 8342–8349; **1997**, *119*, 12026.
[55] Mayer, P.; Morton, T.H. Paper presented at the 46th ASMS Conference on Mass Spectrometry and Allied Topics, Dallas, TX, June, 1999; manuscript in preparation.
[56] Zhao, L.; Chronister, E.L.; Morton, T.H. Unpublished results.
[57] Ausloos, P.; Lias, S.G. *J. Am. Chem. Soc.* **1970**, *92*, 5037–5045.
[58] Cacace, F. *Acc. Chem. Res.* **1988**, *21*, 215–222.
[59] Speranza, M. *Mass Spectrom. Rev.* **1992**, *11*, 73–117.
[60] Doepker, R.D.; Ausloos, P. *J. Chem. Phys.* **1965**, *43*, 3814–3819.
[61] Burns, F.B.; Morton, T.H. *J. Am. Chem. Soc.* **1976**, *98*, 7308–7313.
[62] Fornarini, S.; Muraglia, V. *J. Am. Chem. Soc.* **1989**, *111*, 873–877.
[63] Laguzzi, G.; Angelini, G.; Bucci, R.; Segre, A.L. *J. Phys. Chem.* **1994**, *98*, 6719–6724.
[64] Cacace, F.; Speranza, M. *J. Am. Chem. Soc.* **1979**, *101*, 1587–1589.
[65] Cacace, F.; Chiavarino, F.B.; Crestoni, M.E. *Chem. Eur. J.* **2000**, *6*, 2024–2031.
[66] Babérnics, L.; Cacace, F. *J. Chem. Soc. B* **1971**, 2312–2316.
[67] Audier, H.E.; Leblanc, D.; Mayer, P.S.; Morton, T.H. *European Mass Spectrom.* **2000**, *5*, 419–429.
[68] Morton, T.H. *Radiat. Phys. Chem.* **1982**, *20*, 29–40.
[69] Shaler, T.A.; Morton, T.H. *J. Am. Chem. Soc.* **1991**, *113*, 6771–6779.
[70] Nguyen, V.; Cheng, X.; Morton, T.H. *J. Am. Chem. Soc.* **1992**, *114*, 7127–7132.
[71] Shaler, T.A.; Borchardt, D.; Morton, T.H. *J. Am. Chem. Soc.* **1999**, *121*, 7907–7913.
[72] Shaler, T.A.; Morton, T.H. *J. Am. Chem. Soc.* **1994**, *116*, 9222–9226.
[73] Shaler; T.A.; Morton, T.H. *J. Am. Chem. Soc.* **1989**, *111*, 6868–6870; **1990**, *112*, 4090.
[74] Beckwith, A.L.J.; Eaton, C.J.; Lawrence, T.; Serelis, A.K. *Aust. J. Chem.* **1983**, *36*, 545–556.
[75] Grandinetti, F.; Occhiucci, G.; Crestoni, M.E.; Fornarini, S.; Speranza, M. *Int. J. Mass Spectrom. Ion Processes* **1993**, *127*, 123–135.
[76] Grandinetti, F.; Occhiucci, G.; Crestoni, M.E.; Fornarini, S.; Speranza, M. *Int. J. Mass Spectrom. Ion Processes* **1993**, *127*, 137–146.
[77] Grandinetti, F.; Crestoni, M.E.; Fornarini, S. and Speranza, M. *Int. J. Mass Spectrom. Ion Processes* **1994**, *130*, 207–222.
[78] Aschi, M.; Attiná, M.; Cacace, F. *Chem. Eur. J.* **1998**, *4*, 1535–1541.
[79] Crestoni, M.E.; Fornarini, S.; Lentini, M.; Speranza, M. *Chem. Commun.* **1995**, 121–122.
[80] Crestoni, M.E.; Fornarini, S.; Lentini, M.; Speranza, M. *J. Phys. Chem.* **1996**, *100*, 8285–8294.
[81] Johnson, W.S.; Chenera, B.; Tham, F.S.;. Kullnig, R.K. *J. Am. Chem. Soc.* **1993**, *115*, 493–497.
[82] Johnson, W.S.; Buchanan, R.A.; Bartlett, W.R.; Tham, F.S.; Kullnig, R.K. *J. Am. Chem. Soc.* **1993**, *115*, 504–515.
[83] Eyler, J.R.; Ausloos, P.; Lias, S.G. *J. Am. Chem. Soc.* **1974**, *96*, 3673–3675.
[84] Ausloos, P.; Lias, S.G.; Eyler, J.R. *Int. J. Mass Spectrom. Ion Processes* **1975**, *18*, 261–271.
[85] Redman, E.W.; Johri, K.K.; Lee, R.W.K.; Morton, T.H. *J. Am. Chem. Soc.* **1984**, *106*, 4639–4640.
[86] Redman, E.W.; Johri, K.K.; Morton, T.H. *J. Am. Chem. Soc.* **1985**, *107*, 780–784.
[87] Stams, D.A.; Johri, K.K.; Morton, T.H. *J. Am. Chem. Soc.* **1988**, *110*, 699–706.
[88] Nguyen, V.; Mayer, P.S.; Morton, T.H. *J. Org. Chem.* **2000**, *65*, 8032–8040.
[89] Sorensen, T.S.; Whitworth, S.M. *J. Am. Chem. Soc.* **1990**, *112*, 6647–6651.

MULTIPOLE-BOUND MOLECULAR ANIONS

Robert N. Compton and Nathan I. Hammer

OUTLINE

Advances in Gas-Phase Ion Chemistry
Volume 4, pages 257–305.

ABSTRACT

Recent experiments in the field of negative-ion physics have shown that, in some cases, it is useful to describe the binding of an electron to a molecule as a result of the dominant multipole moment of that molecule (dipole, quadrupole, etc.). These anions are weakly bound and are subject to collisional detachment as well as detachment by modest electric fields. Some general theoretical statements have been made as to the "minimum" dipole moment required to bind an electron to a molecule, but similar statements for the molecular quadrupole are not as well described. In cases where a molecule has a dipole or quadrupole moment of sufficient strength to permanently bind an extra electron, but also has a bound valence anion, the multipole-bound state and valence state can interact to form a coupled system in which the diffuse multipole-bound state can relax into the more localized valence state. In this sense, the dipole (or quadrupole) states can act as entrance channels or "doorway" states to the formation of the more strongly bound valence anion states. The concept of electron molecule multipole expansion is especially useful in describing multiply charged negative ions. The combined monopole (Coulomb repulsion) and polarizability interaction potential of an electron with a molecular negative ion gives rise to a Coulomb barrier for the addition or removal of the extra electron. The Coulomb barrier can act to make a multiply charged anion *more* stable with respect to electron loss than its singly charged parent. This simple picture qualitatively explains many of the prominent features of multiply charged anions.

I. INTRODUCTION

An accurate description of the electron-binding energy for a singly or multiply-charged negative ion represents a formidable challenge to modern molecular quantum theory. Electron correlation and exchange terms must be included in the calculation as well as the employment of a sufficient, often diffuse, orbital basis set in the molecular wave functions. Experimentally, it is now known that most elements in the periodic table have stable ground-state, negative-ion configurations, i.e. positive electron affinities. Notable exceptions are the nitrogen, beryllium, and magnesium atoms. Also, no evidence has yet been reported that atoms such as Hg and Zn can form bound negative ions in the ground state. In addition, the rare gases probably do not form bound ground-state negative ions, although this is not a closed subject. There is some evidence that Xe negative ions may be bound.[1] The nuclear degrees of freedom inherent in molecules present additional problems since large geometry changes often accompany the addition of an excess electron. Positive electron affinities, EA, for molecules can vary from ~zero for some molecules [e.g. EA(NO) = 0.025 eV] to well over 5 eV for the so-called superhalogen molecules [e.g., $EA(AuF_6) > 10$ eV] (for a recent review see Ref. [2]).

It is sometimes convenient to represent the interaction of an electron with an atom, molecule, or negative ion in terms of its static and dynamic electrostatic energies. If we consider only the dipole and quadrupole terms, the long-range (outside the electron cloud) potential of interaction between an electron and a molecule, or molecular negative ion, is conveniently expressed as

$$V(r) = \frac{ne^2}{r} - \frac{e\mu}{r^2} P_1(\vec{r} \cdot \vec{R}) - \frac{eQ}{r^3} P_2(\vec{r} \cdot \vec{R}) - \frac{\alpha e^2}{2r^4}, \tag{1}$$

where μ is the dipole moment and Q is the electric quadrupole moment of the molecule. $P_1(\vec{r} \cdot \vec{R})$ and $P_2(\vec{r} \cdot \vec{R})$ are the first and second order Legendre polynomials. The last term in the equation represents the polarizability attraction where α is the polarizability of the molecule or negative ion. The first term represents the Coulomb repulsion energy for the case of the interaction between an electron and an anion with a charge ne.

In recent years, there has been growing interest in the physics of bound states of the electron–molecule system which results primarily from the molecular multipole moments explicit in equation 1. Since only atoms in degenerate states can have a dipole moment (e.g., H), the quadrupole is generally the first non-zero electric moment for atoms. Ground state molecules can be described by a single multipole moment (a dipole moment for say HF), or the quadrupole moment (e.g., CO_2) or a combination of dipole, quadrupole, etc. moments. The induced polarizability always contributes to the electron–molecule interaction to a lesser or greater extent. There are now many examples of dipole-bound molecular negative-ion states. In addition, there is a small but expanding list of molecules in which the electron binding can be primarily attributed to the molecular quadrupole moment. Of course, in both of these cases the polarizability of the molecule serves to increase the binding energy of the negative ion. The interaction of an electron with a negative ion represents a particularly interesting case in which the combined repulsive monopole and attractive polarizability term, i.e., $ne^2/r - \alpha e^2/2r^4$, gives rise to a "Coulomb barrier" in the electron/anion interaction potential. The addition or removal of any of the excess electrons is dominated by this Coulomb barrier.

The Milliken oil-drop experiment amply demonstrates that multiply charged *droplets* can exist as stable entities. Accepting this fact, we pose the question, "How small can the droplet be? i.e., can an atom or molecule become multiply negatively charged?" In an attempt to address this question, it is instructive to consider a variation of one of Milliken's droplets—an electrically conducting sphere. From classical electrostatics,[3] the force acting on an electron at an external distance r from the center of an insulated conducting sphere of radius a containing n excess electronic charges is given by

$$F(r) = \frac{e}{r^2}\left\{ne - \frac{ea^3(2r^2 - a^2)}{r(r^2 - a^2)^2}\right\} \quad r > a \tag{2}$$

At large distance, the force exhibits the expected ne^2/r^2 long-range Coulomb repulsion. At intermediate distances, the polarization attraction force, a^3e^2/r^3, begins to contribute and eventually dominate. The important point to notice is that the force becomes attractive (negative energy) at very small distances (i.e. near the surface of the sphere). Figure 1 shows the interaction potential derived from equation 1 for an electron "attracted" to a singly or doubly-charged anion. From this we conclude that, classically, the sphere can be highly negatively charged. Note also that the potential contains a "Coulomb barrier" at distances close to the sphere which

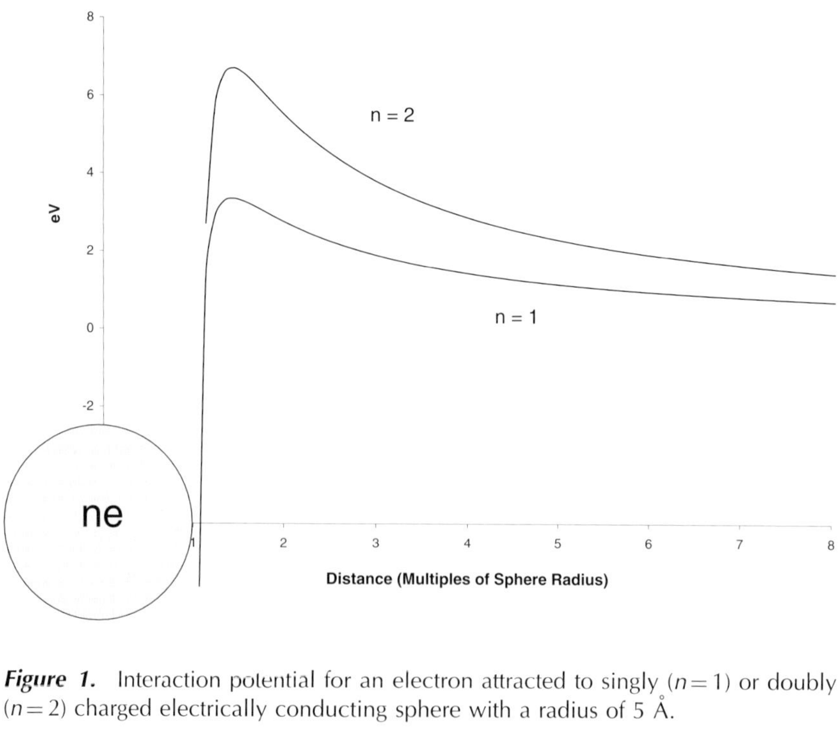

Figure 1. Interaction potential for an electron attracted to singly ($n=1$) or doubly ($n=2$) charged electrically conducting sphere with a radius of 5 Å.

effectively increases the "stability" (i.e., the Coulomb barrier gets larger) with increased charging. At the atomic level this picture, of course, changes. As the radius of the sphere becomes smaller, the system has to be treated quantally. The uncertainty and exclusion principles will ultimately provide the answer to this question through eigenvalue solutions of the Schrödinger equation.

Bohr[4] first considered the question of doubly charged negative ions of the hydrogen atom in the second of his famous three-part masterpiece (Trilogy), "On the Constitution of Atoms and Molecules." Bohr used the symbol N $(n_1, n_2, \ldots)$ to refer to a planar ring of electrons rotating around a nucleus of charge +Ne. The total amount of energy W released by the formation of this Bohr atom is denoted by $W\,[(\mathrm{N}(n_i, n_2, \ldots)]$. The hydrogen atom energy is thus represented as $W_o[1(1)]$ and was calculated to be 0.043 stat volt or 12.9 eV. Similar calculations for H^- gave $W[1(2)] = 1.13\ W_o$, which predicts a bound state for H^-. H^{-2} corresponds to $W[1(3)] = 0.54\ W_o$. From these results, Bohr concluded, "Since $W[1(3)]$ is only 0.54 W_o compared to 1.13 W_o for H^-, a hydrogen atom cannot be expected to be able to acquire a double negative charge." A number of authors have considered the fundamental question of the maximum number of negative particles (fermions or bosons) of charge −e that can be bound to an atomic nucleus of charge $+Ze$.[5–12] As some of the results depend upon statistics of the particles, we consider here only fermions. The theorem of Lieb[10,12] seems most appropriate and predictive for our interests and actually applies to fermions and bosons in a magnetic field as well. Lieb's theorem states: an upper bound for the maximum number (N_c) of negative particles of charge −e that can be bound to an atomic nucleus of charge +Ze is $N_c < 2Z$. Lieb's general theorem is more inclusive and can be applied to molecules as well. For the hydrogen atom, $Z = 1$ and $N < 2$. Thus, Lieb's theorem states that a bound state of H^{-2} cannot exist in the gas phase. Levy-Leblond[13] had earlier demonstrated the nonexistence of bound H^{-2}. Beck and Nicolaides[14] similarly showed that there are no bound states of H^{-2}, including other possible metastable configurations such as the $H^{-2}(2p^3)\ 4S^\circ$ excited state.

For completeness, reference is made to a comment (Footnote 1 in Ref. 15) by Avron, Herbst and Simon,[15] "There is unfortunately a very limited number of rigorous results on Coulombic binding energies, so that there is no rigorous proof that He^- (or even H^{-2}) has no binding." This paper also treats the formation of negative ions in magnetic fields. For example, they predict that He^- is stable in very strong magnetic fields. These authors show that for a one-electron Hamiltonian, the lowest state for which the electron can become unbound has an energy $B/2m$ due to the zero-point energy of a Landau oscillator. Since the energy of the ground state must go up in a magnetic field,[16] we have the rigorous bound $E(B) \leq B/2m$.

Such effects are very small (i.e. for the hydrogen atom $m = 1$, and $E(B)$ is of the order 10^{-7} B °K (with B in gauss)). This result sounds a cautionary note concerning the observation of negative ions (and their stability) where magnetic fields are involved. Many negative ion sources or mass spectrometers (e.g. FTMS) involve strong magnetic fields. Atoms with very small electron affinities or even slightly unbound (shape resonances) anions or dianions may become long-lived or even stable as a result of such large fields.

II. BRIEF REVIEW OF ELECTRON BINDING TO ELEMENTARY PARTICLES

Before proceeding to discuss multipole-bound anions, let us consider electron binding to elementary particles, a process which has a fundamental connection to dipole-bound negative ions. This field has progressed rapidly in recent years and has not been reviewed or included in conventional discussions of negative ions. There are lessons to be learned concerning the effects of mass on electron binding which are analogous to the mass effects on the Rydberg constant. Let us begin by expanding our definition of atomic negative ions to include such esoteric species as the positronium negative ion, $e^-e^+e^-$. The discovery of electron–positron pair production by electromagnetic radiation in 1932 (Anderson[17]) together with the relativistic spinning electron theory of Dirac[18] prompted Wheeler[19] to consider the question of binding in the case of polyelectrons such as positronium, Ps (e^+e^-), the positronium negative ion, Ps ($e^-e^+e^-$), Ps_2, etc. The result for Ps was trivial, i.e., because of the mass differences, the binding energy for Ps is one-half that of the "infinitely" heavy hydrogen atom. Although the accuracy of the resulting calculation was insufficient to predict binding for the positronium molecule, Ps_2, Wheeler was able to show that the positronium negative ion, Ps^-, would be bound. Modern calculations support these conclusions. In 1983, Bhatia and Drachman[20] used elaborate wave functions composed of 200 Hylleraas-type vectors to provide three more significant figures to the earlier binding energy calculation obtained by Kolos, Roothan, and Jack.[21] Bhatia and Drachman obtained a binding energy for Ps of 0.326537 eV. One should note that this is less than half the binding energy of H^- (Pekeris[22]). A summary of fifteen different calculations of the binding energy of Ps can be found on page 539 of the review by Abdel-Raouf,[23] entitled "Positronium Molecules: Their Existence, Formation and Annihilation."

The positronium negative ion, Ps^-, (e^- e^+ e^-) was first observed by Mills in 1981.[24] The method of observation involved detecting the decay of Ps^-

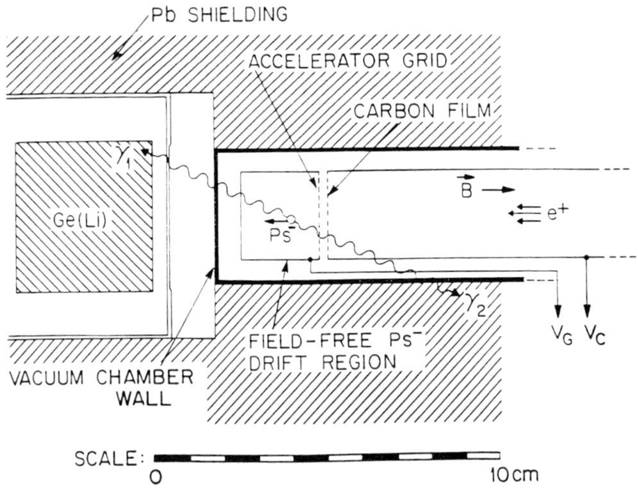

Figure 2. Experimental apparatus for detecting the positronium negative ion, Ps^-.

into two γ-rays The experimental apparatus is shown in Figure 2. A beam of positrons from a 15-MCi source of Co and a Cu(111)+S moderator is guided by a magnetic field into a thin carbon target (37 ± 13 Å thick). The incident positron "picks up" two electrons from the sea of electrons in the carbon (graphite) foil. The conversion efficiency for e^+ into Ps^- is found to be a maximum ($2.8 \pm 0.3 \times 10^{-4}$) at ~400 eV e^+ energy. The Ps^- ions emerging from the carbon foil are then immediately accelerated from 500 to 4,500 eV by applying a positive voltage on a grid located 2.5 mm from the carbon film. Positrons transmitted through the film are turned back where they move with constant velocity. Some of the Ps^- from the carbon foil are accelerated into the drift region. Ps^- decays primarily by 2-γ emission with an annihilation rate nearly equal to that of the spin-averaged Ps decay. By observing the Doppler-shifted annihilation γ-energy spectrum for five different acceleration voltages (12 h data accumulation for each energy), Mills was able to observe the Ps^- 2-γ decay which was Doppler-shifted (19.3 to 39.1 keV) from the 511 keV positronium decay peak. The Doppler-shifted peak corresponded to a mass-to-charge ratio of 3.01114 m/e, clearly demonstrating the existence of the $e^- e^+ e^-$ positronium negative ion. In a later experiment using the same apparatus, Mills[25] measured the number of Ps^- ions reaching a grid under different acceleration conditions in order to determine the total decay rate for Ps^- as $\Gamma = 2.09(9) \times 10^9\ s^{-1}$. To first order, the decay of Ps occurs by 2-γ annihilation. The annihilation rate Γ is proportional to the electron density at the positron $\langle\delta(r_{13})\rangle$ and is given by

$$\Gamma = 2\pi\alpha^4\left(1+\sum_{i=1}^{3} n_i\right)\langle\delta(r_{13})\rangle \tag{3}$$

The quantities n_i are small and account for radiative corrections to the free-particle cross section (n_1), 3-γ annihilation (n_2) and bound-state and relativistic wave-function effects (n_3). Calculations of the electron density at the positron $\langle\delta(r_{13})\rangle$ by Ho[26] and by Bhatia and Drachman[27] give decay widths (Γ) of 2.0842, 2.0858 and 2.0861 $\times 10^9$ s^{-1} using $(1+\sum_{i=1}^{3} n_i)=0.996824$. An earlier calculation by Ferrante[28] resulted in $\Gamma = 1.992 \times 10^9$ s^{-1}. The branching ratio for 3-γ annihilation relative to that for 2-γ annihilation is expected to be the same as that for a positron in a metal and is small ($\sim 2.6 \times 10^{-3}$). The branching ratio for 3-γ annihilation into a 4/3 mc^2 eV photon and a 2/3 mc^2 eV electron is proportional to $\alpha^4 \approx 10^{-9}$ (see Mills[24]).

Mills[25] has also considered the possible existence of a metastable excited state of Ps$^-$ analogous to the 2p^2 ^{3}P^e state of H$^-$. He calculated the Coulomb binding energy for two identical fermions of mass m_1 and a like charge interacting with an oppositely charged spinless particle of mass m_2 for the lowest energy even parity $L=1$ configuration. Figure 3 summarizes these results, showing the dissociation energy, ε, of the 2p^2 ^{3}P^e state of H$^-$ like ions relative to the parent $n=2$ level of the neutral atom versus the relative mass $m_2/(2m_1+m_2)$ for two different numbers of terms ($n=22$ and $n=70$) in the wave functions. Although the picture might change with the inclusion of a larger number of terms, it appears that Ps 2p^2 ^{3}P^e ($m_2/M=1/3$) is not bound. However, note that the ^{3}P^e states of e$^-$ μ^+ e$^-$ and H$^-$ are predicted to be bound. The fact that Mills's calculation for H$^-$ ($m_2/M=1$) gives $\varepsilon = 0.00948$ eV as compared to the more accurate value of 0.00965 eV of Bhatia[30] who used many more terms (90 versus 70), gives some validity to the conclusions in Figure 3: namely H$^-$, e$^-$ p$^+$ e$^-$, H_2^+, and p$^+$ μ^- p$^+$ all have bound 2p^2 ^{3}P^e states relative to the $n=2$ level, but Ps$^-$ probably does not.

Bhatia and Drachman[31] have also examined the binding energies for three charged particles in an attempt to clarify the ^{1}S^e binding energies in relation to the reduced masses. Among other things, the ^{3}P^e state of Ps$^-$ again was found to be unbound in agreement with Mills's[25] prediction. Botero and Greene[32] had earlier considered the three body mass-dependent problem and predicted a single strong shape resonance in the photo-detachment spectrum similar to that for H$^-$. Petelenz and Smith[33] have calculated the binding energies for positronium and muonium negative

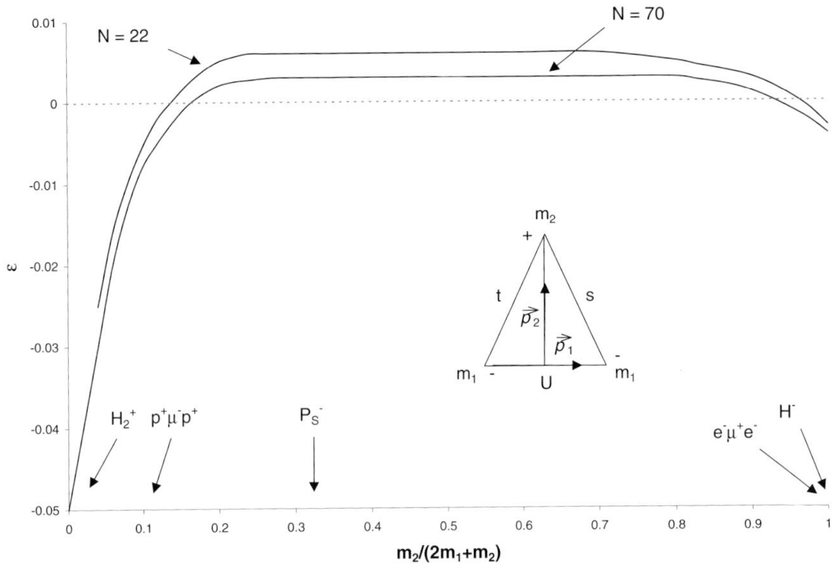

Figure 3. Dissociation Energy, ε, of the $2p^2\ ^3P^e$ state of H^- like ions relative to the parent $n=2$ level of the neutral atom versus the relative mass $m_2/\ (2m_1+m_2)$ for two different numbers of terms ($N=22$ and $N=70$) in the wave functions.

ions using the integral transform method with the result that $E\ (Ps^-)=-0.262005070$ and $E\ (\mu^-)=-0.525054810$ a.u. (Ry). Haftel and Mandelzweig[34] used nonvariational hyperspherical calculations for Ps^- to obtain $E\ (Ps^-)=-0.26200486$ a.u. (Ry). Ho[35] has also calculated bound singlet S states for $e^-\ e^+\ e^-$ (Ps^-) below the $n=2$ and $n=3$ positronium thresholds. The experimental difficulties of detecting such resonances in view of the lifetimes of ortho- ($1.4\ \times 10^{-7}$ s) and para- (1.25×10^{-10} s) positronium are discussed. Very recently, Ho and Bhatia,[36] have calculated a $^1P^\circ$ shape resonance above the $n=2$ Ps threshold at $-0.12434(3)$ Ry with a width of 0.00045(3) Ry, using the complex–coordinate relation method with Hylleraas-type wave functions. They also report $^3P^\circ$ shape resonances lying above the $n=3$ and $n=7$ thresholds. Finally, Kvitsinsky et al.[37] have applied the Faddeev approach to calculate the singlet and triplet scattering lengths for e^- - Ps scattering at zero energy. These authors point out that such cross sections could provide a guide for experiments of fundamental interest, including tests of charge conjugation in purely leptonic systems.

Negative muonium ions, μ^- ($e^-\mu^+e^-$), were reported by two groups in 1986 and 1987. Harshman, et al.[38] and Kuang et al.[39] both used double electron pickup by μ^+ ions passing through thin foils in order to produce μ^-. In the Harshman et al.[38] experiment, energetic muons (μ^+) were injected into thin moderators of LiF, quartz and copper. μ^- ions ($e^-\mu^+e^-$) were observed to exit the surfaces at low energy (<10 eV). The efficiencies for μ^- production were on the order of 1.5×10^{-7} per incident μ^+ for LiF and copper but much lower ($\sim4\times10^{-8}$) for quartz (SiO_2). The apparent low rate for SiO_2 may have been due to surface charging of the quartz. In a second experiment, Kuang et al.[39] observed the production of μ^- ions as μ^+ ions are passed through Al, Au, and Be foils. Somewhat higher yields ($\sim5\times10^{-5}$ μ^-/μ^+) were reported in these experiments than that reported by Harshman et al. (-1.6×10^{-7} μ^-/μ^+). Kuang et al.[39] also performed Monte Carlo calculations for μ^+ traveling through metal foils and obtained good agreement between the measured and calculated yields.

The availability of observable yields of μ^- ions provides hope that further studies of spectroscopy, binding energy, and decay mechanisms might be possible. As mentioned above, Mills[24] has predicted that the $e^-\mu^+e^-$ ion in the $2p^2\ {}^3P^e$ state is bound relative to the $n=2$ neutral μ^+e^- species. Thus, it is possible that some small fraction of the observed μ^- ions are in this state. Further progress in the area of Ps^- or μ^- physics might be conducted using charge exchange of e^+ or μ^+ in alkali vapors. Muonium or positronium negative ions are essentially pure leptonic "atomic" negative ions and could be used for sensitive tests of quantum electrodynamics (QED).

III. BRIEF HISTORY OF ELECTRON BINDING TO A MOLECULAR DIPOLE AND MOLECULAR QUADRUPOLE

The discovery of dipole-bound negative ions has an interesting history and is related to the discussion above concerning electron binding to fundamental particles. Turner[40] has published a detailed account of the theoretical quest for the "minimum dipole moment required to bind an electron." The story begins with theoretical considerations of the capture of negative mesons in matter by Fermi and Teller[41] and later by Wightman.[42] The title of the Fermi–Teller paper was "The Capture of Negative Mesotrons in Matter." Wightman was also interested in the slowing down of μ^- and π^- mesons in hydrogen. In his treatment, an electron was considered bound to the fixed dipole formed by the μ^- and proton. The paper by Wightman contains a calculation of the critical separation of charges +e and −e that just permits the binding of an extra electron and emphasizes that "The existence and size of the critical radius for our problem was pointed out by Fermi and Teller." Both of these groups found a critical radius of 0.639 a_o. Unaware of these calculations, a flurry of theoretical studies[43–46] in 1967–1968 converged on the minimum dipole moment ($D_{min} = 0.639$ ea$_o$ or 1.625 Debye) for electron binding to a stationary dipole. Later authors calculated the binding energy as a function of the dipole moment. These results are summarized in Figure 4. The dotted line and the solid line represent the values by Turner et al.[47] and Wallis[48] et al., respectively. The critical dipole moment is shown as a vertical dashed line. One notices the weak dependence of the binding energy as a function of dipole moment near threshold. The binding energy for many molecules such as those shown in Figure 4 ($D \leq 1$ ea$_o$) is well below the rotational energy for most molecules. Many studies in the 1960's attempted to examine the importance, or even detect the existence, of D_{min} from electron swarm experiments.[49–53] However, Garrett[54] showed that, for a freely rotating dipole, the D_{min} will depend upon the moment of inertia and the length of the dipole and increases with rotational quantum number. This observation, together with the difficulty of coupling the incident electron momentum and energy to the molecule through a diffuse state in which the extra electron has little interaction with the core, makes the effect of dipole states difficult to observe in electron-scattering experiments. It is clear from Figure 4, that a molecule with a sufficiently large dipole moment could permanently bind an extra electron as was shown more rigorously by Crawford and Garrett.[55]

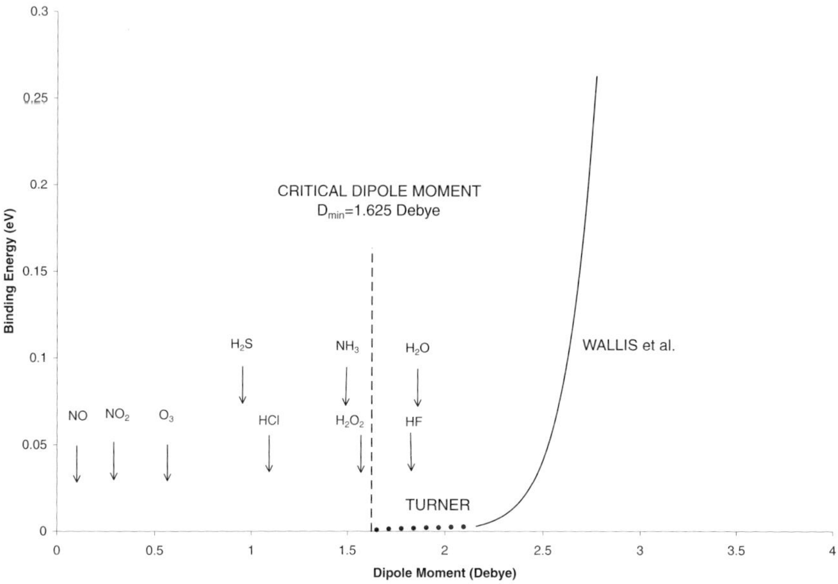

Figure 4. Binding energies of various molecules as a function of dipole moment.

The long-range dipole potential allowed Crawford and Garrett to neglect other effects of the molecular electrons to show that molecules whose dipole moments are above ~2 Debye should permanently bind an extra electron. The inclusion of polarization in the consideration would only increase the binding. The predictions of Crawford and Garrett were borne out by the first observation of a dipole-bound anion, the acetonitrile (CH_3CN^-) anion.[56,57] This anion was produced by charge transfer from an excited Rydberg rare gas atom, where the Rydberg-excited rare gas atoms were produced by electron-impact excitation. Dipole-bound anions were unknown at that time. It was observed also that CH_3CN did not attach slow free electrons, and theoretical considerations indicated that CH_3CN should not exhibit a valence-bound anion state. From these observations, it was suggested that CH_3CN^- exists in a very diffuse state, much like that of a Rydberg state and that Rydberg charge exchange, in this case, was different from free-electron attachment in that the Rydberg electron gently changes centers-of-force from the ion to the dipole during the "collision." In this paper,[57] it was suggested that both CH_3CN and CH_3NO_2 formed dipole-bound anions as a result of their rather large dipole moments (3.89 and 4.2 Debye, respectively). Initial attempts to reproduce these results for CH_3CN using tunable lasers to highly excite an alkali atom to specific n and ℓ states by R. Hill (Stanford Research Institute, private communication) were unsuccessful. These negative results caused one of the present authors (RNC) considerable consternation at that time. However, we now know that the charge transfer from the Rydberg excited atom to the dipole is a "resonance" type process (i.e., only atoms in *certain* quantum states undergo charge transfer), as shown by the group of Schermann. Specifically, Desfrancois et al.[58] have provided direct evidence for dipole-bound anions in a series of elegant experiments showing a narrow n (principal quantum number) dependence in the Rydberg charge transfer rate with molecules having dipole moments above the "critical dipole moment" of ~2.5 D. Furthermore, this group used electric field detachment to demonstrate that these anions were weakly bound and that the wave function describing the extra electron is diffuse. The field detachment thresholds were used to determine electron affinities for many of the polar molecules. The measured affinities fit well with the dipole model. Figure 5 shows the "resonance" like nature of the Rydberg-dipole charge-transfer process by plotting the dipole-bound anion signal versus the effective principal quantum number of the Rydberg atom for a series of polar molecules.

Figure 6 also summarizes the measured binding energies for dipole states versus the molecular dipole moment which adds to the data reported in the seminal papers by Desfrancois et al.[58] The Schermann group[59] has provided a comprehensive review of ground-state dipole-bound anions studied

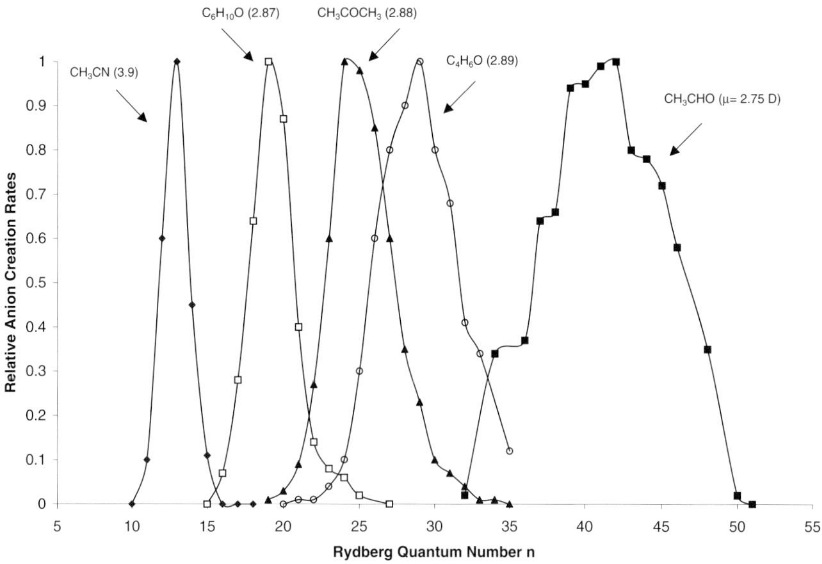

Figure 5. Relative Rydberg atom charge transfer rates as a function of the effective principal quantum number for a series of polar molecules (dipole moment given in Debye).[59]

by Rydberg charge transfer. The reader is referred to these papers for a more complete description of dipole states.

From a historical perspective, the first unambiguous observation of dipole states came from the group of Kit Bowen (Johns Hopkins) in 1990, who studied the important water dimer dipole anion.[60] The Bowen group has also studied ground state dipole-bound anions produced by "electron attachment" under high-pressure nozzle-jet expansion condition.[61] They have also used photodetachment photoelectron spectroscopy to determine electron affinities for a number of the molecules shown in Figure 4.

Of equal historical significance are the contributions of the groups of John Brauman (Stanford) and Carl Lineberger (Colorado). A number of free-radicals are known to exhibit both dipole-bound and more tightly bound conventional (valence) anions. The groups of Brauman[62] and Lineberger[63] have reported very narrow resonance features in the photodetachment spectrum corresponding to rotationally excited shape and Feshbach resonances for many of these dipole-bound radical anions.

The group of Mark Johnson (Yale)[64] has produced dipole-bound anions from photodissociation of the iodine atom/acetone and iodine atom/acetonitrile (CH_3CN) neutral clusters. Dipole-bound anions have been

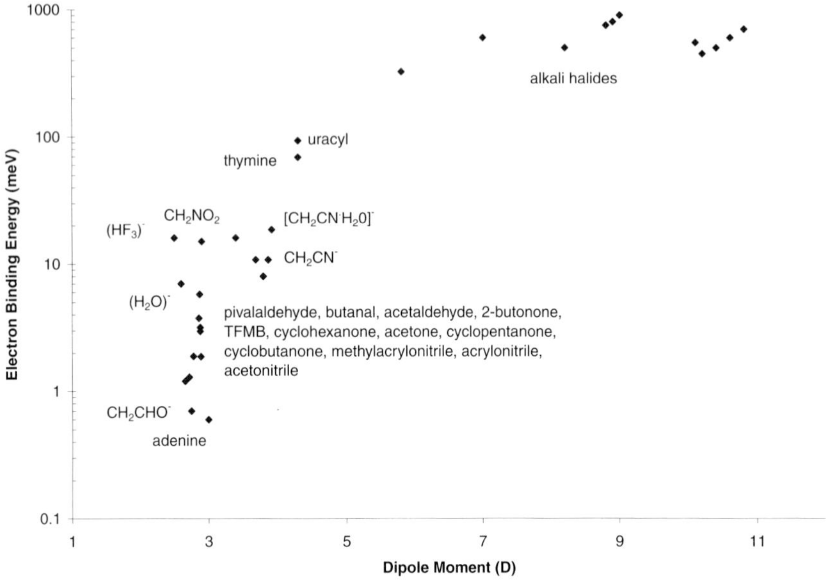

Figure 6. Electron binding energy (electron affinity) for a collection of polar molecules and clusters.

produced by Hashemi and Ellenberger[65] through dissociative electron attachment to clusters such as $(CH_3CN)_n$.

Although electron binding to a molecular dipole is now reasonably well understood theoretically and well documented experimentally, the binding of an electron to a molecular quadrupole is somewhat less certain. The existence of quadrupole-bound anion states was first considered by Jordan and Liebman[66] who predicted a large electron affinity for $(BeO)_2$ (0.65 eV) believed to be due to the quadrupole moment. Gutowski and Skurski have also performed calculations for $(BeO)_2^-$ at the coupled-cluster level with the result that $(BeO)_2^-$ in the D_{2h} ground state may be considered as a quadrupole-bound anion.[67] However, these conclusions concerning quadrupole binding as in the case of $(BeO)_2^-$ has been questioned by Gustev et al.[68] Using the criterion that quadrupole-binding should be approximately as diffuse as that for the dipole-bound anion, they could find no evidence of a diffuse bound state in $(BeO)_2^-$, rather, they find an even more tightly bound valence type anion (EA = 0.9 eV). Their results for potassium chloride dimers, however, do show indications of a quadrupole-bound anion. The rhombic ground-state dimer $(KCl)_2$, has a large quadrupole moment and the corresponding $(KCl)_2^-$ has a single weakly bound state whose properties are similar to that of a dipole-bound anion.

Other theoretical treatments of the possible binding of an excess electron to a molecule possessing a quadrupole moment have also appeared.[69–71] As a result of the shorter range for the quadrupole potential, it is not possible to make a general statement as to the "critical quadrupole moment" necessary to bind an extra electron as has been done in the case of the long-range dipole field. In fact, since the quadrupole potential falls off as $1/r^3$, the singularity at $r=0$ assures binding (negative energy) for a point quadrupole for any magnitude of Q. Since the quadrupolar effects are operative at distances close to the atom or molecule, the details of the atomic or molecular structure contribution must be included, particularly the induced polarizability. Prasad, Wallis and Herman[69,70] calculated the binding energy of an electron to a finite linear electric quadrupole (Q) in two configurations: one which has two positive charges each of charge +q, symmetrically placed about a negative charge of −2q (A) and the other case for the charges reversed (B). Their calculations predict that the minimum quadrupole $Q_{\min}$ to bind an electron is $Q_{\min}$ (A) = 21.0 a.u. and $Q_{\min}$ (B) = 2.66 a.u. Of course, a given molecule may possess both dipole, quadrupole and higher multipole moments and will always exhibit a certain degree of polarizability. Thus the quadrupole moment in these cases will play a role in the binding of electrons to molecules which possess a permanent dipole moment. In some molecules for which the dipole moment is almost large enough to produce a "dipole-bound" state, the quadrupole moment (along with polarizability) may tip the balance and provide the additional energy necessary for positive binding.

There exists presently a small number of experimental measurements of quadrupole-bound negative ions. The first clear evidence for these species came from studies of Rydberg charge transfer to *para*-dinitrobenzene.[72] *Para*-dinitrobenzene has a quadrupole moment of sufficient strength to support a quadrupole-bound state, but it also has a bound valence anion state. The Rydberg charge-transfer cross section exhibits a "resonance" like feature as a function of the principal quantum number n (or (n^*) reminiscent of the dipole features. Abdul-Carime and Desfrancois[73] have recently provided convincing experimental evidence for quadrupole-bound states. Diffuse anions of the succinonitrile molecule and the formamide dimer were studied using Rydberg charge exchange and field attachment. These nonpolar molecules possess very large quadrupole moments and serve as a better example of quadrupole binding than *para*-dinitrobenzene since these molecules are not expected to possess a valence-bound anion state. The authors also used a simple electrostatic model to predict the stability of electrons bound to multipolar potentials.

Historically, the first experimental evidence available for electron binding to a molecular quadrupole comes from an analysis of data presented for the series of molecules CO_2, COS, and CS_2. Harth, Ruf and Hotop[74] were the first to report a very interesting dependence for the CS_2^- production cross section for collisions as a function of n^* with Ne(ns) and Ne(nd) high Rydberg atoms. The cross section exhibited a steep rise at low n^*, a characteristic maximum for n^* around 18 and a steep decrease toward higher n^* (see Figure 7). In addition, Dunning et al.[75] and Carman et al.[76] reported an n^* dependence in the Rydberg charge-transfer cross section for the CS_2 molecule which is also reminiscent of the features exhibited by dipole-bound states. Dunning et al. further showed that the CS_2^- anion was easily field ionized by a weak electric field, again characteristic of a weakly bound dipole state. In an attempt to explain these observations, Compton, Dunning and Nordlander[77] performed calculations which suggested that electron binding to the quadrupole field of CS_2 ($Q = +3.3$ au) might account for these results. The absence of weakly bound states and no pronounced peak in the n^* charge-exchange dependence with COS added further credence to this proposal (Q (COS) $= -0.2$ au). However,

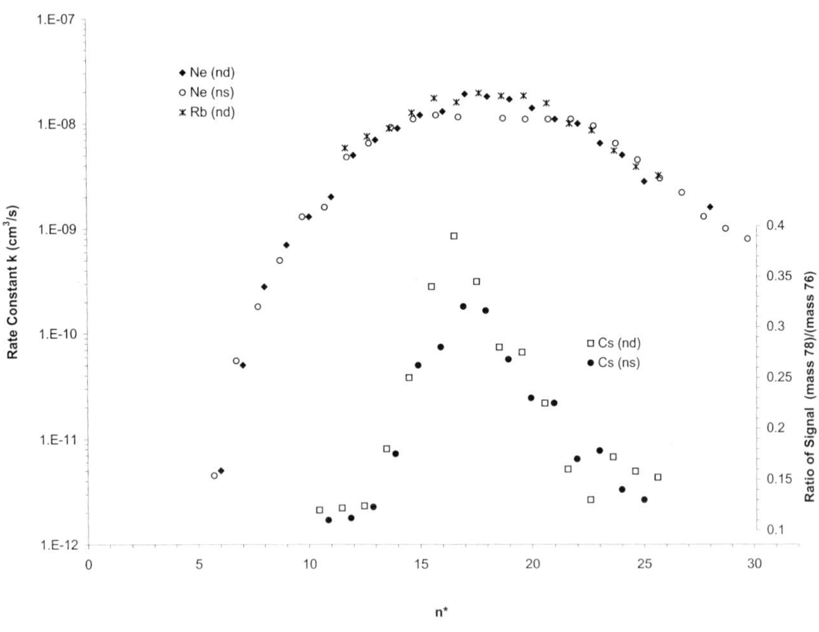

Figure 7. Plot of the rate constant for Ne (*ns, nd*)[74] and Rb(*nd*) (present work) with CS_2 molecules leading to CS_2^- anions as a function of effective quantum number (logarithmic scale) along with a plot of the ratio of $^{32}SC^{34}S^-$ to $^{32}SC^{32}S^-$ ion signal.[76]

given the results of Prasad et al.[69,70] on the binding of an electron to a negative quadrupole moment, CO_2 should have also formed a quadrupole-bound anion ($Q = -3.1$ a.u.). Anions were not reported for CO_2 in Ref. 75 or 76. However, it should be pointed out that these anions were not specifically under investigation and could have eluded detection as a result of the extreme fragility or narrow n dependence. Also somewhat on the negative side, Gutsev, Bartlett and Compton[78] have recently reported calculations for the binding energies for CO_2^-, COS^- and CS_2^-. These calculations were not definitive as to the formation of quadrupole-bound states.

Rydberg charge exchange reactions with CS_2 show another most unusual feature. Carmen et al.[76] reported a pronounced isotope effect in the Rydberg electron transfer reaction between Rb** and $^{32}SC^{32}S$ and $^{32}SC^{34}S$. For a narrow range of (n^* near $n^* = 17$, the rate constant for CS_2^- formation was found to depend upon the isotropic composition of the molecule, producing a negative ion isotope ratio (mass 78 to mass 76 amu) up to 10 times larger than the natural isotope ratio of CS_2. This dramatic effect is shown in Figure 7 along with absolute Rydberg charge exchange rates of Harth et al.[74] and our relative reaction rates for Rb(nd) collisions. Carman et al.[76] considered nuclear spin and statistics to show that only even-J rotational levels are allowed for the symmetric $^{32}SC^{32}S$ species whereas all quadrupole states are allowed for the asymmetric $^{32}SC^{34}S$ molecules. If one evokes quadrupole binding to explain the narrow n^* dependence and the diffuse nature of some of the anions formed in this region of n^* (see Ref. 75), it is reasonable to suggest the presence of a greater density of rotational states for $^{32}SC^{34}S^-$ and a concomitant greater integrated ion yield. This idea could be tested using other symmetric quadrupolar molecules or perhaps polar symmetric top molecules.

Abdul-Carime and Desfrancois[73] also considered quadrupole binding with specific reference to CS_2^-. They considered only the quadrupole and polarization terms

$$V = -\frac{Q}{4r^3}(3\cos^2\theta - 1) - \frac{1}{2r^4}(\alpha_\perp + (\alpha_= - \alpha_\perp)\cos^2\theta) \quad r > r_o \qquad (4)$$

where r and θ are the electron position in cylindrical coordinates and Q is the diagonal element of the molecular quadrupole tensor along the z axis and $\alpha_=$ and $\alpha_\perp$ represent the molecular polarizability tensor elements along (=) and perpendicular ($\perp$) to the z axis. These authors used reasonable parameters in the assumed potentials to predict a binding energy of $\sim$ 5 meV. They further predicted that the Rydberg Electron Transfer (RET) should maximize around $n^* = 20$. Broadening seen in the

RET peak around $n^* = 20$ is attributed to interaction of the quadrupole-bound state with the known valence anion state at -0.9 eV.[79]

A brief discussion of dipole-and quadrupole-bound states was recently included in a review of neutral/neutral charge transfer reactions.[80] In summary, although there exists some experimental and theoretical evidence for quadrupole-bound states, their existence is not as certain as the case dipole-bound states.[73,80]

Finally, we described a remarkable observation of negative-ion formation following the collision of *two* Rydberg excited sodium atoms. Ciocca et al.[81] have reported the first measurements of negative Na^- ions formed by collisions between two Rydberg excited Na** atoms i.e.

$$Na^{**} + Na^{**} \rightarrow Na^+ + Na^- \tag{5}$$

The Na** atoms were produced by two-step (pulsed) laser excitation $(3s - 3p + n\ell)$ of sodium vapor($\sim 10^{11}$ to 10^{12} Na/cm^3). Although initial excitation occurs to *ns* and *nd* Rydberg states, rapid collisional ℓ- mixing leads to a statistical distribution of ℓ states, particularly at high n. The principal quantum number n was varied from 7 to 40. At low n, the density dependence of Na versus the power of the laser producing the $n\ell$ states $(3p - n\ell)$ was approximately linear, indicating that reactions of Na* with ground-state Na atoms resulted in $Na^- + Na^+$ ions in their ground state. This mechanism was originally suggested by Lee and Mahan.[82] Cheret and Barbier[83] have also studied Rb^- formation from collisions between Rb(*6d*) and ground state Rb atoms.

At high n, Ciocca et al.[81] found the laser-power dependence for Na^- formation to be quadratic, suggesting the reaction mechanism (5) above. If the product Na^- ion is in its ground state (excited Na^+ can be dismissed on energetic grounds), then reaction 5 is ~ 5.5 eV exothermic. This led Ciocca et al. to suggest three possible explanations for this rather remarkable observation.

(1) The excess energy is shared as kinetic energy between the recoiling $Na^+ \rightarrow \ldots Na^-$ ions. This would require the $Na^{**} + Na^{**}$ system to proceed to the $Na^+ + Na^-$ channel through a series of potential energy curve crossings in the region of 5–10 eV. This is a plausible mechanism which could be proven by measurement of the Na^- or Na^+ ion kinetic energy (~ 2.7 eV).

(2) A second possibility involves radiative ion-pair formation i.e.,

$$Na^{**} + Na^{**} \rightarrow Na^+ + Na^- + h\nu\ (\sim 5.5\ \text{eV}) \tag{6}$$

Ciocca et al.[81] made a brief attempt to detect the ~ 5.5 eV photons but the results were inconclusive.

(3) A final mechanism involves the formation of Na^- ions in a doubly excited state. This mechanism involves a reaction in which the electron on one ion core switches to another Rydberg core producing a "planetary negative ion." The corresponding Na^- ion would be an electronically excited Feshbach resonance with a lifetime that might exceed a few microseconds. Since the lifetimes of most Rydberg states generally well exceed this time, such a mechanism is plausible. One could also speculate that the extra electron is bound to an excited sodium atom which possesses a large quadrupole moment (quadrupole-bound anion).

Whatever mechanism proves to be responsible for reaction 5, it will indeed be interesting. The possibility of an electron bound to a highly excited sodium atom is of particular interest and is not the first such suggestion. Experimental studies of molecular anions consisting of a closed-shell cation core with two "Rydberg-like" outer electrons, has been evoked following experimental studies of Coe et al.[84] and theoretical calculations of Cardy et al.[85] and Ortiz.[86] Specifically these observations suggested that NH_4^- might exist as a stable or metastable species such as tetrahedral NH_4^+ with two diffuse electrons "orbiting" about the ion core. Further calculations by Gutowski et al.[87] showed that many such anions (e.g., NH_4^-, H_3O^-, H_2F^-, NeH^- and others) may exist in long-lived metastable states whose decay to underlying neutral-molecule states requires the ejection of one electron and the simultaneous "shake-down" redistribution of the second orbital electron. Theoretical and experimental studies of atomic "planetary" Rydberg anions would offer a convenient model system.

IV. EXPERIMENTAL TECHNIQUES

A. Rydberg Charge-Exchange Methods for Producing Dipole and Quadrupole-Bound Anions

In this section, we discuss the methods employed in our group which have been used to produce highly excited Rydberg states for collision studies. First let us discuss the Rydberg state. The ground and excited states of an atom are termed Rydberg states if the energy levels can be described as a quasi-hydrogenic "one-electron" atom, i.e., provided that the energy levels relative to the ground state follow the Rydberg formula

$$E_{n,\ell} = \mathrm{IP_A} - \mathrm{R_A}/n^{*2} \tag{7}$$

Where IP_A represents the ionization potential of the atom, A, R_A is the Rydberg constant for the atom and n^* is the effective principal quantum number ($n^* = n - \delta_\ell$,) with δ_ℓ being the ℓ-dependent quantum defect and n is the principal quantum number). The magnitude of the quantum defect depends upon the degree to which the "one electron" interacts with its "core." For example, the s and p states for the sodium atom have quantum defects of 1.35 and 0.85, respectively, and δ_ℓ is small for $\ell \geq 2$. The dependence of the quantum defect with increasing ℓ can be seen by considering the limiting case in which the wave functions of the core and the outer electron do not overlap, i.e., for angular momentum, ℓ, greater than some value, ℓ_o, so that the interaction potential is mainly due to the induced dipole (α_d/r^4) and quadrupole contribution ($-\alpha_Q/r^6$). The quantum defect then becomes[88–91]

$$\delta_\ell(\ell > \ell_o) \cong 3\alpha_d/4\ell^5 + 35\alpha_Q/16\ell^9 \tag{8}$$

From this expression, one can see that the quantum defect decreases rapidly with increasing angular momentum. Likewise, the classical trajectory of a high n and ℓ Rydberg electron becomes more like that of the orbit of the planets around the sun (a hollow planetary atom). This analogy with the hydrogen atom allows one to describe the energy levels of high-Rydberg atoms in a familiar manner and to examine the "scaling" of different properties with the effective principal quantum number. Table I presents some properties of "hydrogenic" Rydberg atoms and their "scaling" with principle quantum number, n.

In these experiments, the highly excited Rydberg states are produced by one photon excitation of np states or two-photon excitation of ns and nd

Table I. Selected properties of high-Rydberg atoms as a function of the effective principal quantum number, n^*. The ionization potential of the atom is designated as IP_A, a_o denotes the radius of the first Bohr orbit (5.29173×10^{-9} cm), r_n is the mean radius, v_m is the rms velocity of the Rydberg electron, τ_n is the period for electronic motion, and E_{n^*} is the binding energy of the electron in the state n^*. Note that E_{n^*} is equal to $IP_A - E_{n,\ell}$.

Property	*n-dependence*	$n^*=1$	$n^*=10$	$n^*=100$
$\langle r_n \rangle$ (cm)	$n^{*2}a_0$	5.3×10^{-9}	5.3×10^{-7}	5.3×10^{-5}
ν_n (cm/s)	ν_0/n^*	2.2×10^8	2.2×10^7	2.2×10^6
τ_n (s)	$n^{*3}\tau_1$	1.5×10^{-16}	1.5×10^{-13}	1.5×10^{-10}
E_{n^*} (eV)	R_A/n^{*2}	$R_A = 13.6$	$R_A \times 10^{-2}$	$R_A \times 10^{-4}$

Note that the values for $n^* = 1$ correspond to the hydrogen atom (no quantum defect) and IP = 13.6 eV.

states. In some cases, two lasers of different frequencies are used to produce resonantly enhanced two-photon excitation of *ns* and *nd* states (we are assuming that the *n* here corresponds to n^* discussed above). The excitation "steps" are shown below pictorially in Figure 8.

The effects of an external electric field on high Rydberg states (field ionization) is shown in Figures 9 and 10. In this study a Continuum (Sunlite) OPO laser was used to excite *ns* and *nd* Rydberg states of rubidium. A certain time after the laser pulse, a 5 μs voltage pulse of magnitude V is applied to the parallel plate accelerator grids separated by a distance $d = 13$ mm, (field strength $E = V/d$). As the field is increased, the "effective" ionization potential of the atom is lowered such that Rydberg atoms with electronic energy above the potential maximum shown in Figure 9 are field ionized when the field is turned on. The field-modified potential in one dimension (i.e., in the electric field direction) is given by

$$V(r) = -\frac{e}{r} - Er \tag{9}$$

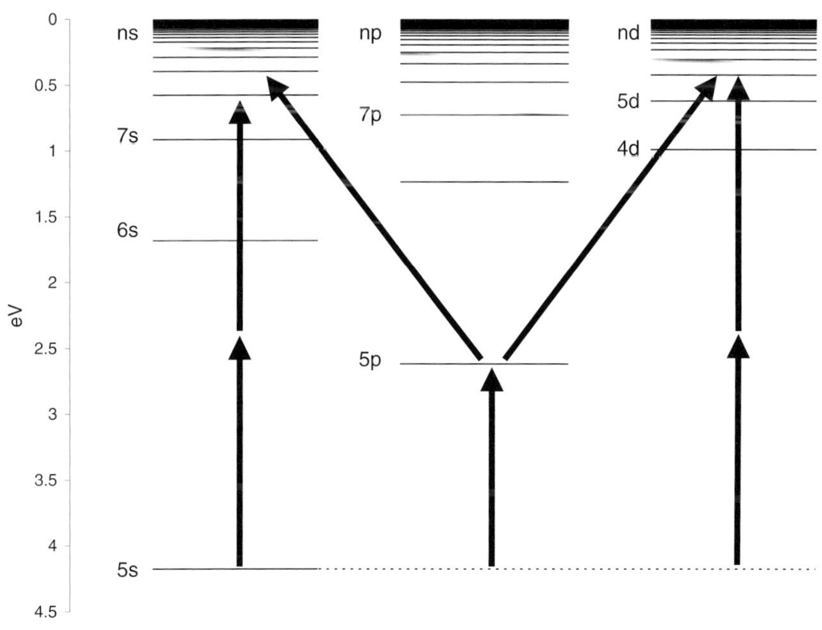

Figure 8. Excitation scheme for producing one- and two-photon excited *ns, np* and *nd* states using the rubidium alkali atom as an example.

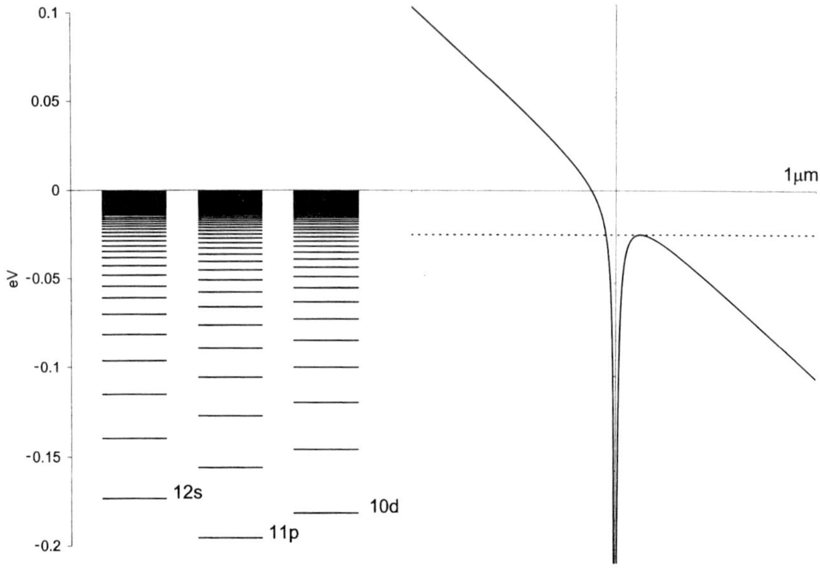

Figure 9. Electric field modified potential energy diagram for rubidium showing the onset of field ionization of high Rydberg states.

where e is the absolute value the electron charge. One can find the potential maximum, $V_{\max}$, (critical potential) as a function of electric field from

$$V_{\max} = -2(eE)^{1/2} \tag{10}$$

Rydberg states with energy above this value escape over the barrier and those below $V_{\max}$ can only tunnel through the potential barrier. Thus ions result from either three-photon ionization or field ionization. The *ns* series was predominantly seen through field detachment, whereas the *nd* series could be readily seen through either mechanism. One will notice that there is not a completely abrupt jump in progressing from three-photon ionization to field ionization. Presumably some degree of tunneling is occurring during the 5 μs width of the draw-out pulse for those states lying just below $V_{\max}$. Field ionization of dipole and quadrupole states can also occur in the same manner if the bound state is above the electric field modified potential of the anion. The combined potentials of a dipole-bound anion and a quadrupole anion in a uniform electric field, respectively, can also be written as

$$V(r) = -\frac{\mu}{r^2} - Er \tag{11}$$

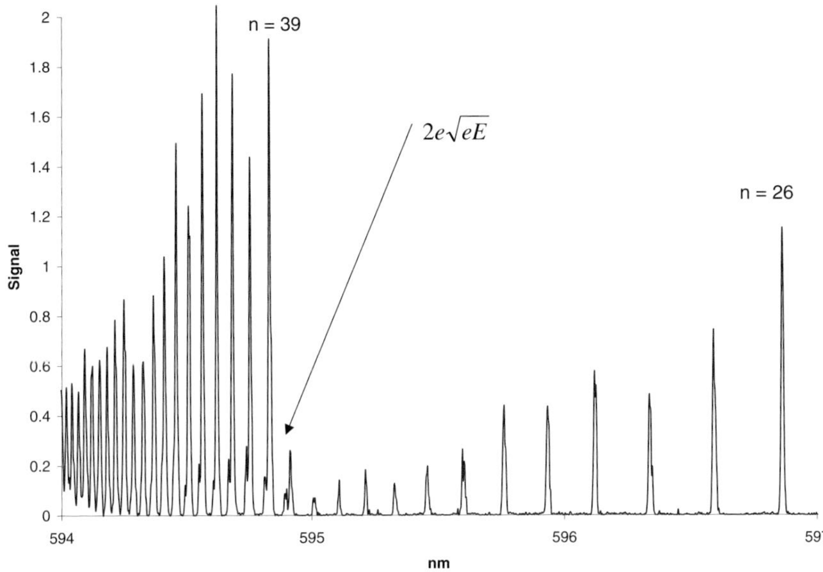

Figure 10. Rydberg Series for rubidium showing field ionization at wavelengths below the field ionization threshold and 2 + 1 REMPI above it. Field ionization in this instance begins at $n = 38$, which is a result of an applied electric field of 15,000 V/m.

$$V(r) = -\frac{Q}{r^3} - Er \tag{12}$$

where μ is the molecular dipole moment and Q is the molecular quadrupole moment. The resulting critical potentials for field detachment of dipole and quadrupole-bound anions are, respectively,

$$V_{\max,\text{dipole}} = -3\left(\frac{\mu E^2}{4}\right)^{1/3} \tag{13}$$

$$V_{\max,\text{quad}} = -4\left(\frac{QE^3}{27}\right)^{1/4} \tag{14}$$

Dipole-bound anions are characterized both by their small electron affinities and diffuse nature. Laser photodetachment of dipole-bound anions leave the neutral molecule in the same (or approximately so) vibrational state as that of the anion. The diffuse nature of the dipole-bound anions allow for field-detachment as discussed above. Field-detachment has conventionally been carried out by applying a strong electric field in the direction of the ion motion in a time-of-flight mass spectrometer and observing the onset of fast neutral atoms as a function of electric field. We have recently devised another experimental scheme for studying electric field detachment. It is found that application of an increasing electric field in the ion draw-out region of the time-of-flight mass spectrometer also results in the abrupt disappearance of negative ions. This is shown in Figure 11 where the acetone dipole-bound negative ion signal ($\mu = 2.88$ Debye, EA $= 2.65$ meV)[59] is observed to promptly disappear at fields above ~ 500 V/cm and totally disappears at ~ 1 kV. Signals from stable negative ions (e.g. SF_6^-) or higher dipole moment molecular anions are constant over this electric field range. The field detachment shown in Figure 11 agrees exactly with that reported by Desfrancois, et al.[59] using field detachment of fast ions although the ion residence time in our experiment is much longer. After some analysis of the data, electron affinities can also be deduced from these measurements. The simplest analysis of the field detachment threshold shown in Figure 11 using equation 13 predicts an electron affinity of 1.4 ± 0.1 meV which agrees with the original study reported by

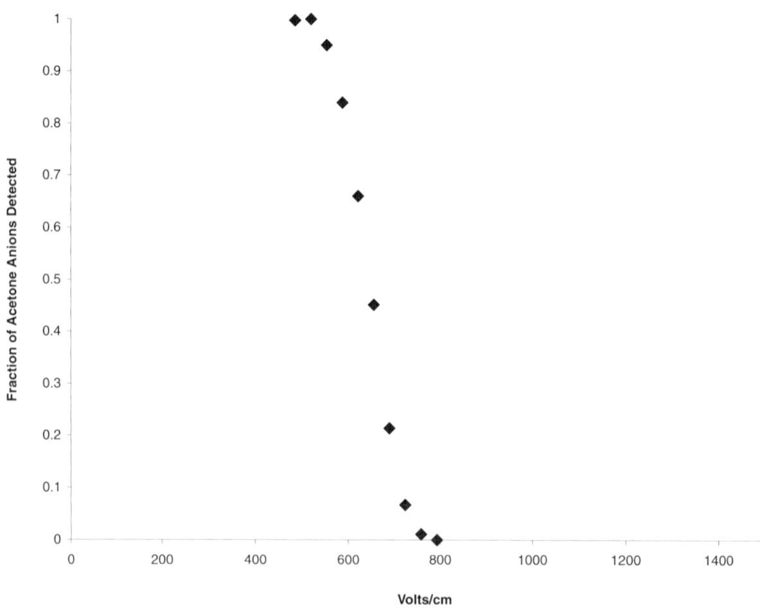

Figure 11. Acetone negative ion ($(CH_3)_2CO^-$) signal as a function of ion draw-out voltage in the TOF mass spectrometer. The disappearance of signal is attributed to electric field detachment.

Desfrancois, et al.[58] Desfrancois has also considered tunneling through the barrier which involves the time the anion spends in the electric field. From this analysis he arrived at an electron affinity of 2.65 meV.[58] Electron tunneling through the narrow potential barrier in the dipole-bound case is clearly important but difficult to treat exactly. Tunneling is expected to be even more important for field detachment of quadrupole-bound anions. Finally, our data shown in Figure 11 emphasize the importance of weak electric fields in the acceleration (and detection) of low dipole moment multipole-bound anions.

Our experimental apparatus used to study RET reactions is shown schematically in Figure 12. In a particular study,[92] cesium or potassium atoms in an effusive beam were excited to *ns* and *nd* Rydberg levels by two-photon absorption from a pulsed, Nd:YAG-pumped dye laser using the dyes DCM for Cs and R6G for K. The *np* Rydberg states were obtained by one-photon absorption of the frequency-doubled output of a dye laser. The excited alkali beam was crossed at 90° with a pulsed supersonic beam containing a dilute mixture of CCl_4 and SF_6 in helium. The nozzle jet produced pulses of sample molecules often mixed with rare gas that were introduced into the collision region through a skimmer which was pumped by a Roots blower backed by a fore pump. The beams were crossed in a

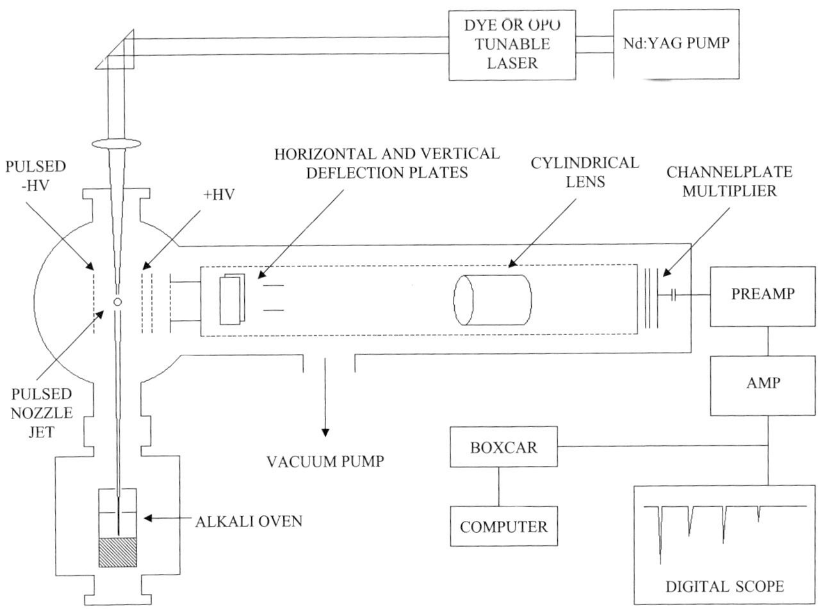

Figure 12. Schematic diagram of Rydberg electron transfer experimental apparatus.

field-free region between two parallel mesh grids in a Wiley–McLaren-type ion source of a time-of-flight mass spectrometer (TOFMS). The positive or negative ions produced by RET were extracted into the TOFMS by applying a voltage pulse to one of the grids at a time t_r (typically 0.1 to 5 μs) after the laser pulse. The ions were detected with a dual channelplate electron multiplier, and the mass spectra were recorded with a digital oscilloscope.

As an illustration of the reaction of high Rydberg atoms with electrons attaching to molecules, we shall discuss the results for SF_6^- in terms of n^*. SF_6^- signals were measured as a function of n^* using a gated boxcar integrator to monitor the ion signal as the laser was scanned through the Cs Rydberg series. For a given ℓ, the two-photon excitation rate should scale approximately as $(n^*)^{-3}$. After making small corrections for variations of laser power with wavelength and Cs** radiative lifetimes, it was found that the SF_6^- signal varies as $(n^*)^{-3}$ for $n^* \geq 25$. We also found that the ratios of signals I (*nd*)/I (*ns*) agree well with calculated ratios of two-photon excitation rates for the *ns* and *nd* state of Cs. It was therefore concluded that $k_{SF_6^-}$ is independent of n and ℓ for $n^* \geq 25$, consistent with the results of Dunning and co-workers for Xe (*nf*),[93] Rb (*ns*, nd),[94] and K (*nd*)[95] Rydberg atoms. Rate constants for SF_6^- formation were thus obtained by multiplying the measured signal intensities by $(n^*)^3$ and normalizing the results at large n^* to $k_{SF_6^-} = 4 \times 10^{-7}$ cm^3 s^{-1}, as measured by Dunning and co-workers. For $n^* \leq 25$ we found that $k_{SF_6^-}$ falls significantly below the value observed at large n^*. Similar results have been previously reported for K(*nd*) and Na(*np*) Rydberg atoms. Zollars et al.[95] attributed this decrease in ion production to electrostatic interactions between the nascent negative ion and the positive Rydberg core. At lower n^*, the negative ion is formed in closer proximity to the core, with a correspondingly stronger Coulombic attraction between them. Thus as n^* decreases, fewer ion pairs possess sufficient kinetic energy to separate and be detected. Using a simple model based upon these assumptions, Zollars et al. were able to successfully fit their experimental data. To further test this interpretation, we measured the n^* dependence of $k_{SF_6^-}$ for the *nd* states of Cs at two different collision energies. Nozzle-jet expansion using rare gas buffer gases was used to vary the collision energy. For higher-energy collisions, a larger fraction of the nascent ion pairs are able to overcome their Coulombic attraction and separate, yielding larger values for $k_{SF_6^-}$. Figure 13 summarizes the formation of SF_6^- as a function of n^* and collision energies. In addition, typical wavelength scans of the Cl^- and SF_6^- signals are shown in Figure 14 for the region $20 \leq n^* \leq 27$. It can be seen even from this figure that the ratios of signals obtained for the *d* states to those obtained for the *s* states, I(*nd*)/I(*ns*), are different for the two molecules.

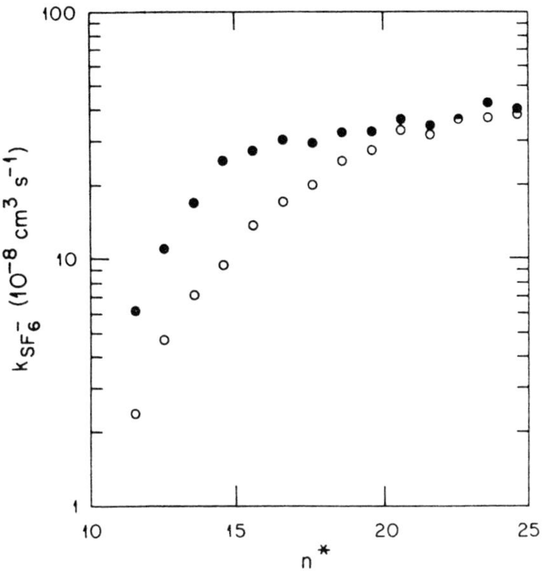

Figure 13. Apparent Rate Constants for production of SF_6^- for collisions between Cs (*nd*) and SF_6 with collisional energies of ~0.15 eV (filled circles) and ~0.05 eV (open).

B. Methods of Producing Multiply Charged Negative Ions

Multiply charged negative ions have been produced through a number of mechanisms employing a variety of experimental methods. We outline some of these mechanisms below and provide illustrations with experimental methods used in our studies at the University of Tennessee and the Oak Ridge National Laboratory:

(1) Studies of positive ion collisions with surfaces have reported the formation of gas phase doubly charged anions. One of the first complete reports of doubly charged anions came from studies of carbon cluster dianions (C_n^{-2}) produced by the collisions of high energy Cs^+ ions with graphite.[96] This *sputtering* of negative ions is not fully understood at this time, but could involve the attachment of an electron to singly-charged anions in the gas phase. The singly-charged anions could be produced by a variety of mechanisms.

(2) Laser desorption of negative ions (LDNI) from surfaces sometimes leads to the formation of doubly charged negative ions. Notable among these studies is the formation of fullerene dianions such as

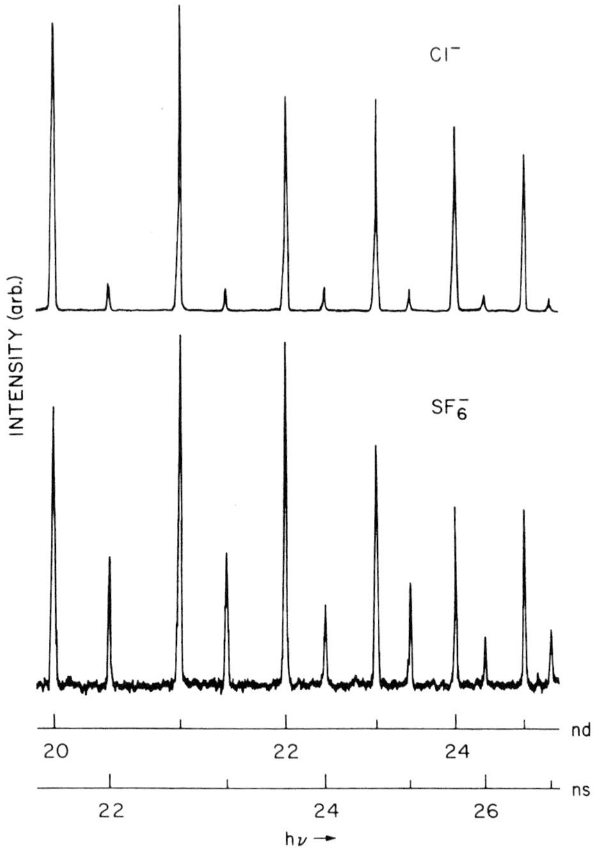

Figure 14. Wavelength Scans of Cl^- and SF_6^- intensities for collisions of Cs (*ns, nd*) with CCl_4 and SF_6.

C_{60}^{-2} (see Refs. 97 and 98). To date, laser desorption has only produced dianions and no more highly charged anions. The actual mechanisms leading to dianions in these studies is not certain. Surface ionization was first proposed,[98] but in light of the following mechanism (electron attachment) this proposal is now less attractive. We should also point out that LDNI has only been observed in cases where strong magnetic fields are present (FTMS) suggesting that the B field confines the fast electrons in the vicinity of the singly charged negative ions and allowing for a greater probability of attaching a second electron.

(3) The step-wise attachment of two electrons to the $C_{60}F_{48}$ molecule *in the gas phase* has been demonstrated.[99] This mechanism involves the

attachment of an incident electron to the preexisting gas phase anion *over* the top of the Coulomb barrier described in Section IV. A. Dianions formed in this manner are metastable (i.e., the total energy is positive) and can decay by autodetachment. In many cases, the molecular degrees of freedom are sufficiently large that the lifetime is sufficiently long enough to be studied in a mass spectrometer, as in the case of $C_{60}F_{48}^{-2}$. Ultimately, infrared radiative decay could also make the gas-phase anions stable. Step-wise attachment of two electrons was also studied for the C_{84} fullerene.[100] In this study, the lifetime for autodetachment over the top of the Coulomb barrier i.e.

$$e + C_{84}^{-*} \rightleftarrows C_{84}^{-2**} \tag{15}$$

was measured and found to compare favorably with vibrational autodetachment rate constants employing quasiequilibrium theory.

(4) Boltalina et al.[101] have recently reported the observation of dissociative charge exchange of singly charged anions at high collision energy (50 keV) with a number of target gases that results in doubly charged anions, e.g.

$$C_{60}F_{35}^{-} + CH_4 \rightarrow C_{60}F_{34}^{-2} + F + CH_4^{+} \tag{16}$$

Tuinman and Compton[102] recently performed a variation of this experiment to study the nondissociative reaction

$$C_{60}F_{36}^{-} + CH_4 \rightarrow C_{60}F_{36}^{-2} + CH_4^{+} \tag{17}$$

Figure 15 shows a schematic diagram of the ZAB mass spectrometer used in these studies. Singly charged negative ions are produced by nondissociative electron attachment in the source S followed by mass (B) and energy (E) analyses. Ions formed after collisions with gases (e.g. CH_4) in CC3 are mass analyzed with a quadrupole mass spectrometer. Figure 16 shows the primary ion mass spectrum (top) along with the doubly charged anions formed through collisions (bottom). Also shown is the energy threshold for the formation of $C_{60}F_{36}^{-2}$. One will note that the apparent center-of-mass threshold is well above that expected from the calculated energetics of the reaction assuming a Coulomb barrier of ~1.5 eV. This would indicate that this method does not offer promise for the measurements of second electron affinities. We also note that collisions of C_{84}^{-} with CH_4 did not produce observable yields of C_{84}^{-2}, making the method less promising as a general method for producing multiply charged negative ions.

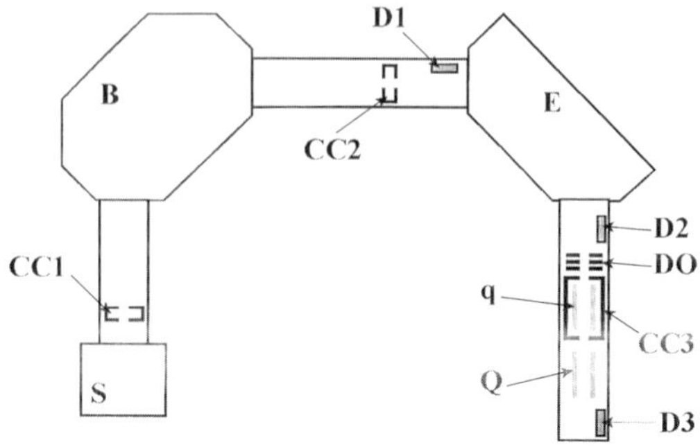

Figure 15. Schematic diagram of the ZAB mass spectrometer where S is the electron attachment ion source, CC1, CC2, and CC3 are collision cells, D1, D2, and D3 are ion detectors, B is an analyzing magnet, E is an electric field energy analyzer, q is the quadrupole field, Q is the quadrupole mass filter, and DO is ion optics.

(5) Electrospray ionization appears to be the most general method currently used to produce multiply charged negative ions. A detailed review of electrospray is given by Klassen et al.[103] and a brief review of this method, adequate for the purposes of this discussion, is given by Scheller et al.[2] The negative-ion electrospray apparatus employed at the University of Tennessee is shown in Figure 17. Tuinman and Compton have used such an apparatus to study the structures of gas phase $(C_{60})_n(CN)_m^{-3}$ trianions from reactions of C_{60} with NaCN in solution.[104] The reaction mixture is directly infused into the apparatus with a source temperature of 90°C and capillary and cone (nozzle to skimmer) voltages of 2250 and 30 V, respectively. They were able to identify numerous species with the general structure $(C_{60})_n(CN)_n(OH)_u(OOH)_v^{x-}$ that originated from oxygen being introduced from either dissolved O_2 or water. They found that, when $n>1$ and $x>1$ these species were composed of building blocks such as $C_{60}(CN)_2^-$ which could be separated into individual components by collision-induced dissociation.

V. EXAMPLES OF DIPOLE-BOUND ANIONS AND LIKELY EXAMPLES OF QUADRUPOLE-BOUND ANIONS

The Rydberg Electron Transfer technique (RET) has proven to be an excellent method for selectively producing fragile ground state dipole-bound

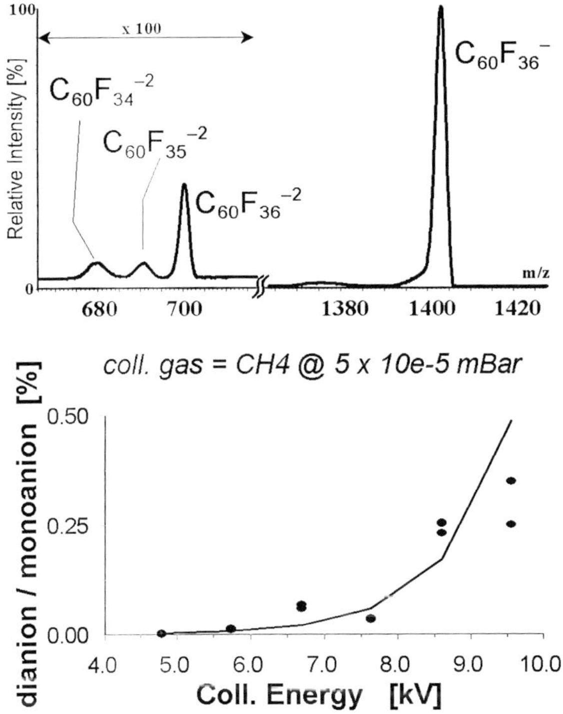

Figure 16. Top — Mass-analyzed ion kinetic energy spectrum (MIKES) of the precursor ion $C_{60}F_{36}^-$ acquired at 10 keV collisons energy with the collision gas. Bottom — the intensity ratio of the product dianion ($C_{60}F_{36}^{-2}$) to the monoanion precursor ($C_{60}F_{36}^-$) as a function of collision energy.

anions.[59] This method uses atoms excited to high Rydberg states by means of tunable lasers. These Rydberg atoms collide with thermal or supersonic beams of polar molecules or clusters resulting in the transfer of their diffuse electron, giving birth to anions which are stabilized against the reverse process, autodetachment, by the presence of the Rydberg ionic core. A crucial point of the RET method is that charge-exchange reaction rates exhibit pronounced maxima in the cross sections due to the necessity of favorable orbital "matching" for efficient electron transfer to polar molecules. These maxima appear at either specific values of the quantum number, n, or the effective quantum number, n^*, (see Section IV. A.) which decrease with increasing dipole moment. The RET signature for the production of weakly bound anions is the presence of a narrow peak at $n_{\max}$ or $n^*_{\max}$ in the RET n-dependence. This Rydberg selectivity in anion

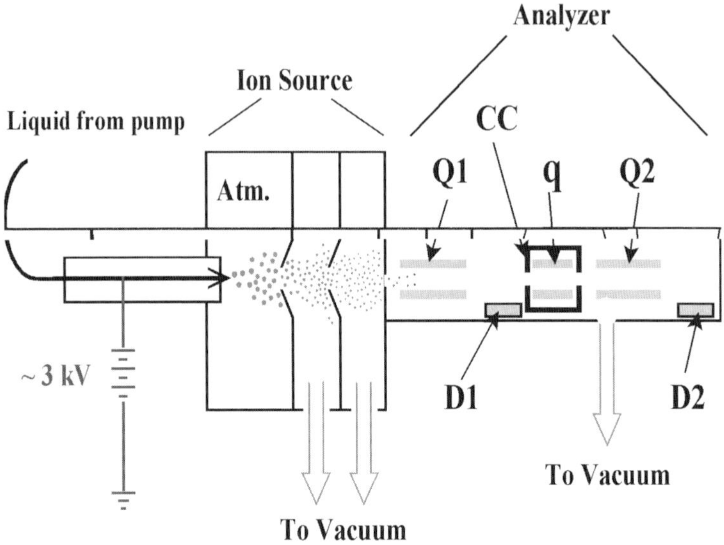

Figure 17. Negative ion electrospray apparatus where Q1 and Q2 are quadrupole mass spectrometers, D1 and D2 are ion detectors, CC is a gas collision cell, and q is a quadrupole ion focussing lens.

production can be understood within the framework of a multiple ionic-covalent crossing model.[59] An empirical law, which has been tested with a large number of dipole-bound anions (and a few quadrupole-bound anions), directly relates the corresponding electron affinity (EA) to the experimentally determined n_{max} value and is given by

$$\mathrm{EA(eV)} = \frac{23}{(n_{\max})^{2.8}} \tag{18}$$

The signature for creation of valence anions ("free electron" attachment), on the other hand, is the observation of a smooth variation of the RET n-dependencies over a wide range of n corresponding to high n-values.

Charge exchange reactions between excited Rydberg atoms and pulsed supersonic beams of helium seeded with nitromethane were studied both at the Oak Ridge National Laboratory using Cs(ns, nd) Rydberg atoms and at the University of Paris-Nord using Xe(nf) Rydberg atoms.[105] The negative ions were mass-analyzed by time-of-flight spectroscopy ($m/\Delta m = 100$). Monomer anions $CH_3NO_2^-$ as well as cluster anions $(CH_3NO_2)^-_{2,3}$ were observed under usual conditions. Figure 18 shows the

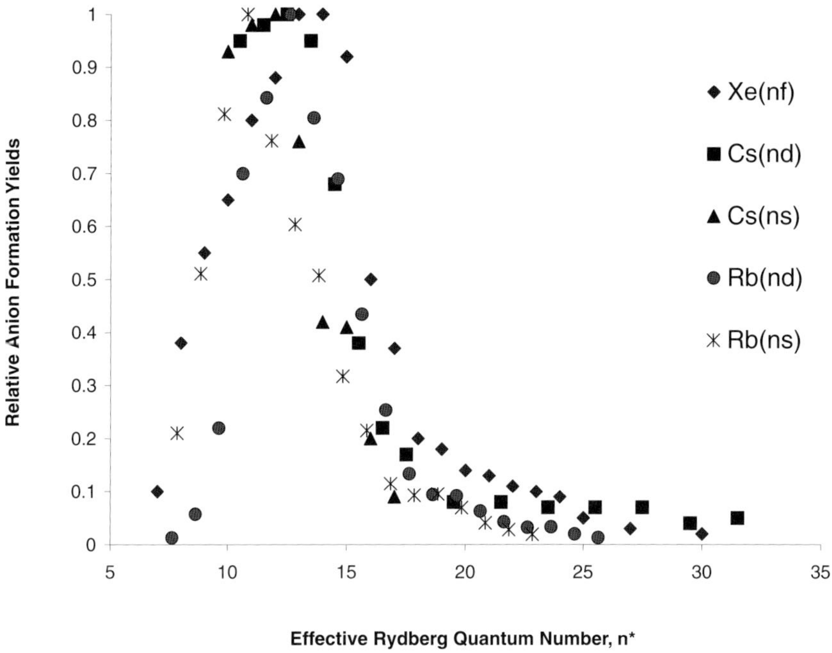

Figure 18. Relative rate constants for the production of $(CH_3NO_2)^-$ anions in electron transfer reactions between laser excited Cs(*ns*), Cs(*nd*), Xe(*nf*), Rb(*nd*), and Rb(*ns*) Rydberg atoms and nitromethane molecules.

experimentally measured $(CH_3NO_2)^-$ yield as a function of the effective principal quantum number, n^*, for collisions between Rb(*ns*, *nd*), Cs(*ns*, *nd*), and Xe(*nf*) Rydberg atoms and nitromethane molecules entrained in a supersonic jet. The maxima for all of these species occur at around $n^* = 13$ and there is very little difference observed of the angular momentum, ℓ, of the Rydberg atom, with the core, Rb^+, Cs^+ or Xe^+. It is instructive to consider the dynamic interaction between the adiabatic-dipole and valence anion states in nitromethane. Figure 19 represents a simplified two-dimensional picture of the adiabatic potential energy surface for $(CH_3NO_2)^-$. In this picture, the $(CH_3NO_2)^-$ state at the equilibrium geometry of the neutral is dominated by the dipole field. As the molecule is distorted to the "bent" equilibrium geometry of the valence state, the excess electron density begins to shift to the $-NO_2$ end of the molecule. For small angular tilt of the $-NO_2$ group, it is assumed that the energy level increases, producing a small barrier to the formation of the ground valence state. Thus, in this picture, the dipole state is a metastable state whose lifetime is

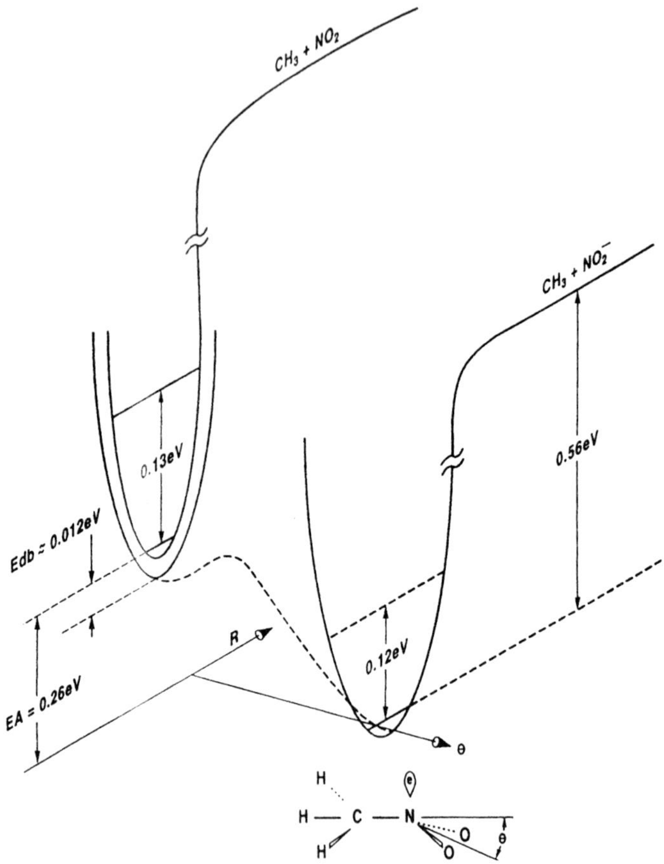

Figure 19. Potential energy surfaces for CH_3NO_2 and $(CH_3NO_2)^-$ anions. The surface representing the dipole-bound anion lies (~0.012 eV) below that of CH_3NO_2 and the valence-bound anion lies 0.26 eV below the neutral.

determined by vibrational coupling with the anion ground state. Free electrons can temporarily attach into the valence anion state via nuclear excited Feshbach resonances as well as into dipole-supported shape and Feshbach resonances in the continuum. The system is then a superposition of dipole and valence states eventually exiting again into the continuum. In this case, the dipole states act as a separate doorway to the eventual formation of a stable valence anion. We should also point out that the attachment of free electrons to CH_3NO_2 produces a short-lived negative ion ($\tau < 10^{-6}$ s). However, the presence of the Rydberg ion core stabilizes the anion into a negative energy state (infinite lifetime).

Nitrobenzene, *para*-dinitrobenzene (*p*DNB) and *meta*-dinitrobenzene (*m*DNB) anions have both diffuse multipole-bound and valence-anion states. For nitrobenzene and *m*-dinitrobenzene, the diffuse states are believed to correspond to dipole-bound anions whereas *p*DNB anions are considered to be primarily described as quadrupole-bound states. Nitrobenzene anions (NB^-) in both valence and dipole-bound states were examined using RET spectroscopy. *Para*-dinitrobenzene (zero dipole moment and a large quadrupole moment) and *meta*-dinitrobenzene (large dipole moment and a small quadrupole moment) were also studied using RET.[106] Figure 20 shows the *n*-dependence of the reaction rates for charge transfer between nitrobenzene and atoms of cesium, Cs(*ns*, *nd*), and xenon, Xe(*nf*), in high Rydberg states. The RET *n*-dependence curve exhibits a prominent peak at an effective principal quantum number $n^*_{max} \cong 11$ and does not appear to depend upon the alkali ion core or the angular momentum of the Rydberg electron. This result is consistent with the formation of a dipole-bound anion and the corresponding electron affinity deduced from an empirical formula given above where $EA_{db} = 28$ meV. In order to test the nature of the orbital occupied by the excess electron of the NB^- anions, field detachment measurements were performed while tuning the laser for production of Xe(11*f*) or Xe(25*f*) Rydberg atoms. In both cases, no detachment of the NB^- anions was observed. It was concluded from the absence of field-detachment of the dipole-bound anions (for $n^* = 11$), that there exists a strong coupling of the diffuse dipole state to the valence state. Nitrobenzene has a large dipole moment of 4.22 D and a rather small quadrupole moment component along

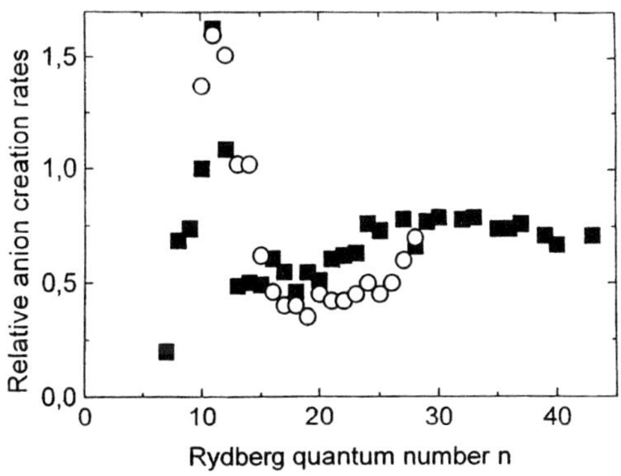

Figure 20. The *n*-dependence of the relative rate constant for the formation of nitrobenzene anions in collisions of Xe(*nf*) (squares) and Cs(*ns, nd*) (circles) laser-excited atoms with nitrobenzene.

the dipole axis of only −9 a.u. Its parallel and perpendicular polarizabilities are estimated to be 17.3 A^3. and 11.6 A^3, respectively. From these values and using an electrostatic model[59] (see Section IV. Experimental Techniques), an approximate value of the dipole-bound electron affinity was found to be 30 ± 10 meV.

The study of RET electron attachment to *p*DNB was prompted by the anticipated observation of quadrupole-bound anions. *p*DNB appears to be a good candidate since its dipole moment is zero, and the components of its quadrupole moment are equal to $Q_{xx} = +45$ a.u.; $Q_{yy} = -59$ a.u.; $Q_{ZZ} = +14$ a.u. The parallel and perpendicular polarizabilities were evaluated to be 22.0 A^3 and 16.6 A^3, respectively. From the simple electrostatic model, the approximate value of the quadrupole-bound electron affinity was predicted to be 23 ± 15 meV. The RET *n*-dependence for the production of *para*-dinitrobenzene ions $p\mathrm{DNB}^-$, shown in Figure 21, displays a peak around $n^* = 11 - 12$ suggesting the existence of a weakly bound anion. The corresponding experimental electron affinity deduced from the above empirical relationship is about 25 meV. The increase of the anion-creation rate between $n^* = 20$ and $n^* = 25$ is similar to that observed for several other anions created by Rydberg electron attachment to aromatic systems. This may be due to Rydberg electron transfer into π orbitals but, until now, no quantitative explanation has been given for this observation. No field-detachment has been observed for $p\mathrm{DNB}^-$ anions produced by RET from Xe(12*f*) atoms. This result and the rather large width of the RET *n**-dependence indicate that the diffuse state is strongly coupled to a valence state. Indeed, it is known that NO_2-containing disubstituted benzene derivatives strongly

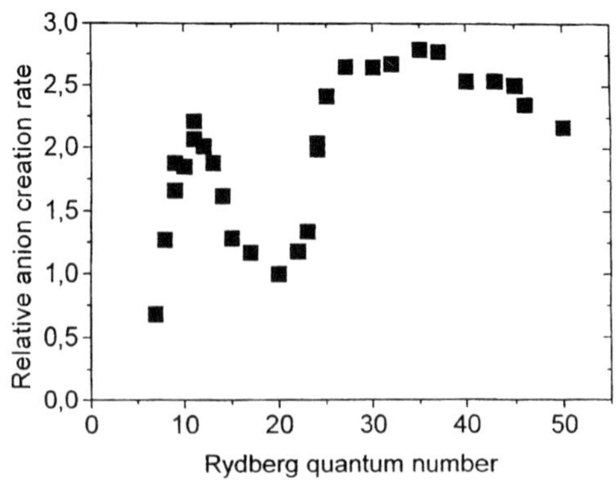

Figure 21. The *n*-dependence of the relative rate constant for the formation of 1, 4-dinitrobenzene anions ($p\mathrm{DNB}^-$) in collisions of *p*DNB with Xe(*nf*) atoms.

attach epithermal electrons and form long-lived valence negative ions. By comparison to the dipole features, it is suggested that the pDNB$^-$ anions formed under Rydberg charge transfer for $n^* = 10$–15 are a result of transfer into a quadrupole-bound state. In this connection, a recent theoretical study by Gutowski and Skurski[107] finds that pDNB does not support a quadrupole-bound anion state but only a valence-anion state. The quadrupole moment tensor reported by that group is however close to the one reported here. A further possibility therefore is that a short-lived quadrupole state is formed followed by rapid intramolecular conversion to the bound valence state. In this sense, the quadrupole state is a doorway state to the formation of bound valence anions. This issue will only be resolved through further experimental and theoretical studies.

Meta-dinitrobenzene has a large dipole moment ($\mu = 4.29$ D), very close to that of nitrobenzene, and a quadrupole moment of $Q = +18$ a.u. Its parallel and perpendicular polarizabilities are estimated to be 21.2 A^3 and 17 A^3, respectively. From the electrostatic model, the predicted electron affinity of mDNB is 105 ± 25 meV, corresponding to a peak in the RET curve at around $n^* = 7$. The curve for mDNB (not shown) is very different from the RET curve corresponding to pDNB. The broad peak at $n^* = 11$–12 is not present and is replaced by a very small peak at $n^* = 8$, close to the predicted value for the dipole-bound mDNB$^-$ anion. The anions reported here are not observed to undergo field detachment. Again, this observation is attributed to the coupling of these diffuse states with the ground valence anion state.

We now briefly discuss two other studies of RET to molecules for which, in retrospect, one may relate their properties to multipole-bound anion states. The studies involve hydrogen iodide and bare carbon clusters. The scattering and attachment of low-energy electrons by hydrogen halide molecules, HX (where X = F, Cl, Br, I) have proved a challenging problem in chemical physics for many years. In particular, the processes of dissociative attachment (DA) and vibrational excitation (VE) have received considerable attention. RET to jet-cooled hydrogen iodide (HI) molecules has been studied for alkali atoms excited to *ns* and *nd* Rydberg levels ($9 < n < 40$).[108] Sodium, rubidium, or cesium atoms in a collimated effusive beam were excited to *ns* and *nd* Rydberg levels by two-color, two-photon absorption using two independently tunable dye lasers pumped by a single nitrogen laser. The first dye laser was tuned to the first $^2S_{1/2} \rightarrow {}^2P_{3/2}$ transition of the alkali atom, while the second dye laser was tuned to a selected $^2P_{3/2} \rightarrow ns, nd$ transition (see Figure 14). The excited alkali beam was crossed at 90° with a collimated beam of HI or DI molecules formed by supersonic expansion of pure HI or DI gas from a pulsed nozzle. The beams were crossed in a field-free region between two parallel mesh grids in a Wiley–McLaren–type acceleration region of a 1.5 m linear time-of-flight mass spectrometer (TOFMS), whose axis was orthogonal to both molecular beams. Subsequent to the firing of the lasers used to excite the alkali atoms, collisions were allowed to proceed in the field-free region for a given time, after which the positive or negative ions produced by Rydberg electron transfer were extracted into the TOFMS by applying a voltage pulse to one of the grids. The ions were detected with a dual microchannelplate electron multiplier and the mass spectra were averaged and recorded. Relative rate constants for ion formation as a function of Rydberg level were then extracted from the spectra after making corrections for variations in laser power with wavelength, radiative lifetimes of the alkali excited states, and two-photon excitation probabilities. For Rydberg atoms with $n^* > 13$, only dissociative electron transfer yielding I^- ions was observed. Thus the RET can be described as a free electron attachment. For $n^* < 13$, it was observed that, as n^* decreased, a peak at mass 128 steadily increased relative to the signal for I^- (mass 127), which was attributed to the HI^- ion. The ratio of ion signals, $R = S[HI^-(DI^-)]/S[I^-]$, as a function of n^*, is fairly similar for HI and DI and is independent of the alkali atom, as shown in Figure 22. The exact

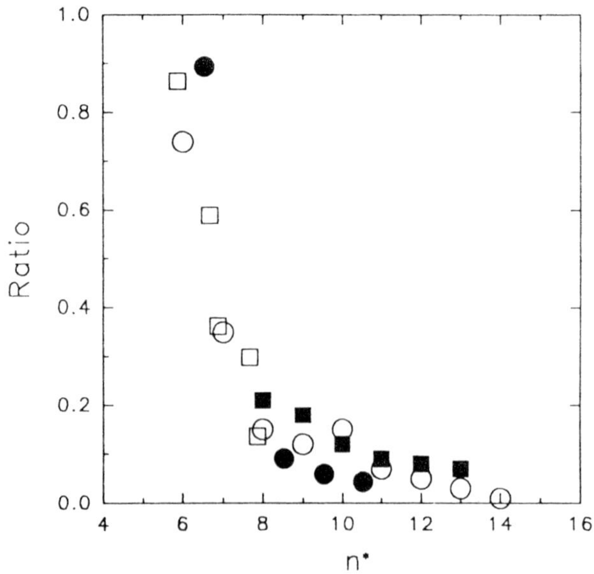

Figure 22. Signal ratio, $R = S[HI^-(DI^-)]/S[I^-]$, versus effective principal quantum number, n^*, obtained for Na(nd) + HI (open circles), Rb(nd) + HI (open squares), Cs (nd) + HI (closed circles), and Na(nd) + DI (closed squares) collisions.

mechanism leading to HI^- formation is not clear. One possibility is that a fraction of the HI^- ions, produced by electron attachment, are stabilized by subsequent collisions with the core ion, before dissociation occurs. It seems plausible that this fraction would increase with decreasing n^*. It is significant, however, that no HI^- ions are observed for $n^* > 13$, even though core interactions are appreciable for $n^* < 23$. This observation suggests that, for $n^* < 13$, the core directly affects the initial process of negative ion formation. The production of HI^- may then be better described in terms of nonadiabatic transitions between ionic and covalent potential energy surfaces. For low values of n^*, the presence of the Rydberg core ion obviously affects the dynamics of negative ion formation. Since the lowest $H(D) + I^-$ limit lies below the rovibrational ground state of HI, any HI^- state correlating with this limit must be stable. Observation of long-lived DI^- ions suggests that the anion state has a potential well deep enough to support vibrational levels for both HI^- and DI^-. Theoretical calculations also show a potential well minimum at long range.[109] The dipole moment of HI at its equilibrium separation, r_e, is only 0.44 Debye and well below the minimum value to produce a dipole-bound state in the gas phase. However, one can raise the question as to whether the dipole moment of HI at r_e would be large enough to affect the HI^- potential at that separation. The occurrence of a diffuse state would account for the larger cross section at large n^*. Partial confirmation of this proposal would come from field detachment studies of HI^- ions. Further theoretical consideration would also be helpful in this regard.

Low-energy electron attachment to neutral C_n clusters ($n < 30$) has also been studied using the RET technique.[110] As has become common in cluster research, inferences about the structures and stabilities of carbon clusters have often centered around so-called "magic numbers" observed in the mass spectral intensity distributions for cluster ions formed by various means. Magic numbers are usually interpreted as indicating clusters of unusual stability. In contrast to positive ions, distributions of negative ions observed for vaporization of graphite are dependent upon the method of generation. For example, laser vaporization of graphite into a vacuum produces C_n^- distributions ($n < 20$) with an even–odd intensity (even > odd) alternation but no distinct magic numbers are apparent. The greater intensities for the even clusters reflect their larger electron affinities. For the RET attachment of electrons, neutral and negatively charged C_n clusters were generated using a laser vaporization nozzle source in which a rotating graphite rod was vaporized using the 532 nm second harmonic of a Nd:YAG laser. The vaporized material was entrained in a pulse of helium or argon gas from a pulsed valve and allowed to cool and condense prior to jet expansion into a vacuum and collimation into a beam. The C_n cluster beam was crossed at 90° with a collimated effusive beam of Rb atoms in a field-free region between two parallel mesh grids in a Wiley–McLaren-type acceleration region of a linear TOFMS whose axis was orthogonal to both molecular beams. Rubidium atoms were excited to *ns* or *nd* Rydberg states by resonant two-color excitation using two independently tunable dye lasers pumped by a single nitrogen laser as before. Collisions between neutral C_n clusters and Rb Rydberg- atoms were allowed to proceed in the field-free region for a specified reaction time after which positive or negative ions formed during the collisions were accelerated into the TOFMS by applying a voltage pulse to one of the grids. By flagging various combinations of the lasers and the Rb beam, C_n^- distributions were obtained for negative ions produced either directly in the source or by collisions with Rb atoms in the ground or Rydberg states (see Ref. [104]). Figure 23 shows C_n^- distributions obtained for ions formed by Rydberg electron transfer from the *30d* state of Rb. "Magic numbers" are clearly seen at $n = 5$, 10, 12, 16, and 18. One rationalization of these magic numbers is that these molecules have exceptionally large electron attachment cross sections. Another possibility is that these magic clusters possess large enough multipole moments, particularly quadrupole moments, that would accommodate bound states. Unfortunately, these experiments were not performed for various n^* values. Such data could be used to test this hypothesis. RET to form quadrupole-bound anions would be expected to maximize for n^* appropriate to the size of the quadrupole moment. In fact, if peaks in the n^* cross section are observed, these data would be used to estimate quadrupole moments for carbon clusters (of unknown geometries, however).

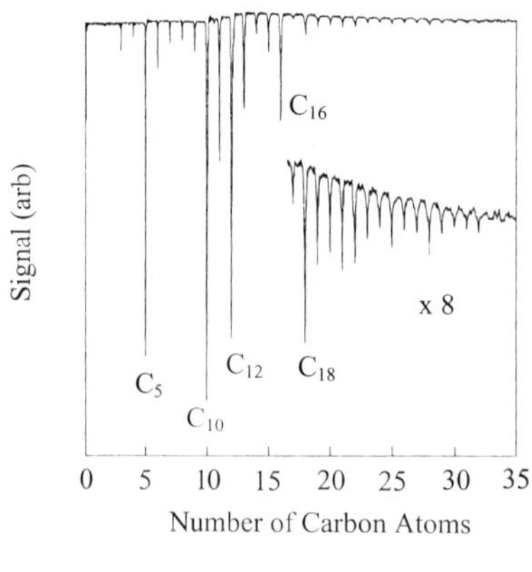

Figure 23. Time of Flight Mass Spectrum obtained for C_n^- ions produced by Rydberg electron transfer from the *30d* state of Rb to C_n neutrals.

Whitehouse and Buckingham[111] have recently measured the atomic quadrupole moment of graphite and made a correlation with the moment by simply adding up the quadrupole moments of all the carbon atoms. It is interesting to note that the quadrupole moment of benzene is roughly six times the quadrupole moment of the individual carbon atom.[112] Thus one can *postulate* that the ions observed in Figure 23 are the result of quadrupole-bound electrons to C_n clusters present in the laser ablation plume. Possible polycyclic structures for the carbon clusters observed in the negative ion spectrum produced by RET are

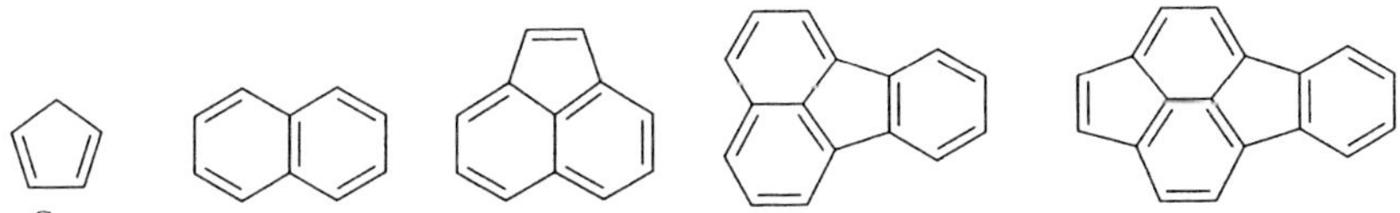

C_5 C_{10} C_{12} C_{16} C_{18}

It is interesting to note that electron attachment to acenaphthelene ($C_{12}H_8$) and fluoranthene ($C_{16}H_{10}$) have been observed.[113] RET measurements as a function of n^* might allow one to distinguish between electron attachment directly into valence bound anion states or electron attachment via mediation through quadrupole-bound anions. If these polycyclic species are present in the plume, one would expect both anion states to play a role. We further point out that these species have been proposed (and detected) as intermediates in fullerene formation.[114] Under tight focusing of the laser beam during ablation of graphite in vacuum, it is possible to produce C_{60}^+ and C_{70}^+ ions.[115] Presumably, a small "channel" was created which keeps the expanding carbon atoms and clusters together long enough for fullerene formation.

Recently, Gutowski et al.[116] have examined negative ions of $(MgO)_n$ clusters both experimentally and theoretically. Photoelectron spectra of $(MgO)_n^-$ ($n=1$–5) showed an unusual decrease from $n=1$ to $n=4$ (1.6, 1.1, 0.7, 0.7 eV, respectively) followed by an increase at $n=5$ (0.98 eV). This trend was attributed to a component of the electron-binding energy which could be identified with the long-range electron-molecule interaction. The monomer and pentamer produce a dipolar ($1/r^2$) long-range potential, while the dimer and trimer produce a quadrupolar ($1/r^3$) potential and a ($1/r^4$) octopolar long-range potential for the tetramer. These binding energies are considerably larger than those found for diffuse dipole-bound anion states discussed previously. Unfortunately, based upon the discussion earlier for the case of $(BeO)_2^-$ quadrupole-bound anions and the contradictory results of Gutsev, Jena, and Bartlett,[68] these states may be more appropriately described as valence bound anions.

VI. MULTIPLY CHARGED NEGATIVE IONS: THE COULOMB BARRIER

Schauer et al.[117] published the first clear and unambiguous observation of doubly charged anions in the gas phase. In these experiments, ion bombardment of graphite was found to produce C_n^{-2} ($n = 7$–28). This result was quickly confirmed by three other groups.[118–122] In addition, small doubly-charged anions of silicon and carbon clusters, SiC_n^- ($n = 6$, 8, and 10) have also been produced by sputtering a SiC surface with a 14.5 keV Cs^+ ion beam.[121] References to theoretical studies on these dianion clusters are found in the paper by Gnaser.[121] Not much is known about the relative stabilities of these dianions. However Mathr et al.[122] have used a multicoincidence time-of-flight technique which provides evidence for symmetric fission of dianions of carbon into C_n^-–C_n^- ion-pairs.

Perhaps the simplest model for the formation of multiply charged anions is the charged conducting sphere depicted in Figure 1. Accordingly, there is considerable recent interest in the multiple charging of metal clusters. In one study, gold cluster dianions of gas-phase gold clusters have been reported.[123,124] Yannouleas and Landman[125] have theoretically considered the stability and decay channels of Au_n^{-2} and Ag_n^{-2}. They find that autodetachment dominates fission as the mode of decay at a finite temperature. Earlier they had studied sodium cluster dianions.[126] These calculations show pronounced shell-effects in the second electron affinities illustrating the failure of the simple electrostatic model of charging a conducting sphere as depicted in Figure 1. However, the Coulomb barrier calculated with such simple assumptions probably is a reasonably good approximation. The importance of the Coulomb barrier in multiply charged anions is emphasized below.

Historically, the first reference to the Coulomb barrier for multiply charged negative ions related to the occurrence of what was believed to be AuF_6^{-3}. The occurrence of an intense, single-isotope, negative-ion signal at an m/z corresponding to 103 ± 1 amu/z from gold wires "burning" in fluorine gas was tentatively attributed to AuF_6^{-3}. Gold and fluorine each have only one naturally occurring isotope. The occurrence and long lifetime of this assumed anion was attributed[127] to a giant shape resonance as a result of the expected Coulomb barrier. Unfortunately, more accurate mass spectroscopic studies[128] show that this ion's mass is 103.0 ± 0.07 m/z and is most probably the "impurity" ion $Na_2F_3^-$.

More recently, the occurrence of long-lived C_{60}^{-2} anions is attributed to the Coulomb barrier.[2,98] Later, electron attachment to C_{84}^- to produce a highly vibrationally excited C_{84}^{-2} anion was observed to decay by auto detachment over the top of the Coulomb barrier.[99] In a number of cases, the doubly charged anions are observed to be more stable with respect to the loss of either excess electron than their singly charged parent anion. The Coulomb barrier in these experiments was illustrated by considering the interaction potential for an electron with a singly-charged sphere of radius a and dielectric constant k,

$$V(r) = \frac{e^2(k-1)a^3}{2(k+2)r^2(r^2-a^2)} + \frac{e^2}{r} + \frac{\ell(\ell+1)\hbar^2}{2mr^2} \tag{19}$$

where the last term represents the centrifugal potential for an approaching (or exiting) electron with angular momentum$[\ell(\ell+1)]^{1/2}\hbar$. Figure 24 shows this potential for C_{84} assuming $a^3 = 80$ Å^3 and $k \approx 4.4$. Electron attachment to C_{84}^- is expected to exhibit broad resonances in the electron attachment cross section corresponding to angular momentum shape resonances beginning at the top of the Coulomb barrier at ~1.4 eV.

All of this discussion assumes the occurrence of a Coulomb barrier of the type shown in Figures 1 and 24. Lai-Sheng Wang and his group

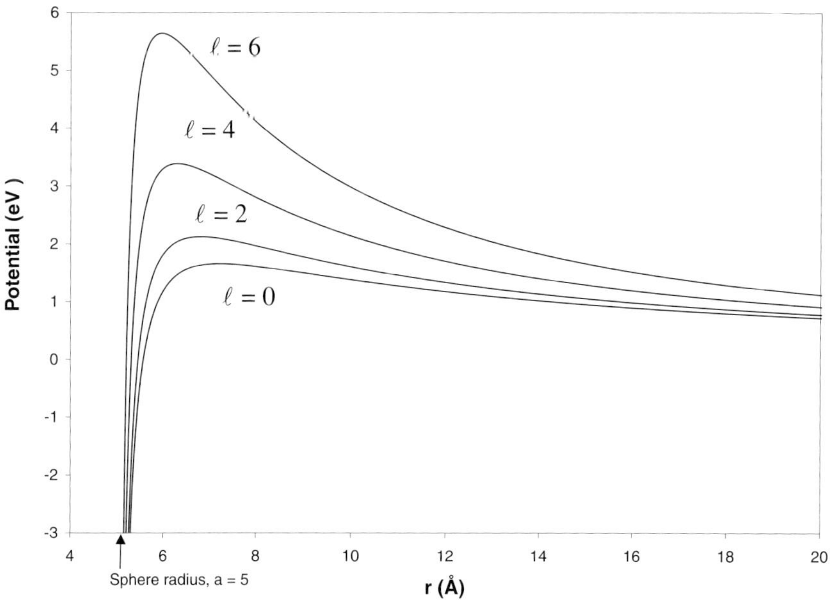

Figure 24. Interaction potential for an electron attracted to singly charged dielectric sphere (k=4.4) with a radius of 5 Å computed for a number of angular momentum states.

have published a series of studies which provide an estimate of the magnitude of the Coulomb barrier.[129–136] They employ electrospray ionization to produce sufficiently intense beams of doubly charged anions in order to obtain photodetachment electron spectra. They estimate the Coulomb barrier to be ~1.9–2.5 eV. In addition, they obtain positive second electron affinities for many molecules. In one case,[134] they reported a long-lived dianion with a negative second electron-binding energy similar to that inferred for C_{60}^{-2}. Electron tunneling through the Coulomb barrier was reported for $O_2C(CH_2)_nCO_2^{-2}$ ($n = 3$–6).[133]

VII. DIPOLE-BOUND MULTIPLY CHARGED MOLECULAR ANIONS

J. Simons and his group at the University of Utah have performed a number of calculations on multiply charged anions which involve dipole states. They studied the possibility of binding two electrons to a fixed finite point dipole (limit of large q and small R) and found a critical moment for electron binding to approach a value below 2 Debye.[137] Interestingly, in the point dipole limit, the value was found to approach the same as that for the single electron binding (1.625 D). Dianionic species involving mixed dipole and valence anion states have also been considered.[138] More recently, this group suggested that the nonpolar HCN–HCCH–NCH molecule could possibly bind two extra electrons through binding to each of the oppositely directed dipoles.[139] This molecule also has a large quadrupole moment, but the binding was better described as that due to binding to the two "isolated" dipoles.

In another similar study, the Simons group has considered the possibility of binding two excess electrons by a molecule possessing two polar ends, each of which is capable of binding an extra electron.[140] Dianion states for $(LiCN)_2$, HC≡CH and $(NCLi)_2$ in the nearly degenerate $^3\Sigma_g^+$ and $^1\Sigma_u^+$ (linear) states are found to be bound by ~0.8 eV with respect to the monoanion. These are vertical detachment energies and the dianions are electronically stable at the equilibrium geometry of its monoanionic daughter by ~0.79 eV and neutral granddaughter by ~0.75 eV.

At present, there is no firm experimental evidence for dipole-bound dianions. However, it is instructive to consider the results of multiply charged anions of fullerene derivative clusters discussed in Section VI. Tuinman and Compton[104] reported the formation of doubly charged anions of $C_{60}(CN)_2$ dimers, $[C_{60}(CN)_2]_2^{-2}$, in electrospray ionization. The dipole moment of the $C_{60}(CN)_2$ molecule in which the two CN radicals are bonded to two adjacent carbon atoms in C_{60} is expected to be large

(PC Model gives ~6 Debye) and one might expect the dimer to orient in the geometry of two opposing dipoles, i.e., $_2(NC)C_{60}$–$C_{60}(CN)_2$. We suggest that this system has an excellent chance of falling into the category of a dipole-bound cluster dianion. This system will most likely have a "valence bound" dianion state as well. The $C_{60}(CN)_2^{-2}$ anions observed are most likely in the valence anion bound state, but this system is expected to have a dipole-bound dianion as well. In this respect, it should be stated that many multiply charged anions may have excited dipole-bound states in which one or more electron could be considered bound to the "local" dipole moment in the sense described by the Simons group.

ACKNOWLEDGMENTS

RNC is indebted to colleagues Jean Pierre Schermann, Charles Desfrancois, Howard Carman, Eph Klots, Al Tuinman, and Kit Bowen for their collaboration over the years. We thank Ray Garrett for a critical reading of the manuscript and for commenting on the theory of dipole states. NIH thanks the National Science Foundation (CHE-9981945) for financial support.

REFERENCES

[1] Haberland, H.; Kolar, T.; Reiners, T. *Phys. Rev. Lett.* **1989**, *63*, 1219.
[2] Scheller, M. K.; Compton, R. N.; Cederbaum, L. S. *Science* **1995**, *270*, 1160.
[3] Jackson, J. D. *Classical Electrodynamics*, 2nd Ed.; John Wiley & Sons, Inc., 1975, p 59.
[4] Bohr, N. "On the Constitution of Atoms and Molecules. Part II: Systems Containing Only a Single Nucleus," *Philosophical Magazine*, **1913**, *26*, 476–502.
[5] Ruskai, M. B. *Commun. Math. Phys.* **1982**, 82, 457.
[6] Sigal, I. M. *Commun. Math. Phys.* **1982**, *85*, 309.
[7] Ruskai, M. B. *Commun. Math. Phys.* **1982**, *85*, 325.
[8] Benguria, R.; Lieb, E. H.; *Phys. Rev. Lett.* **1983**, *50*, 1771.
[9] Hill, R. N. In *Mathematical Problems in Theoretical Physics*; Osterwalden, K., Ed.; Springer-Verlag: New York, 1979, p 52–56.
[10] Lieb, E. H. *Phys. Rev. Lett.* **1984**, *52*, 315.
[11] Brownstein, K. R. *Phys. Rev. Lett.* **1984**, *53*, 907.
[12] Lieb, E. H. *Phys. Rev.* **1984**, *29*, 3018.
[13] Levy-Leblond, J. M. *J. Phys.* **1971**, B4, L23; **1972**, *B5*, 720.
[14] Beck, D. R.; Nicolaides, C. A. *Chem. Phys. Lett.* **1978**, *59*, S2S.
[15] Avron, J.; Herbst, I.; Simon, B. *Phys. Rev. Lett.* **1977**, *38*, 1068.
[16] Simon, B. *Phy. Rev. Lett.* **1976**, *36*, 1083.
[17] Anderson, C. D. *Phys. Rev.* **1932**, *41*, 405.
[18] Dirac, P. A. M. In *The Principles of Quantum Mechanics, 3rd ed.*; Oxford: New York, 1947, Chap Xi.
[19] Wheeler, J. A. In *Annals of the New York Academy of Sciences, XLVIII*, 1946, p 219.
[20] Bhatia, A.; Drachman, R. *J. Phys. Rev. A* **1983**, *28*, 2523.
[21] Kolos, W.; Roothan, C. C. J.; Jack, R. A. *Reviews of Modern Physics* **1960**, *32*, 178.
[22] Pekeris, C. L. *Phys. Rev.* **1962**, *126*, 1470.

[23] Abdel-Raouf, M. A. *Fortschrift Physih* **1988**, *36*, 7521.
[24] Mills, A. P., Jr. *Phys. Rev. Lett.* **1981**, *46*, 717.
[25] Mills, A. P., Jr. *Phys. Rev. Lett.* **1983**, *50*, 671.
[26] Ho, Y. K. J. *J. Phys. B* **1983**, *16*, 1503.
[27] Bhatia, A. X.; Drachman, R. J. *Phys. Rev. A* **1987**, *35*, 4051.
[28] Ferrante, G. *Phys. Rev. Lett.* **1968**, *170*, 76.
[29] Mills, A. P., Jr. *Phys. Rev. Lett.* **1981**, *24*, 3242.
[30] Bhatia, A. X. *Phys. Rev. A*, **1987**, *2*, 1667.
[31] Bhatia, A. X.; Drachman, R. J. *Phys. Rev. A* **1987**, *35*, 4051.
[32] Botero, J.; Greene, C. H. *Phys Rev. Lett.* **1986**, *56*, 1366.
[33] Petelenz, P.; Smith V.H., Jr. *Phys. Rev A* **1987**, *36*, 5125.
[34] Haftel, M. I.; Mandelzweig, V. B. *Phys. Rev. A* **1989**, *39*, 2813.
[35] Ho, Y. K. J. *Phys. Rev. A* **1979**, *19*, 2347.
[36] Ho, Y. K. J.; Bhatia, A. K. *Phys. Rev. A* **1997**, *47*, 1497.
[37] Kvitsinsky, A. A.; Carbonell, J.; Gignoux, C. *Phys. Rev. A* **1992**, *46*, 1310.
[38] Harshman, D. R.; Warren, J. B.; Beveridge, J. L.; Kendall, K. R.; Kiefl, R. F.; Oram, C. J.; Mills, A. P., Jr.; Crane, W. S.; Rupaal, A. S.; Turner, J. H. *Phys. Rev. Lett.* **1986**, *56*, 2850.
[39] Kuang, Y.; Arnold, K. P.; Chmely, F.; Eckhause, M.; Hughes, V. W.; Kane, J. R.; Siketell; Kim, D. H.; Kumar, K.; Lu, D. C.; Ni, B.; Matthias, B.; Orth, H.; Putlitz, G.; Zu, Schaefer, H. R.; Souder, P. A.; Woodle, K. *Phys. Rev. A* **1987**, *35*, 3172. and *Phys. Rev. A* **1989**, *39*, 6109.
[40] Turner, J. E. *American Journal of Physics* **1977**, *45*, 758.
[41] Fermi, E.; Teller, E. *Phys. Rev.* **1947**, *72*, 406.
[42] Wightman, A. S. *Phys. Rev.* **1949**, *77*, 521.
[43] Turner, J. E.; Fox, K. *Phys. Rev. Lett.* **1966**, *23*, 547.
[44] Mittleman, M. H.; Meyerscough, V. P. *Phys Lett.* **1966**, *23*, 545.
[45] Levy-Leblond, J. M. *Phys. Rev.* **1967**, *153*, 1.
[46] Brown, W. B.; Roberts, R. E. *J. Chem. Phys.* **1967**, *46*, 2006.
[47] Turner, J. E.; Anderson, V. E.; Fox, K. *Phys. Rev.* **1968**, *174*, 81.
[48] Wallis, R. F.; Herman, R.; Milnes, H. W. *J. Mol. Spectrosc.* **1960**, *4*, 51.
[49] Hurst, G. S.; Stockdale, J. A.; O'Kelly, L. B. *J. Chem. Phys.* **1963**, *38*, 2572.
[50] Stockdale, J. A.; Christophorou, L. G.; Turner, J. E.; Anderson, V. E. *Phys. Lett.* **1967**, *25A*, 510.
[51] Levy-Leblond, J. M.; Provost, J. P. *Phys. Lett.* **1967**, *26A*, 104.
[52] Takayanagi, K.; Itikawa, Y. *J. Phys. Soc. Jpn.* **1968**, *24*, 160.
[53] Christophorou, L. G.; Pittman, D. *J. Phys. B* **1969**, *2*, 71.
[54] Garrett, W. R. *Mol. Phys.* **1972**, *24*, 465.
[55] Crawford, O. H. *Mol. Phys.* **1971**, *20*, 585; Garrett, W. R. *Phys. Rev.* **1971**, *3*, 961; Crawford, O. H.; Garrett, W. R. *J. Chem. Phys.* **1977**, *66*, 4968.
[56] Sigiura, T.; Arakawa, A. In *Proceedings of the International Conference on Mass Spectroscopy*; Univ. of Tokyo: Tokyo, 1970, p 848.
[57] Stockdale J. A. D.; Davis, F. J.; Compton, R. N.; Klots, C. E. *J. Chem. Phys.* **1974**, *60*, 4279.
[58] Desfrancois, C.; Khelifa, N. A.; Lisfi, A.; Schermann, J. P.; Eaton, J. G.; Bowen, K. H. *J. Chem. Phys.* **1991**, *95*, 7760; Desfrancois, C.; Baillon, B.; Schermann, J. P.; Arnold, S. T.; Hendricks, J. H.; Bowen, K. H. *Phys. Rev Lett.* **1992**, *72*, 48; Desfrancois, C.; Abdoul-Carime, H.; Khelifa, N.; Schermann, J. P. *Phys. Rev. Lett.* **1994**, *73*, 2436; Desfrancois, C. *Phys. Rev. A* **1995**, *51*, 3667.
[59] Desfrancois, C.; Abdoul-Carmine, H.; Schermann, J. P. *International Journal of Modern Physics B* **1996**, *10*, 1339.
[60] Coe, J. V.; Lee, G. H.; Eaton, J. G.; Arnold, S. T.; Sarkas, H. W.; Bowen, K. H.; Ludewigt, C.; Haberland, H.; Worsnop, D. R. *J. Chem. Phys.* **1990**, *92*, 3980.

[61] Hendricks, J. H.; Lyapustina, S. A.; de Clercq, H. L.; Snodgrass, J. T.; Bowen, K. H. *J. Chem. Phys.* **1996**, *104*, 7788.
[62] Brinkman, E. A.; Berger, S.; Marks, J.; Brauman, J. I. *J. Chem. Phys.* **1993**, *99*, 7586 and others cited therein.
[63] Mullins, A. S.; Murray, K. K.; Schultz, C. P.; Lineberger, W. C. *J. Phys. Chem.* **1993**, *97*, 10281 and others cited therein.
[64] Dessent, C. E. H.; Bailey, C. G.; Johnson, M. J. Chem. Phys. **1995**, *102*, 6335.
[65] Hashemi, R.; Illenberger, E. *J. Phys. Chem.* **1991**, *95*, 6402.
[66] Jordan, K. D.; Liebman, J. F. *Chem. Phys. Lett.* **1979**, *62*, 143.
[67] Gutowski, M.; Skurski, P. *Chem. Phys. Lett.* **1999**, *303*, 65.
[68] Gustev, G. L.; Jena, P.; Bartlett, R. S. *J. Chem. Phys.* **1999**, *111*, 504.
[69] Ambika Prasad, M. V. N.; Wallis, R. F.; Herman, R. *Phys. Rev. B* **1989**, *40*, 5924.
[70] Ambika Prasad, M. V. N.; Wallis, R. F.; Herman, R. *Solid State Communications* **1991**, *77*, 973.
[71] Gutsev, G. L.; Bartlett, R. J.; Compton, R. N. *J. Chem. Phys.* **1998**, *108*, 6756.
[72] Desfrancois, C.; Periquet, V.; Lyapustina, S. A.; Lippa, T. P.; Robinson, D. W.; Bowen, K. H.; Nonaka, H.; Compton, R. N. *J. Chem. Phys.* **1999**, *111*, 4569.
[73] Abdul-Carime, H.; Desfrancois, C.; *Eur. Phys. J. D.* **1998**, *2*, 149.
[74] Harth, K.; Ruf, M. W.; Hotop, H. *Z. Phys. D. At Mol. Clusters* **1989**, *14*, 149.
[75] Kalamaries, A.; Walker, C. W.; Smith, K. A.; Dunning, F. B. *J. Chem. Phys.* **1988**, *89*, 7226.
[76] Carman, H. S., Jr.; Klots, C. E.; Compton, R. N. *J. Chem. Phys.* **1990**, *92*, 5751; Carman, H. S. Jr.; Klots, C. E.; Compton, R. N. "Negative-Ion Formation by Rydberg Electron Transfer" In *Isotope Effects in Gas-Phase Chemistry* ACS Symposium Series 502 Kaye, J. Ed.; 1992.
[77] Compton, R. N.; Dunning, F. B.; Nordlander, P. *Chem. Phys. Lett.* **1996**, *253*, 8.
[78] Gutsev, G.; Bartlett, R.; Compton, R. N. *J. Chem. Phys.* **1998**, *108*, 6756.
[79] Oakes, J. M.; Ellison, G. B. Tetrahedron **1968**, *42*, 6263.
[80] Compton, R. N. In *The Role of Rydberg States in Spectroscopy and Photochemistry*; Sandorfy, C., Ed.; Kluwer Academic Publishers, 1999, p 419–436.
[81] Ciocca, M.; Allegrini, M.; Arimondo, E.; Burkhardt, C. E.; Garrer, W. P.; Leventhal, J. J. *Phys. Rev. Lett.* **1986**, *56*, 704.
[82] Lee, Y. T.; Mahan B. H. *J. Chem. Phys.* **1965**, *42*, 2893.
[83] Cheret, M.; Barbier, L. *Phys. Rev. A* **1984**, *30*, 1132.
[84] Coe, J. V.; Snodgrass, J. T.; Freidhoff, C. B.; McHugh, K. M.; Bowen, K. H. *J. Chem. Phys.* **1985**, *83*, 3169.
[85] Cardy, H.; Larrieu, C.; Dargelos, A. *Chem. Phys. Lett.* **1986**, *131*, 5076.
[86] Ortiz, J. V. J. *Chem. Phys.* **1987**, *87*, 3557.
[87] Gutowski, M.; Simons, J.; Hernandez, R.; Taylor, H. L. *J. Phys. Chem.* **1988**, *92*, 6179.
[88] Drake, G. W. F. *Atomic, Molecular, and Optical Physics Handbook*; Springer-Verlag: New York, 1996.
[89] Gallagher, T. *Rydberg Atoms*; Cambridge University Press: Cambridge, 1983.
[90] Stebbings, R. F.; Dunning, F. B. *Rydberg States of Atoms and Molecules*, Cambridge University Press: Cambridge, 1983.
[91] Dunning, F. B.; Stebbings, R. F. "Experimental Studies of Thermal-Energy Collisions of Rydberg Atoms with Molecules" in *Rydberg States of Atoms and Molecules*, Cambridge University Press: Cambridge, 1983.
[92] Carman, H. S., Jr.; Klots, C. E.; Compton, R. N. *J. Chem. Phys.* **1989**, *90*, 2580.
[93] Foltz, G. W.; Latimer, C. J.; Hildebrandt, G. F.; Kellert, F. G.; Smith, K. A.; West, W. P. *J. Chem. Phys.* **1977**, *67*, 1352.
[94] Zollars, B. G.; Higgs, C.; Lu, F.; Walter, C. W.; Gray, L. G.; Smith, K. A.; Dunning, F. B.; Stebbings, R. F. *Phys. Rev. A* **1985**, *32*, 3330.

[95] Zollars, B. G.; Walter, C. W.; Lu, F.; Johnson, C. B.; Smith, K. A.; Dunning, F. B. *J. Chem. Phys.* **1986**, *84*, 5589.

[96] Schauer, S. N.; Williams, P.; Compton, R. N. *Phys. Rev. Lett.* **1990**, *65*, 625.

[97] Limbach, P. A.; Schweinhard, L.; Cowan, K. A.; McDermott, M.; Marshall, A. G.; Coe, J. V. *J. Am. Chem. Soc.* **1991**, *113*, 6795.

[98] Hettich, R. L.; Compton, R. N.; Ritchie, R. L. *Phys. Rev. Lett.* **1991**, *67*, 1242.

[99] Jin, C.; Hettich, R. L.; Compton, R. N.; Tuinman, A.; Derecski-Kovacs, A.; Marynick, D. S.; Dunlap, B. I. *Phys. Rev. Lett.* **1994**, *73*, 2821.

[100] Compton, R. N.; Tuinman, A. A.; Klots, C. E.; Pederson, M. R.; Patton, D. C. *Phys. Rev. Lett.* **1997**, *78*, 4367.

[101] Boltalina, O. V.; Hvelplund, P.; Larsen, M. C.; Larson, M. O. *Phys. Rev. Lett.* **1998**, *80*, 5101.

[102] Tuinman, A.; Compton, R.N. (To be published).

[103] Klassen, J. S.; Ho, Y.; Blades, A. T.; Kebarle, P. K. *Adv. Gas-Phase Ion Che.* **1998**, *3*, 255–318.

[104] Tuinman, A.; Compton, R.N. *J. Phys. Chem. A* **1998**, *102*, 9791.

[105] Compton, R. N.; Carman, H. S., Jr.; Desfrancois, C.; Abdoul-Carmine, H.; Schermann, J. P.; Hendricks, J. H.; Lyapustina, S. A.; Bowen, K. H. *J. Chem. Phys.* **1996**, *105*, 3472.

[106] Desfrancois, C.; Periquet, V.; Lyapustina, S. A.; Lippa, T. P.; Robinson, D. W.; Bowen, K. H.; Compton, R.N. *J. Chem. Phys.* **1999**, *111*, 4569.

[107] Gutowski, M.; Skurski, P. *Chem. Phys, Lett.* **1999**, *303*, 65.

[108] Carman, H. S., Jr.; Klots, C. E.; Compton, R. N. *J. Chem. Phys.* **1993**, *99*, 1734.

[109] Chapman, D. A.; Balasubramanian, K.; Lin, S. H. *Phys. Rev. A.* **1988**, *38*, 6098.

[110] Carman, H. S., Jr.; Compton, R. N. *J. Chem. Phys.* **1993**, *98*, 2473.

[111] Whitehouse, D. B.; Buckingham, A. D. *J. Chem. Soc. Faraday Trans.* **1993**, *89*, 1909.

[112] Battaglia, M. R.; Buckingham, A. D.; Williams, J. H. *Chem. Phys. Lett.* **1981**, *78*, 421.

[113] Frey, W. F.; Compton, R. N.; Naff, W. T.; Schweitzer, H. C. *Int. J. Mass Spectrum Ion Phys.* **1973**, *12*, 19.

[114] Chang, T. M.; Naim, A.; Ahmen, S. N.; Goodloe, G.; Sherlin, P. B.; *J. Am. Chem. Soc.* **1992**, *141*, 7603.

[115] Shea, M.; Compton, R. N. unpublished.

[116] Gutowski, M.; Skurski, P.; Li, Xi; Wang, L. S. *Phys. Rev. Lett.* **2000**, *85*, 3145.

[117] Schauer, S. N.; Williams, P.; Compton, R. N. *Phys. Rev. Lett.* **1990**, *65*, 625.

[118] Gnaser, H.; Oechsner, *Nucl. Instrum. Methods Phys. Res. B* **1993**, *82*, 518.

[119] Calabrese, D.; Covington, A. M.; Thompson, J. S. *J. Chem. Phys.* **1996**, *105*, 2936.

[120] Middleton, R.; Klein, J. *Nucl. Instrum. Methods Phys. Res. B* **1997**, *123*, 532.

[121] Gnaser, H. *Phys. Rev.* **1999**, *60*, 2645.

[122] Mathur, D.; Bhardwaj; Rajgara, F.A.; Safvan, C. P. *Chem. Phys. Lett.* **1997**, *277*, 558.

[123] Herlert, A. et al. *Physica Scripta* **1999**, *T80*, 200.

[124] Schweikhard, L.; Herlert, A.; Kruckeberg, S.; Vogel, M. *Philos. Mag.* **1999**, *79*, 1343.

[125] Yannouleas, C.; Landman, U. *Phys. Rev.* **B** *2000*, *61*, R10587.

[126] Yannouleas, C.; Landman, U. *Phys. Rev. B* **1993**, *48*, 8376; *Chem Phys. Lett.* **1993**, *210*, 437.

[127] Compton, R. N. In *Photophysics and Photochemistry in the Vacuum Ultraviolet*, McGlyn, S. P. et al. Eds. Reidel, Dordrecht: 1985, p 261.

[128] Tuinman, A. A.; Lahamer, A. S.; Compton, R. N. *Int. J. Mass Spec.* **2000**, *205*, 309.

[129] Wang, X. B.; Ding, C. F.; Wang, L. S. *Phys. Rev. Lett.* **1998**, *81*, 3351.

[130] Wang, L. S.; Ding, C. F.; Wang, X.B.; Nicholas, J. B. *Phys. Rev. Lett.* **1998**, *81*, 2667.

[131] Ding, C. F.; Wang, X. B.; Wang, L. S. *J. Phys. Chem.* **1998**, *102*, 8633.

[132] Ding, C. F.; Wang, X. B.; Wang, L. S. *J, Chem. Phys.* **1999**, *110*, 3635.

[133] Wang, X. B.; Ding, C. F.; Wang, L. S. *Chem. Phys. Lett.* **1999**, *307*, 391.

[134] Wang, X. B.; Wang, L. S. *Nature* **1999**, *400*, 245.
[135] Wang, X. B.; Wang, L. S. *J. Chem. Phys.* **1999**, *111*, 4497.
[136] Wang, X. B.; Wang, L. S. *J. Phys. Chem.* **2000**, *104*, 5358.
[137] Skurski, P.; Gutowski, M.; Simons, J. *Int. J. Quantum Chem.* **2000**, *76*, 197.
[138] Skurski, P.; Gutowski, M.; Simons, J. *J. Chem. Phys.* **2000**, *111*, 9469.
[139] Gutowski, M.; Skurski, P.; Simons, J. *Inter. J. Mass Spec.* **2000**, *201*, 245.
[140] Skurski, P.; Gutowski, M.; Simons, J. *Chem. Phys. Lett.* **2000**, *322*, 175.

INDEX